# Probabilistic Theory of Ship Dynamics

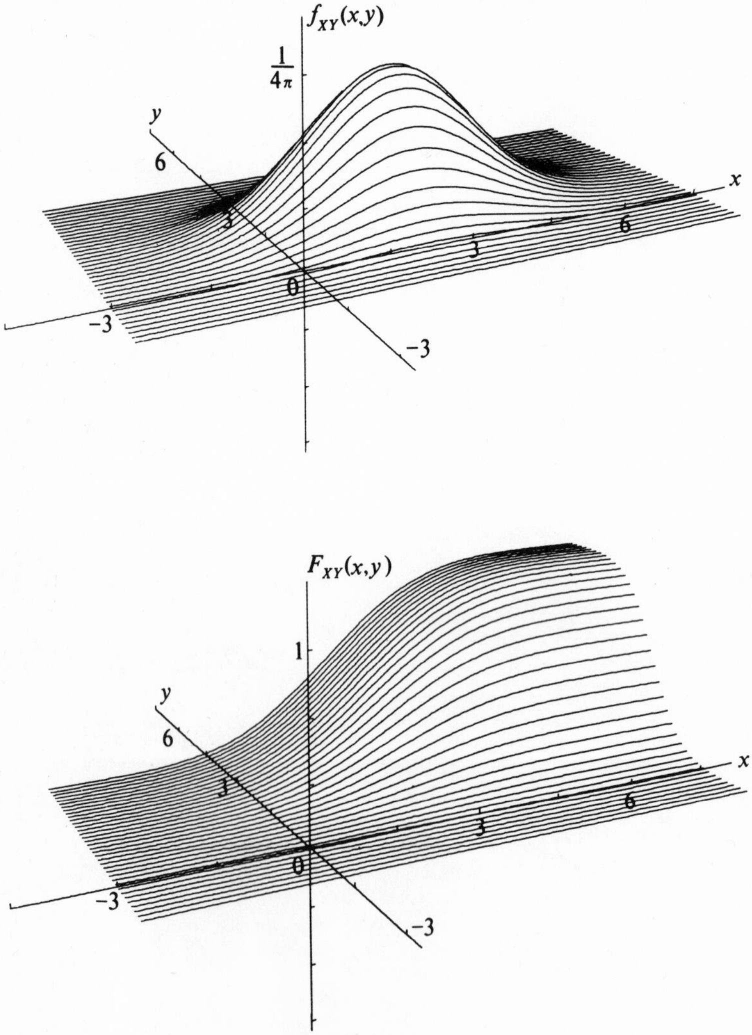


Perspective views of the Gaussian joint probability density function $f_{XY}(x,y)$ and distribution function $F_{XY}(x,y)$. For these surfaces, $\mu_X = 3$; $\mu_Y = 2$; $\sigma_X = 2$; $\sigma_Y = 1$; $\rho_{XY} = 0$.

These drawings were computed by the authors' colleague, the artist Edward Ihnatowicz.

# Probabilistic Theory of Ship Dynamics

W. G. Price
Lecturer in Mechanical Engineering
University College, London.

R. E. D. Bishop
Kennedy Professor of Mechanical Engineering
in the University of London and
Fellow of University College.

CHAPMAN AND HALL
London

A HALSTED PRESS BOOK
JOHN WILEY & SONS
New York

First published 1974
by Chapman and Hall Ltd
11 New Fetter Lane, London EC4P 4EE

Typeset by
Santype Limited, Coldtype Division
Salisbury, Wiltshire
and printed in Great Britain by
T. & A. Constable Ltd.
Edinburgh





**Library of Congress Cataloging in Publication Data**

Price, W. G.
Probabilistic theory of ship dynamics.

"A Halsted Press book."
Includes bibliographical references.
1. Ships–Hydrodynamics. 2. Probabilities.
3. Ocean Waves. 4. Stability of ships. I. Bishop,
R. E. D. joint author. II. Title.
VMI56.B57 623.82 74-911
ISBN 0-470-69733-4

# Preface

Even if the motions of ships in waves are studied on dry land – seen but not shared, so to speak – the analyst still has his problems. He is dealing with random processes having comparatively low dominant frequencies, he requires wave data which in the very nature of things are difficult to collect, he has to make difficult calculations which even for the most thoroughly documented ship would hardly be above suspicion.

Seakeeping calculations have to be made nevertheless. The motions are sometimes of intrinsic interest; it is not sensible to build a costly ship whose foredeck becomes awash in a force 3 breeze. Again, waves have a profound effect on hull resistance, structural dynamics and on handling characteristics. In short, the dynamics of ships in waves is one of those fields where it is necessary to make the best of a bad job and to try to make that best better and better. After all the presence of waves on the sea is the rule rather than the exception; thus waves more than 2·5 m high are observed in the North Atlantic for more than 40% of the time (to take a simple example from the wave atlas).

Of course it might be possible to deal with seakeeping and related topics entirely on the basis of rules. According to the notes on the Beaufort Wind Scale in the Manual of Seamanship, for instance, '(fishing) smacks are advised to seek shelter if possible' in a fresh gale (i.e. in a force 8 wind of 34–40 knots). But this sort of approach is on its way out, though ironically one can see that it may re-appear when criteria have been established in a scholarly manner; unfortunately that will be long after any reader of this book will have lost interest in the problems of this world.

Persuaded of the sheer necessity of mastering the dynamics of a ship in rough sea (and, by clear inference, of reading this book) the naval architect will not wish to tarry over this Preface. If however he will force himself to do so, we will inform him about the layout of material that follows.

Modern analysis of ship dynamics requires the use of a probabilistic theory. This is not, perhaps, the sort of approach that most engineers can immediately apply so we have written Part I of this book as an introduction to probability and random process theory. This is in fact a fairly extensive excursion into statistics, but to write a useful book we felt obliged to make it. If this is in fact something that the reader finds entirely superfluous, well as Sir Despard Murgatroyd would have said

'I blush for my wild extravagances,
But be so kind
To bear in mind',

to which his newly acquired wife (née Mad Margaret) adds

'We were the victims of circumstances'.

Part II deals with waves. It is an attempt to draw together relevant material from physical oceanography in a coherent fashion. Wave dynamics is, of course, the subject of a large literature and it is easy to become engulfed by it. In the process of presenting this material, therefore, we have been at pains to develop the underlying theory from first principles.

At last, in Part III we come to ships. And it is not until we get to this stage that we come to realise what a young subject probabilistic ship dynamics is. Perhaps the main conclusion to be drawn from Part III is that there is an urgent need for more research in this field. Even so we have tried to show the present scope of linear theory and briefly to introduce the concepts of non-linear.

Parenthetically we venture to suggest that the book is only a stone's throw from being an introduction to random vibration theory in general. Although the theory refers specifically to ships – and indeed to the *motions* of ships – the principles discussed are of much wider application.

That a great deal more will be done – will *have* to be done – in probabilistic ship dynamics during the next decade seems certain to us. We hope that this book will help, if only by driving others to try to do better.

Finally, we are indebted to three people for their help and comments. They are Professor D. V. Lindley and Mr. J. E. Conolly who commented on a first draft and Jenny Price who did the typing (and retyping!) with endless patience and good humour.

W.G.P.
R.E.D.B.

# Contents

# 1 Introduction

No more than a glance at the surface of the sea is needed to decide that the behaviour of a marine vehicle in a seaway is a matter for probabilistic (rather than deterministic) analysis. Accordingly the theory of seakeeping that we shall develop rests on probability theory, and the first part of this book is a review of elementary statistics.

The theory of probability is commonly introduced in terms of an 'event'. The event is a particular outcome of a hypothetical experiment or trial. If, on $N$ repetitions of the experiment, the number of times the event $E$ occurs is $r$, then the ratio $(r/N)$ denotes the 'relative frequency' of the event. For increasing values of $N$ the relative frequency fluctuates with a diminishing band of values and is said to exhibit 'statistical regularity'. The 'probability of the event $E$', $P[E]$, which corresponds to the estimate of the ultimate relative frequency is held to be

$$P[E] = \lim_{N\to\infty} \left(\frac{r}{N}\right).$$

This traditional approach is simple, intuitive, and illogical since in practice $N$ can never approach infinity.

A version of statistical regularity derived by Bernoulli in the law of large numbers is

$$\lim_{N\to\infty} P\left[\,\left|\frac{r}{N} - p\right| \geqslant \delta\right] = 0$$

where $p = P[E]$ and $\delta$ is a small quantity. The law states that for $N$ independent repetitions of a random or hypothetical experiment in which there are $r$ occurrences of an event $E$ with probability $p$, then the probability that the relative frequency $(r/N)$ differs from $p$ by more than a small quantity $\delta$, tends to zero as $N$ tends to infinity.

It is important to note that the law of large numbers does not state that the relative frequency tends to a limit as the number of repetitions becomes infinite. In practice, for a large number of repetitions a limit is reached but it is impossible to prove mathematically that the sequence of relative frequency ratios would approach this limit. However, it is possible to show that the probability of a relative frequency lying outside the limits $p + \delta$ and $p - \delta$ tends to zero as the number of repetitions of the experiment tends to infinity.

This classical approach of relative frequency was controversially proposed by von Mises[1] in 1919 as the foundation of a philosophical and rigorous mathematical treatise on probability theory. The mathematical difficulties of the limiting process were overcome by Kolmogorov[2] in 1933; in a more rigorous mathematical approach, he considered probabilities as functions which satisfy *axioms*. We prefer to have as few axioms as possible and choose those likely to yield the largest number of theorems valuable in the real world. Since it is relative frequency of events in the real world which we hope to describe and understand, the axioms selected are suggested by the elementary properties of relative frequency. There is, therefore, little conflict between the consequences of the definition of probability as the limit of an observable relative frequency on the one hand and the axiomatic approach on the other. We shall adopt the rather more modern axiomatic approach in our preliminary review.

The configuration of the surface of the sea is a random process. So too is the motion of a marine vehicle on that surface. Starting from our elementary theory of probability, therefore, we have now to consider the following topics:

1. Random process theory.
2. Measurement of the sea's surface configuration and its specification as a random process.
3. Specification of the relevant properties of a vehicle.
4. Estimation and assessment of the motion of the vehicle (as a random process).

A brief introduction to the relevant theory of random processes is given in this book and in the works of Crandall and Mark[3] and Robson[4], although the latter use probabilities based on infinite numbers of experiments and not on the 'axiomatic' approach that we shall use. This will allow us to appreciate how the remaining three topics can be dealt with *at all*. This introduction constitutes Part I of this monograph. The measurement and specification of the sea's waves as a random process is a matter of oceanography and it is dealt with elsewhere[5,6,7]; we shall merely summarise results in Part II of this book.

When we come to discuss the vehicle, we note that it is capable of (a) performing bodily motions and (b) distorting. Bodily motions are usually investigated under the assumption that distortions may be disregarded and the theory is known as that of 'seakeeping'. There are, however, few marine vehicles where this assumption is valid. The distortion of vehicles by waves is steadily becoming a more serious problem, notably with large tankers and container ships. It must be recognised, however, that both the seakeeping and the random vibration theories are still in their infancy; moreover both topics are somewhat complicated. There can therefore be no question of doing more than merely introducing them in this monograph. We shall largely confine our attention in Part III of this book to seakeeping.

## References

[1] von MISES, R., 1939 (revised 1957), *Probability, Statistics and Truth*, Macmillan, New York.

[2] KOLMOGOROV, A. A., 1954, *Foundations of the Theory of Probability*, Chelsea, New York.

[3] CRANDALL, S. H. and MARK, W. D., 1963, *Random Vibration in Mechanical Systems*, Academic Press, New York.

[4] ROBSON, J. D., 1963, *An Introduction to Random Vibrations*, E.U.P., Edinburgh.

[5] BRAHTZ, J. F. (Ed.), 1968, *Ocean Engineering*, John Wiley, New York.

[6] DEFANT, A., 1961, *Physical Oceanography*, Pergamon Press, New York.

[7] HILL, M. W. (Ed.), 1962, *The Sea*, vol. i, Interscience, New York.

# Part 1 Basic Statistics

# 2 Sets and subsets

A 'set' is a collection of quantities (the 'elements' of the set) all of which possess some relevant distinguishing feature. We shall want to refer to different degrees of completeness of sets, so we speak of

a 'universal set', $\mathscr{S}$, containing *all* the quantities of interest
a 'set', $A$, drawn from $\mathscr{S}$
a 'subset', which is a selection from the set $A$.

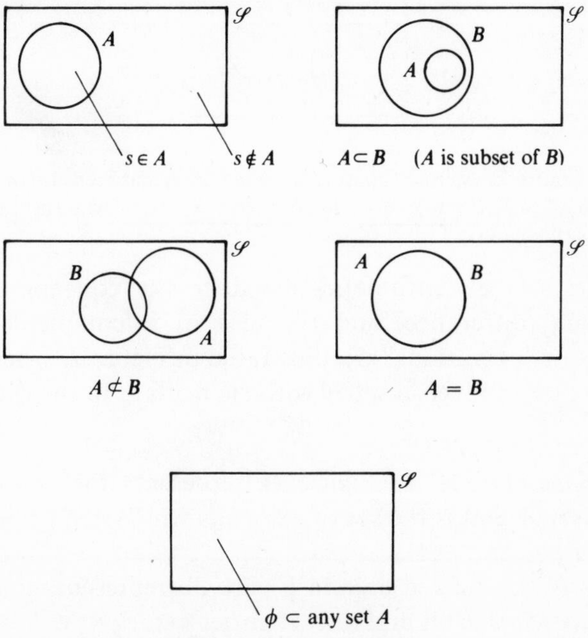


**Figure 1.** Venn diagrams.

If the area within a rectangle represents $\mathscr{S}$, then we may draw Venn diagrams as in Figure 1, where

- $\in$ expresses a relationship between an element and a set; i.e. $s \in \mathscr{S}$ means that the element $s$ 'belongs to' or 'is a member of' or 'falls within' the set $\mathscr{S}$ whereas $\lambda \notin \mathscr{S}$ means that the element $\lambda$ 'does not belong to' the set $\mathscr{S}$.
- $\subset$ expresses a relationship between sets; i.e. $A \subset B$ means the set $A$ is 'completely contained by' the set $B$.
- $\phi$ is a set of no size (i.e. containing no element) – c.f. a null matrix in matrix algebra.

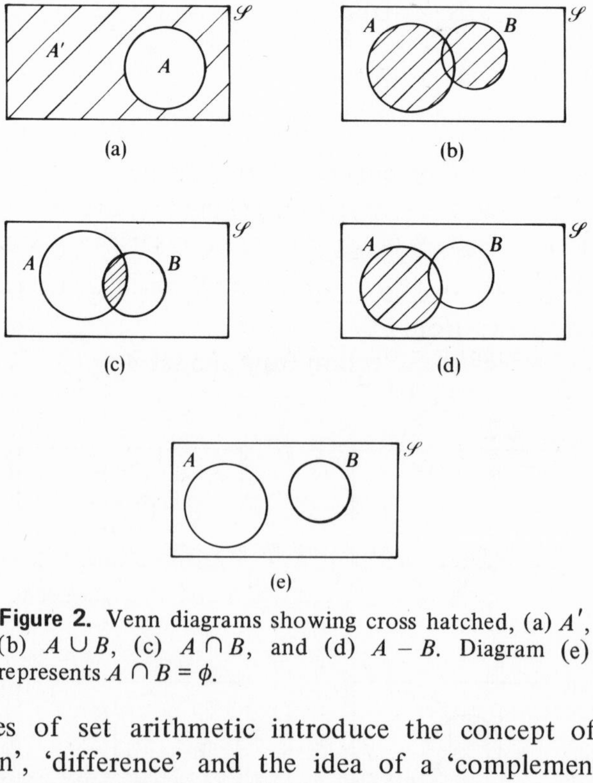


**Figure 2.** Venn diagrams showing cross hatched, (a) $A'$, (b) $A \cup B$, (c) $A \cap B$, and (d) $A - B$. Diagram (e) represents $A \cap B = \phi$.

The rules of set arithmetic introduce the concept of 'union', 'intersection', 'difference' and the idea of a 'complementary set'. These rules or 'operations' are illustrated in Figure 2, where $A$ and $B$ are any sets of a universal set $\mathscr{S}$ with elements $s$. In the order shown, we have:

(a) *complement:* $A'$ in Figure 2a represents the complement of the set $A$ and is the set of elements which do not belong to $A$; i.e. $s \notin A$ or $s \in A'$.

(b) *union:* the shaded area in Figure 2b represents union $A \cup B$; i.e. $s \in (A \cup B)$ if and only if either $s \in A$ *or* $s \in B$ *or* both.

(c) *intersection:* the shaded area in Figure 2c represents the intersection $A \cap B$; i.e. $s \in (A \cap B)$ if and only if $s \in A$ *and* $s \in B$. Notice that if $B$ is the null set, then $A \cap \phi = \phi$. Moreover, when $A$ and $B$ 'do not intersect', i.e. are 'disjoint' or 'mutually exclusive', then $A \cap B = \phi$ as shown in Figure 2e.

(d) *difference:* the shaded area in Figure 2d represents $A - B$, i.e. $s \in (A - B)$ if and only if $s \in A$ and $s \notin B$. This can also be expressed as $s \in (A \cap B')$.

The rules for three sets can be obtained by letting, for example, $A$ be two sets to which the rules have already been applied. If this is repeated enough times we have all sets in the class $\{A_1, A_2, A_3 \ldots\}$ and the operations are stated as follows:

*union:* $s \in (A_1 \cap A_2 \cup A_3 \cup \ldots)$, or $s \in \bigcup_{i=1}^{\infty} A_i$, if and only if $s \in A_i$ for *one or more* $i = 1, 2, 3 \ldots$

*intersection:* $s \in (A_1 \cap A_2 \cap A_3 \cap \ldots)$, or $s \in \bigcap_{i=1}^{\infty} A_i$, if and only if $s \in A_i$ for every $i = 1, 2, 3 \ldots$

Let us illustrate these concepts by means of a simple example. Suppose that all the cards of a pack are set out on a table, face up, and that we are going to choose one; no element of chance enters. The universal set $\mathscr{S}$ consists of the 52 cards, and we might identify two sets:

$A$ consists of all spades, $\qquad$ $B$ consists of all honours.

Then

3 of spades $\in A$, 3 of spades $\notin B$

and neither of the statements

$A \subset B$, $B \subset A$

is true. We may go on thus,

Jack of hearts $\in B$ and $\in A'$.

The union of $A$ and $B$ is such that

(all spades and honour cards of clubs, diamond, hearts) $\in (A \cup B)$.

On the other hand the intersection of $A$ and $B$ is the much smaller subset

(ace, king, queen, jack, ten of spades) $\in (A \cap B)$

while

(9, 8, 7, 6, 5, 4, 3, 2 of spades) $\in (A - B)$.

Had we chosen $B$ as the set consisting of all red honours, (leaving $A$ as before) then, $A$ and $B$ are disjoint sets so that

$$A \cap B = \phi$$

and

$$A - B = A.$$

## 2.1 Sample space

The reason for our interest in set theory is that the universal set $\mathscr{S}$ may be thought of as containing all the possible outcomes of an experiment. In this context $\mathscr{S}$ is a 'sample space' – though another word in the mathematical jargon, *viz.* 'population', is perhaps better. We can distinguish three types of sample spaces. A finite sample space contains a finite number of elements such as $\{1, 2, 3\}$ or $\{a, b, c, \ldots, x, y, z\}$. A countably infinite sample space contains a countably infinite number of elements such as the set of integers. Thirdly, an uncountable sample space contains uncountably many elements such as magnitudes in any given numerical interval.

An 'event' is now associated with a set of elements of $\mathscr{S}$, as indicated in Figure 3. Probability theory is concerned with the likelihood of an event's occurrence.

Whereas the cards in our previous illustration lay face up, let us now suppose them to form a shuffled pack so that the selection of a

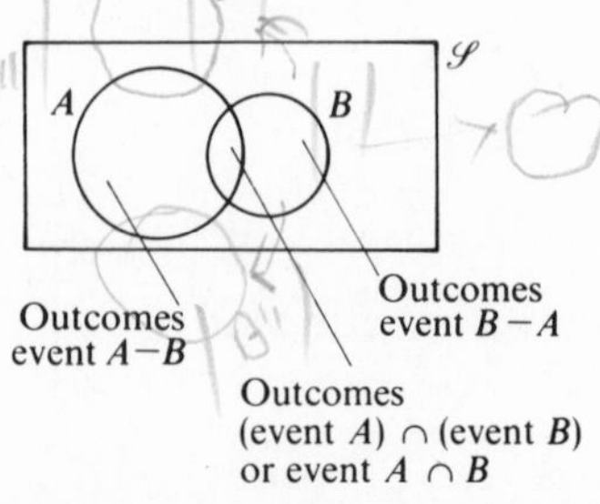

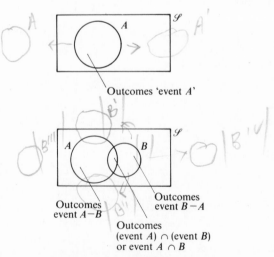

**Figure 3.** Venn diagrams employed to define 'events'.

card introduces the element of chance. The finite sample space is the universal set of 52 denominations. Using our previously defined sets $A$ (all spades) and $B$ (all honours) we draw a card from the pack. If it is:

(a) the 3 of clubs, then no 'event' occurs
(b) the 5 of spades, then the events $A$ and $A - B$ occur
(c) the queen of spades, then events $A$, $B$ and $A \cap B$ occur
(d) the king of diamonds, then events $B$ and $B - A$ occur.

If the location of some point on the sea-bed is fixed *exactly*, we may enquire what the depth of the water is at some specified instant at that point. The depth is determined by the locality, by tides and by irregular waves; and the waves introduce the element of chance. The depth $s$ at that point, and at that instant, is an element in the uncountably infinite universal set of outcomes $\mathscr{S}$ from 0 to infinity in any preselected system of units. Notice that $s$ is most unlikely to be 0 (unless the selected point is at the water's edge) and even more unlikely to be $\infty$; it will have a value that fluctuates, possibly with departures normally less than 10 m from some mean value. Notice, too, that some difficulty attends such a statement as '$s = 25$ m', since a question of accuracy arises – does this mean $s = 25 \text{ m} \pm 0$? The essential difference between a countably infinite and an uncountably infinite sample space is that in the former the possible outcomes are discrete whereas in the latter they form a continuous range.

## 2.2 Probability

The whole of the sample space may be divided up into events $A_1, A_2, \ldots, A_n$ (Figures 4a and 4b). To each event we identify a number $P[A_k]$ which is a measure of the probability of its occurrence. This number is arrived at by hunch or in any other convenient way.

The outcome 'heads' when a coin is tossed can be labelled 'event $A$'. We do not attempt to *prove* that

$$P[A] = \tfrac{1}{2}.$$

It merely seems reasonable.

Following up this idea of 'seeming reasonable' we observe that the probability of a certain event is unity and that of an impossible event is zero. But on the other hand an event with probability 1 is not necessarily certain; nor is an event with probability zero impossible.

The probabilities must comply with certain simple rules:

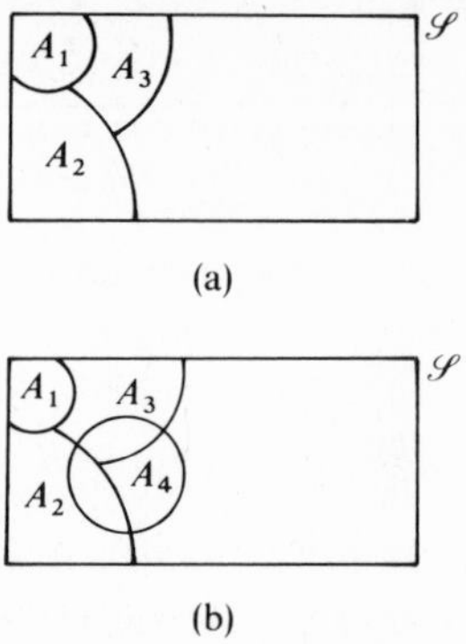


**Figure 4.** Representation in a Venn diagram of (a) mutually exclusive events, and (b) intersecting events.

1) $P[A_k] \geqslant 0$

   for $k = 1, 2, \ldots, n$.

2) $P[\mathscr{S}] = 1$.

3) $P\left[\bigcup_{k=1}^{n} A_k\right] = \sum_{k=1}^{n} P[A_k]$

   if the $A_k$ do *not* intersect (i.e. are mutually exclusive).

At first sight these rules seem insufficient since the $A_k$ may intersect (as in Figure 4b). But if they do,

$$
\begin{aligned}
P[A_i \cup A_j] &= P[(A_i \cap A_j') \cup A_j] \\
&= P[A_i \cap A_j'] + P[A_j]
\end{aligned}
$$

as clarified by Figure 5. But

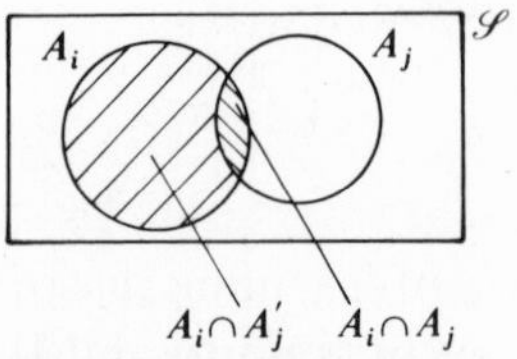


**Figure 5.** Venn diagrams illustrating intersection of the events $A_i$ and $A_j$.

$$A_i = (A_i \cap A_j') \cup (A_i \cap A_j)$$

whence

$$P[A_i] = P[A_i \cap A_j'] + P[A_i \cap A_j]$$

so that we *deduce* the (very reasonable) requirement:

$$P[A_i \cup A_j] = P[A_i] + P[A_j] - P[A_i \cap A_j].$$

### 2.2.1 Conditional probability and independence of events

For the sake of definiteness, consider again the selection of a card from a shuffled pack with

event $A$ = card selected is a spade
event $B$ = card selected is an honour card.

Suppose that a card is drawn and that it is known to be an honour card; an assistant assures us that, say, event $B$ has occurred. We now ask what the probability is that event $A$, too, has occurred. This is an example of a 'conditional probability' and it is written $P[A|B]$.

In this simple example we might argue as follows. As event $B$ has occurred the card may be any one of 20 since there are 5 honour cards in each suit (A, K, Q, J, 10). If event $A$ is to have taken place, the card must be one of 5 of those 20. That is

$$P[A|B] = \frac{5}{20} = \frac{1}{4}.$$

Before we attempt to place this reasoning on a more formal footing, let us first note that conditional probabilities are by no means as unusual as they may appear. Indeed, philosophically there is no such thing as a probability – only a conditional probability, since before we can arrive at a logical measure of probability we have to define a sample space, and that too is subject to some uncertainty. (A coin is not sufficiently biassed, or cards are all sufficiently alike not to upset preconceived ideas on measures of probability . . .). However we shall not follow up this rather highly theoretical line of reasoning.

If $A_i$ and $A_j$ are two events defined in the sample space $\mathscr{S}$ described in the previous section, then the conditional probability of event $A_i$ given $A_j$ is written as $P[A_i|A_j]$. The probability axioms may be written as conditional probability axioms as follows:

1) $P[A_k|A_k] = 1$

for all $k = 1, 2, \ldots, n$.

2) $P\left[\bigcup_{k=1}^{n} A_k | B\right] = \sum_{k=1}^{n} P[A_k | B]$

if, given event $B$, the events $A_k$ are mutually exclusive.

3) $P[A_i | A_j] = \dfrac{P[A_i \cap A_j]}{P[A_j]}$ for $P[A_j] > 0$.

To revert to the card drawing experiment, we note that

$$P[B] = \frac{20}{52},$$

$$P[A \cap B] = \frac{5}{52}.$$

Hence we arrive at our previous result.

We see from the last axiom that if the event $A_j = \mathscr{S}$ then

$$A_i \cap \mathscr{S} = A_i$$

and

$$P[A_i | \mathscr{S}] = \frac{P[A_i \cap \mathscr{S}]}{P[\mathscr{S}]} = P[A_i].$$

Further if, alternatively, $A_i = \mathscr{S}$ then

$$P[\mathscr{S} | A_j] = \frac{P[\mathscr{S} \cap A_j]}{P[A_j]} = \frac{P[A_j]}{P[A_j]} = 1.$$

Again, since $P[A_i \cap A_j] \geqslant 0$ and $P[A_j] > 0$, the conditional probability has a lower bound of zero and an upper bound of unity, i.e.

$$0 \leqslant P[A_i | A_j] \leqslant 1.$$

By interchanging the roles of $A_i$ and $A_j$ it may further be shown that

$$P[A_i \cap A_j] = P[A_i | A_j] P[A_j] = P[A_j | A_i] P[A_i].$$

This last result is sometimes called the 'theorem of compound probabilities'.

The probability of the simultaneous occurrence of three events $A_h$, $A_i$, $A_j$ is obtained by repeated application of the third conditional probability axiom; that is

$$\begin{aligned} P[A_h \cap A_i \cap A_j] &= P[A_h | A_i \cap A_j] P[A_i \cap A_j] \\ &= P[A_h | A_i \cap A_j] P[A_i | A_j] P[A_j] \end{aligned}$$

provided that $P[A_i \cap A_j] > 0$ and $P[A_j] > 0$. In the first application, the simultaneous occurrence of $A_i$ and $A_j$ is considered as a single event, and in the second the event $(A_i \cap A_j)$ is expressed as a product.

To revert once more to the experiment of drawing a card, let event $C$ be *not* a picture card. The event $(A \cap B \cap C)$ is thus that the card drawn is the ace or 10 of spades, i.e. $P[A \cap B \cap C) = 2/52$. The theory shows that

$$P[A \cap B \cap C] = P[A|B \cap C]P[B|C]P[C]$$

$$= \frac{2}{8} \times \frac{8}{40} \times \frac{40}{52}$$

as the reader may verify.

Suppose that the conditional probability $P[A_i|A_j]$ is equal to the probability of the occurrence of the event $A_i$, i.e.

$$P[A_i|A_j] = P[A_i].$$

It follows that

$$P[A_i \cap A_j] = P[A_i]P[A_j]$$

and

$$P[A_j|A_i] = P[A_j].$$

Thus the conditional probability of the event $A_j$, knowing $A_i$ has occurred, is the probability of the occurrence of $A_j$. Knowing the event $A_i$ has occurred has not altered the probability of $A_j$ occurring. If this relationship exists between the events, then $A_j$ is 'statistically independent' of $A_i$.

The card drawing experiment is such that

$$P[A] = \frac{1}{4}; \qquad P[B] = \frac{5}{13}; \qquad P[C] = \frac{10}{13}$$

whereas

$$P[A|B] = \frac{1}{4}; \qquad P[B|A] = \frac{5}{13};$$

$$P[B|C] = \frac{1}{5}; \qquad P[C|B] = \frac{2}{5};$$

$$P[C|A] = \frac{10}{13}; \qquad P[A|C] = \frac{1}{4}.$$

We see that event $A$ is statistically independent of events $B$ and $C$, but that events $B$ and $C$ are not statistically independent of each other. The reason for this is not as straightforward as one would wish.

Note that independence does not imply that $P[A_i \cap A_j] = 0$; nor does it imply that the events $A_i$ and $A_j$ do not intersect, i.e. that $A_i \cap A_j = \phi$. These relationships would imply that $P[A_i|A_j] = 0$ and that is not what independence means.

The concept of statistical independence may be taken a step further. The set of events $A_k (k = 1, 2, \ldots, n)$ are statistically independent if all the $(n - 1)$ conditions

$$P[A_1 \cap A_2] = P[A_1]P[A_2]$$

$$P[A_1 \cap A_2 \cap A_3] = P[A_1]P[A_2]P[A_3]$$

$$\cdots\cdots\cdots\cdots\cdots\cdots\cdots\cdots\cdots\cdots$$

$$P[\bigcap_{k=1}^{n} A_k] = \prod_{k=1}^{n} P[A_k]$$

are satisfied.

However, if

$$P[A_1 \cap A_2, \ldots \cap A_n] = P[A_1]P[A_2]P[A_3] \ldots P[A_n]$$

it does not mean that all the events $A_k$ $(1 \leqslant k \leqslant n)$ are independent. This is only true if all the $(n - 1)$ conditions listed are fulfilled.

For the card drawing experiment $(A \cap B \cap C)$ is the event ace or 10 of spades so $P[A \cap B \cap C] = 2/52$; but this is not given by the product $P[A]P[B]P[C]$ so the set of events are *not* independent.

### 2.2.2 Bayes' Formula

The sample space $\mathcal{S}$ shown in Figure 4a consists of the mutually exclusive events $A_1, A_2, \ldots, A_n$ such that

$$\begin{aligned} &A_i \subset \mathcal{S}, && i = 1, 2, \ldots, n; \\ &P[A_i] \geqslant 0, && i = 1, 2, \ldots, n; \\ &A_i \cap A_j = \phi, && i \neq j = 1, 2, \ldots, n; \\ &A_1 \cup A_2 \cup \ldots \cup A_n = \mathcal{S}. \end{aligned}$$

When these conditions are satisfied $\{A_1, A_2, \ldots, A_n\}$ forms a 'partition' of the sample space $\mathcal{S}$.

Let $A$ be any event in $\mathcal{S}$. It follows that

$$\begin{aligned} A \cap \mathcal{S} = A = A \cap (A_1 \cup A_2 \cup \ldots \cup A_n) \\ = (A \cap A_1) \cup (A \cap A_2) \cup \ldots \cup (A \cap A_n) \end{aligned}$$

which expresses the event $A$ as the union of $n$ mutually exclusive events. By applying the probability and conditional probability axioms we find that

$$P[A] = \sum_{i=1}^{n} P[A \cap A_i] = \sum_{i=1}^{n} P[A|A_i]P[A_i].$$

Also the conditional probability of an event $A_k$ $(1 \leqslant k \leqslant n)$ given event $A$ is

$$P[A_k|A] = \frac{P[A_k \cap A]}{P[A]} = \frac{P[A_k]P[A|A_k]}{\sum_{i=1}^{n} P[A_i]P[A|A_i]}.$$

The last result is known as Bayes' Formula and is obtained by repeated use of the conditional probability axioms.

# 3 Random variable

Consider a sample space $\mathscr{S}$ containing all possible outcomes $s$ of an experiment. It is frequently necessary to give a numerical evaluation of the outcomes. This means that a rule must be framed for assigning *numbers* to the outcomes $s$, thereby specifying a function $X(s)$. This function is a 'random variable' and it may take real values such that

$$-\infty < X(s) < \infty.$$

There is a special case in which the experiment is in the nature of a measurement so that $\mathscr{S}$ contains all real numbers over some range. It is then perfectly admissible to take $X(s) = s$ where $s(\in\mathscr{S})$ is already a continuous variable.

If the $s$ form the elements of $\mathscr{S}$ it is clear that the continuous random variable $X(s)$ may be thought of as an element in an uncountable space. This new space is infinitely large since $X(s)$ may take any real value. An event $A$ in this new sample space may be represented by any segment of a line extending from $-\infty$ to $+\infty$. The event may be represented by a finite range of $X$ (c.f. Figure 6a) or a semi-infinite range (c.f. Figure 6b). If all possible outcomes are to fall within the event $A$ then the whole line from $-\infty$ to $+\infty$ represents $A$.

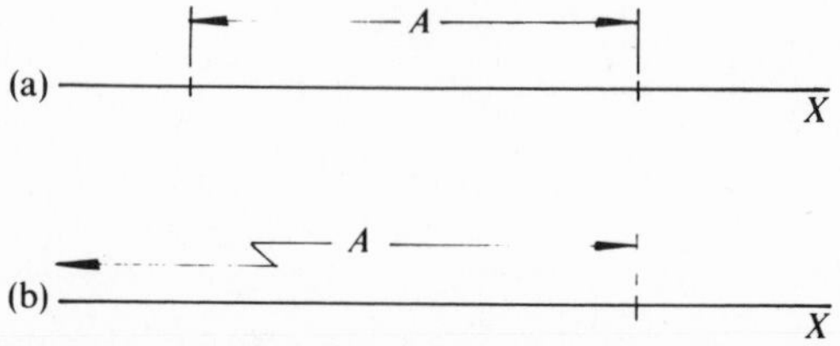


**Figure 6.** An event $A$ with (a) a finite range, (b) a semi-infinite range.

## 3.1 Probability distribution function

The probability distribution function $F_X(x)$ or $F(x)$ of a random variable $X(s)$ is defined for any real number $x$ as the probability of an event

$$P[s : X(s) \leqslant x] = F(x).$$

In words, this means that the random variable $X$, with sample space $s \in \mathscr{S}$ as domain, is a function such that $X(s) \leqslant x$ and the probability of the occurrence of this inequality being satisfied is called the probability distribution function.

If we consider the range of values of the random variable $X$ to be the real line $(-\infty, \infty)$, it follows by the probability axioms that the distribution function has the limiting values

$$\lim_{x \to \infty} F(x) = F(\infty) = P[s : X(s) \leqslant \infty] = 1$$

$$\lim_{x \to -\infty} F(x) = F(-\infty) = P[s : X(s) \leqslant -\infty] = 0.$$

Also, if $x_1$ and $x_2$ are two real numbers such that $x_1 \leqslant x_2$, it follows that $\{s : X(s) \leqslant x_1\} \subset \{s : X(s) \leqslant x_2\}$ and, moreover,

$$F(x_1) = P[x : X(s) \leqslant x_1] \leqslant P[s : X(s) \leqslant x_2] = F(x_2)$$

so that

$$F(x_1) \leqslant F(x_2).$$

Again,

$$P[s : x_1 < X(s) \leqslant x_2] = F(x_2) - F(x_1).$$

Thus the distribution function $F(x)$ or $F_X(x)$ is a non-decreasing monotonic function having values lying in the range $0 \leqslant F(x) \leqslant 1$. Two such curves are shown in Figure 7.

### 3.1.1 Probability density function

A continuous random variable is defined in an uncountable sample space, and the probability that it will take some discrete specific value is generally zero. For this reason, to describe a continuous random variable it is necessary to create a probability density function.

Consider an event in which the continuous random variable $X(s)$ falls in a limited range of values $(x, x + \delta x)$. According to the rules that probabilities must satisfy,

$$P[s : x < X(s) \leqslant x + \delta x] = F(x + \delta x) - F(x) = \delta F(x).$$

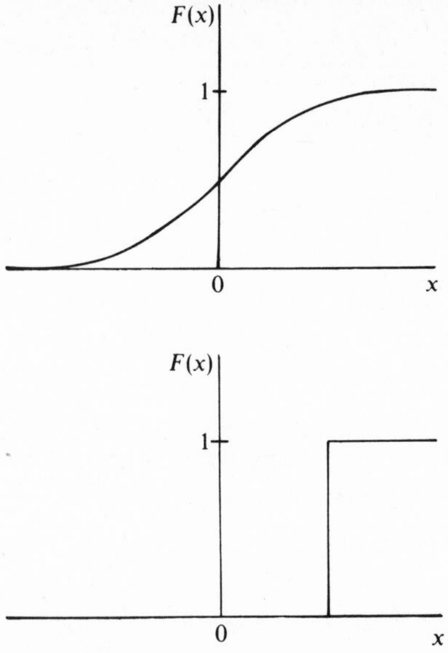


**Figure 7.** Possible forms of a probability distribution function for a single variable.

Assuming that the derivative of the probability distribution function exists, the probability density function $f(x)$ or $f_X(x)$ of the random variable $X(s)$ is defined by

$$f(x) = \lim_{\delta x \to 0} \frac{F(x + \delta x) - F(x)}{\delta x} = \frac{dF(x)}{dx}.$$

Thus the probability that the random variable $X(s)$ lies in the infinitesimal interval $(x, x + \delta x)$ is

$$P[s : x < X(s) \leqslant x + \delta x] = f(x)\, \delta x$$

and is represented by the area of the rectangle of width $\delta x$ under the curve of $f(x)$ at $x$, as shown in Figure 8a. The function $f_X(x)$ or $f(x)$ is the probability density function of the random variable $X(s)$ and, like $F(x)$, it must satisfy certain conditions. In particular:

a) Since the probability distribution function is non-decreasing, a probability density function is non-negative; i.e. $f(x) \geqslant 0$.
b) The total area beneath the probability density function curve is given by

$$\int_{-\infty}^{\infty} f(x) dx = \int_{-\infty}^{\infty} \frac{dF(x)}{dx} dx = F(\infty) - F(-\infty) = 1$$

whereas the area beneath the probability density function curve in an arbitrary range $(a, b)$ is given by

$$\int_a^b f(x)dx = \int_a^b \frac{dF(x)}{dx}\, dx = F(b) - F(a) = P[s : a < X(s) \leqslant b]\,.$$

This denotes the probability of the random variable $X(s)$ falling within the range $(a, b)$. If we let the range have limits $a = -\infty$ and $b = x$, then

$$\int_{-\infty}^{x} f(x')dx' = F(x) - F(-\infty) = F(x)\,.$$

Figures 8a and 8b show valid probability density functions. In general, the probabilities $P[s : X(s) \leqslant x]$, $P[s : x < X(s) \leqslant x + \delta x]$, etc., are usually abbreviated to $P[X \leqslant x]$, $P[x < X \leqslant x + \delta x]$, etc.

## 3.2 Two random variables

Some experiments have pairs of outcomes – suits and values, heave and pitch, . . . – so that the sample space contains pairs of elements. We shall assume that the random variables $X$ and $Y$ are measured simultaneously and defined for all pairs of real numbers $(x, y)$

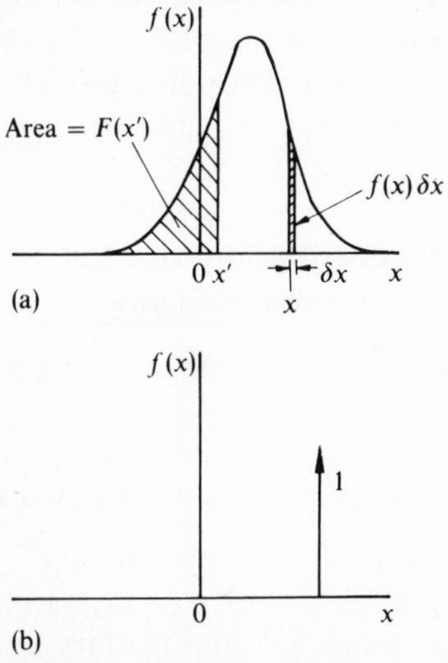


**Figure 8.** Possible forms of a probability density function for a single variable. The areas beneath the curve in diagram (a) represent probabilities.

respectively. The events $A(x)$ and $B(y)$ are defined on the same sample space $\mathcal{S}$, $s \in \mathcal{S}$, to be

$$A(x) = [s : X(s) \leqslant x],$$
$$B(y) = [s : Y(s) \leqslant y].$$

The random variables may be defined for the member pair $(x, y)$, such that they lie in the range

$$-\infty < X(s) < \infty,$$
$$-\infty < Y(s) < \infty.$$

Suppose that we wish to discuss (a) the vertical displacement of the centre of mass of a ship in rough sea from the horizontal plane in which it would lie in a flat calm (we may refer to this roughly as a displacement of 'heave', though this is not strictly correct as that term is associated with body axes and not inertial axes) and (b) the tangent of the angle that the longitudinal axis in the water plane section makes with the horizontal (we shall refer to this quantity, again loosely, as 'pitch'). These two quantities may be denoted by $X(s)$ and $Y(s)$ respectively. The outcomes $s$ are no longer simple measurements – though $X(s)$ and $Y(s)$ are – and they can only easily be thought of in physical terms as pairs of outcomes $X(s)$ and $Y(s)$. That is to say, $\mathcal{S}$ is the sample space containing all the uncountably infinite combinations of heave and pitch.

We may now proceed to define joint probability distribution and density functions. These serve to relate pairs of random variables of this sort.

### 3.2.1 Joint probability distribution function

The joint probability distribution function

$$P[s : X(s) \leqslant x \cap Y(s) \leqslant y] = F_{XY}(x, y),$$

or

$$P[A(x) \cap B(y)] = F_{XY}(x, y),$$

denotes the probability of the events $A(x)$ and $B(y)$ occurring together. This function $F(x, y)$ or $F_{XY}(x, y)$ may be represented by a surface of which a somewhat rudimentary example is sketched in Figure 9. A more general shape would have plateaux at the values 0 and 1 with a less abrupt cliff from one to the other.

If the ranges of values of the random variables $X(s)$ and $Y(s)$ extend from $-\infty$ to $\infty$ it follows from the probability axioms that

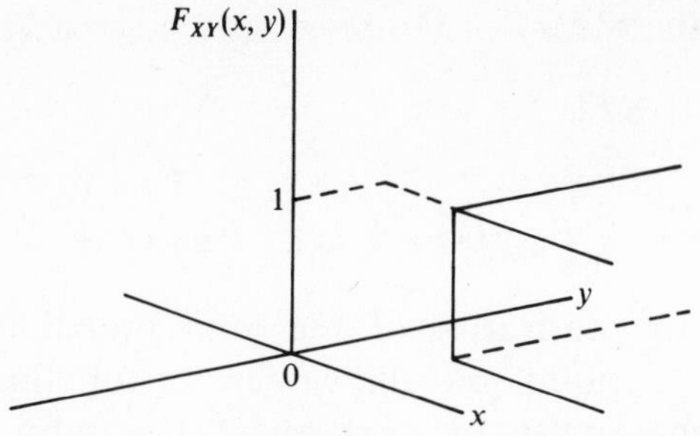


**Figure 9.** A possible form of a probability distribution function for two variables.

a) since $A(\infty) = \{s : X(s) \leqslant \infty\} = \mathscr{S}$ and $A(-\infty) = \{s : X(s) \leqslant -\infty\} = \phi$, etc., it follows that

$$F_{XY}(\infty, \infty) = P[A(\infty) \cap B(\infty)] = P[\mathscr{S}] = 1,$$
$$F_{XY}(-\infty, -\infty) = P[A(-\infty) \cap B(-\infty)] = P[\phi] = 0,$$
$$F_{XY}(\infty, y) = P[B(y)] = F_Y(y),$$
$$F_{XY}(-\infty, y) = P[\phi] = 0.$$

b) The joint probability distribution function $F_{XY}(x, y)$ increases monotonically in both variables. For arbitrary $h \geqslant 0$ and $k \geqslant 0$, we may express the probability

$$P[X \leqslant x + h \cap Y \leqslant y]$$
$$= P[x < X \leqslant x + h \cap Y \leqslant y] + P[X \leqslant x \cap Y \leqslant y]$$

or

$$P[x < X \leqslant x + h \cap Y \leqslant y] = F_{XY}(x + h, y) - F_{XY}(x, y) \geqslant 0$$

from the probability axioms defined in section 2.2. Similarly

$$P[X \leqslant x \cap y < Y \leqslant y + k] = F_{XY}(x, y + k) - F_{XY}(x, y) \geqslant 0$$

and by combining these results we find that

$$P[x < X \leqslant x + h \cap y < Y \leqslant y + k] = F_{XY}(x + h, y + k) - F_{XY}(x + h, y) - F_{XY}(x, y + k) + F_{XY}(x, y) \geqslant 0.$$

### 3.2.2 Joint probability density function

The 'events', $A(x)$ and $B(y)$ say, referred to in the definition of $F$ are semi-infinite ranges of the random variables $X(s)$ and $Y(s)$. They may equally well represent limited ranges and it is in this form that we shall now regard them. Thus let the events be

$$A(x) = [s : x < X(s) \leqslant x + \delta x],$$
$$B(x) = [s : y < Y(s) \leqslant y + \delta y].$$

It follows from the theory of the previous section that

$$\begin{aligned} P[A(x) \cap B(y)] &= P[s : x < X \leqslant x + \delta x \cap y < Y \leqslant y + \delta y] \\ &= F_{XY}(x + \delta x, y + \delta y) - F_{XY}(x + \delta x, y) \\ &\quad - F_{XY}(x, y + \delta y) + F_{XY}(x, y) \geqslant 0. \end{aligned}$$

Assuming that the derivative of the joint probability distribution function exists, the joint probability density function $f_{XY}(x, y)$, or $f(x, y)$, of the random variables $X(s)$ and $Y(s)$ is defined by

$$f_{XY}(x, y) = \lim_{\substack{\delta x \to 0 \\ \delta y \to 0}} \frac{F_{XY}(x + \delta x, y + \delta y) - F_{XY}(x + \delta x, y) - F_{XY}(x, y + \delta y) + F_{XY}(x, y)}{\delta x \, \delta y}$$

$$= \frac{\partial^2 F_{XY}(x, y)}{\partial x \, \partial y}$$

by the theory of differentiation.

Thus the probability that the random variable $X(s)$ lies in the range $(x, x + \delta x)$ and simultaneously $Y(s)$ lies in the range $(y, y + \delta y)$, is given by

$$P[s : x < X \leqslant x + \delta x \cap y < Y \leqslant y] = f_{XY}(x, y)\delta x \delta y \geqslant 0.$$

For a *single* random variable $X(s)$, the probability density function at a point is proportional to the probability that the random variable $X(s)$ lies within a small interval containing the point. The area under the curve corresponding to the interval, represents the probability of $X(s)$ falling within the interval. Similarly, $f_{XY}(x, y)$ is proportional to the probability that the random variable $X(s)$ lies within a small interval containing $x$, and simultaneously $Y(s)$ lies within a small interval containing $y$. The volume above the area of the rectangle carved out by these intervals represents the probability of $X(s)$ and $Y(s)$ falling simultaneously within the rectangle.

The total volume enclosed by the surface representing the joint probability density function is

$$\int_{-\infty}^{\infty}\int_{-\infty}^{\infty} f_{XY}(x, y) dx\, dy = P[s : -\infty < X(s) \leqslant \infty \cap -\infty < Y(s) \leqslant \infty]$$

$$= 1$$

whereas

$$\int_{-\infty}^{\infty} f_{XY}(x, y)dx = \int_{-\infty}^{\infty} \frac{\partial^2 F_{XY}(x, y)}{\partial x \partial y} dx$$

$$= \frac{d}{dy} [F_{XY}(\infty, y) - F_{XY}(-\infty, y)]$$

which, as the results of section 3.2.1 show, reduces to

$$\frac{dF_Y(y)}{dy} = f_Y(y).$$

A rudimentary density function of this sort is shown in Figure 10. More generally, the function is represented by a 'hillock' rather than a concentrated peak. But, whatever its form, $f_{XY}(x, y)$ is non-negative and the surface representing it has unit volume beneath it.

### 3.2.3 Independence of two random variables

Our definition of $f_{XY}(x, y)$ for the events $A(x)$ and $B(y)$ defined in their limited ranges may be expressed in the form

$$P[A(x) \cap B(y)] = f_{XY}(x, y)\delta x \, \delta y.$$

By analogy with our previous argument concerning conditional probability, we may reasonably write

$$P[A(x)|B(y)] = \frac{P[A(x) \cap B(y)]}{P[B(y)]} = \frac{f_{XY}(x, y)\delta x \, \delta y}{f_Y(y)\delta y}$$

for $P[B(y)] > 0$.

In the limiting form, as $\delta x, \delta y \to 0$, the probability of the continuous random variable $X(s)$ being in the infinitesimal interval

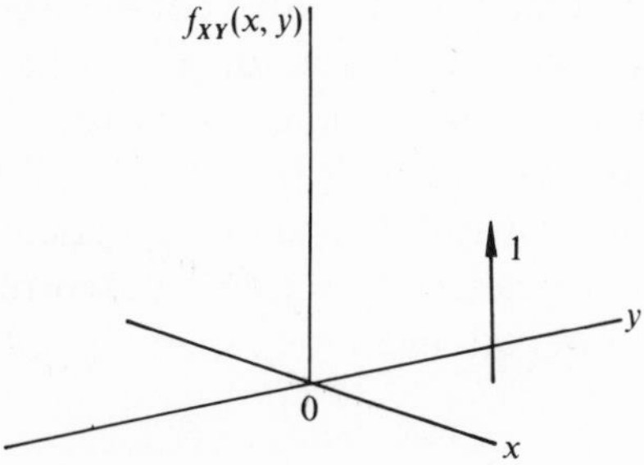


**Figure 10.** A possible form of a probability density function for two variables.

$(x, x + \delta x)$ is conditional on $Y(s)$ being in the infinitesimal interval $(y, y + \delta y)$. On this basis we define a conditional probability density function $f(x|y)$ such that this probability is $f(x|y)\delta x$; hence

$$f(x|y) = \frac{f_{XY}(x, y)}{f_Y(y)}$$

for $f_Y(y) > 0$, and it follows that

$$f_X(x) = \int_{-\infty}^{\infty} f_{XY}(x, y)dy = \int_{-\infty}^{\infty} f(x|y)f_Y(y)dy.$$

If $A(x)$ and $B(y)$ are independent events

$$P[A(x)|B(y)] = P[A(x)] = f_X(x)\delta x$$

which gives the result

$$f_{XY}(x, y) = f_X(x) f_Y(y).$$

## 3.3 Transformation of probability density function

Suppose that we are given a random variable $X$, with a probability density function $f_X(x)$, and that a new random variable $Y$ is defined which is related to $X$ in some way

$$Y = r(X),$$

such that the inverse relationship (or 'transformation') is

$$X = t(Y).$$

It is assumed that the random variables have a one-to-one correspondence to each other, so that, for each value of $X$ there exists only one value of $Y$. For example, we admit $Y = aX$ or $Y = a + bX$ but not $Y = X^2$ since two different values of $X$ can then produce the same value of $Y$. We need to determine the probability density function $f_Y(y)$ of the new random variable $Y$.

As a consequence of the one-to-one correspondence, every event of $Y$ occurring in the range $(y, y + \delta y)$ is dependent on the event $X$ occurring in the interval $(x, x + \delta x)$. In probabilities, this may be stated as

$$P[y < Y \leqslant y + \delta y] = P[x < X \leqslant x + \delta x],$$

which in terms of probability density functions gives

$$f_X(x)\delta x = f_Y(y)\delta y.$$

It follows that $f_Y(y) = f_X(x)/(\delta y/\delta x)$ and, in the limit as $\delta x$ tends to zero this becomes

$$f_Y(y) = \frac{f_X(x)}{\left|\dfrac{dy}{dx}\right|}.$$

Since the probability density functions $f_X(x)$, $f_Y(y)$ are always positive, we must take care to avoid the negative sign of $(dy/dx)$ if it arises. Thus the probability density function of the random variable $Y$ is given by

$$f_Y(y) = \left.\frac{f_X(x)}{\left|\dfrac{dy}{dx}\right|}\right|_{x = t(y)}.$$

We may extend this result to the transformation of density functions with several random variables, each having a one-to-one correspondence. Suppose, for example, that we have a joint probability density function $f_{X_1X_2}(x_1, x_2)$ and that we wish to find the joint probability density function $f_{Y_1Y_2}(y_1, y_2)$ that this would imply under the transformations

$$y_1 = y_1(x_1, x_2),$$
$$y_2 = y_2(x_1, x_2).$$

Evidently we now require that

$$f_{Y_1Y_2}(y_1, y_2)\delta y_1 \delta y_2 = f_{X_1X_2}(x_1, x_2)\delta x_1 \delta x_2$$

and we must relate $\delta y_1 \delta y_2$ to $\delta x_1 \delta x_2$.

There is a one-to-one correspondence of points under the surfaces $f_{X_1X_2}(x_1, x_2)$ and $f_{Y_1Y_2}(y_1, y_2)$, as indicated in Figure 11 which

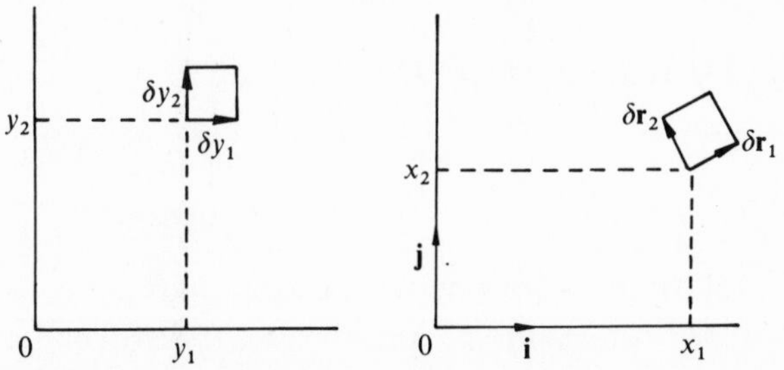


**Figure 11.** Transformation of elemental areas when viewing two two-variable probability density functions from above.

is to be imagined as showing the two humped surfaces from above. When an element $\delta y_1 \delta y_2$ is defined under $f_{Y_1 Y_2}(y_1, y_2)$ we can imagine an element $\delta \mathbf{r}_1 \delta \mathbf{r}_2$ determined thereby under $f_{X_1 X_2}(x_1, x_2)$. To the first order, if **i** and **j** are unit vectors along the axes $0x_1$, $0x_2$, we have

$$\delta \mathbf{r}_1 = \left(\frac{\partial x_1}{\partial y_1} \delta y_1\right) \mathbf{i} + \left(\frac{\partial x_2}{\partial y_1} \delta y_1\right) \mathbf{j}$$

$$\delta \mathbf{r}_2 = \left(\frac{\partial x_1}{\partial y_2} \delta y_2\right) \mathbf{i} + \left(\frac{\partial x_2}{\partial y_2} \delta y_2\right) \mathbf{j}$$

because $\delta \mathbf{r}_1$ corresponds to $\delta y_1$ and $\delta \mathbf{r}_2$ to $\delta y_2$.

The area between the small vectors $\delta \mathbf{r}_1$ and $\delta \mathbf{r}_2$ is given by $\delta \mathbf{r}_1 \times \delta \mathbf{r}_2$. It is found to be

$$\left[\frac{\partial x_1}{\partial y_1}\frac{\partial x_2}{\partial y_2} - \frac{\partial x_1}{\partial y_2}\frac{\partial x_2}{\partial y_1}\right] \delta y_1 \delta y_2 \quad \text{or} \quad |J| \delta y_1 \, \delta y_2$$

where the Jacobian determinant is defined as

$$|J| = \begin{vmatrix} \dfrac{\partial x_1}{\partial y_1} & \dfrac{\partial x_1}{\partial y_2} \\[2ex] \dfrac{\partial x_2}{\partial y_1} & \dfrac{\partial x_2}{\partial y_2} \end{vmatrix}.$$

It follows that

$$f_{Y_1 Y_2}(y_1, y_2) \, \delta y_1 \, \delta y_2 = f_{X_1 X_2}(x_1, x_2) \, \|J\| \, \delta y_1 \, \delta y_2$$

where the modulus $\|J\|$ of $|J|$ is used to ensure that the joint probability density function is positive. It is to be remembered, of course, that $\|J\|$ must be evaluated when $X_1 = x_1(y_1, y_2)$ and $X_2 = x_2(y_1, y_2)$. Thus we have, finally,

$$f_{Y_1 Y_2}(y_1, y_2) = f_{X_1 X_2}(x_1, x_2) \, \|J\| \, \Bigg|_{\substack{x_1 = x_1(y_1, y_2) \\ x_2 = x_2(y_1, y_2)}}.$$

Suppose that there is no one-to-one correspondence between the random variables. For example, consider the transformation

$$Y = X^2 \qquad -\infty < x < \infty$$

where for each event $Y$ occurring in the interval $(y, y + \delta y)$, the event $X$ occurs either in the interval $(x, x+\delta x)$ *or* in $(-x, -(x+\delta x))$. Since the two events of $X$ are mutually exclusive, by the probability axioms defined in section 2.2, it follows that

$$P[y < Y \leqslant y+\delta y] = P[x < X \leqslant x+\delta x] + P[-x < X \leqslant -(x+\delta x)]$$

and

$$f_Y(y)\delta y = \{f_X(x) + f_X(-x)\}\,\delta x.$$

The probability density function of the random variable $Y$ reduces to

$$f_Y(y) = \left.\frac{f_X(x) + f_X(-x)}{\left|\dfrac{dy}{dx}\right|}\right|_{x=\sqrt{y}}$$

That is,

$$f_Y(y) = \frac{f_X(\sqrt{y}) + f_X(-\sqrt{y})}{2\sqrt{y}} \qquad \text{for } y > 0$$

$$= 0 \qquad \text{for } y < 0.$$

Suppose that a new random variable is obtained from the sum of known random variables. Consider, for example, the transformation,

$$Y = X_1 + X_2$$

Suppose that the value of $X_2$ is known to fall in the range $(x_2, x_2+\delta x_2)$. Then

$$P[y < Y \leqslant y+\delta y | x_2 < X_2 \leqslant x_2+\delta x_2]$$
$$= P[x_1 < X_1 \leqslant x_1+\delta x_1 | x_2 < X_2 \leqslant x_2+\delta x_2]$$

where $\delta x_1$ and $\delta y$ are related. Now

$$P[y < Y \leqslant y+\delta y | x_2 < X_2 \leqslant x_2+\delta x_2]$$
$$= \frac{P[y < Y \leqslant y+\delta y \cap x_2 < X_2 \leqslant x_2+\delta x_2]}{P[x_2 < X_2 \leqslant x_2+\delta x_2]}$$
$$= \frac{f_{YX_2}(y, x_2)\delta y \delta x_2}{f_{X_2}(x_2)\delta x_2}.$$

By the same token

$$P[x_1 < X_1 \leqslant x_1+\delta x_1 | x_2 < X_2 \leqslant x_2+\delta x_2]$$
$$= \frac{f_{X_1X_2}(x_1, x_2)\delta x_1 \delta x_2}{f_{X_2}(x_2)\delta x_2}.$$

It follows that

$$f_{YX_2}(y, x_2) = \frac{f_{X_1X_2}(x_1, x_2)}{\left|\dfrac{dy}{dx_1}\right|}$$

in the limit as $\delta x_1 \to 0$. Moreover, since

$$x_1 = y - x_2$$

we have

$$\left|\frac{dy}{dx_1}\right| = 1$$

whence

$$f_{YX_2}(y, x_2) = f_{X_1X_2}(x_1, x_2).$$

The $x_2$ variable may be eliminated by integration. We see that

$$f_Y(y) = \int_{-\infty}^{\infty} f_{YX_2}(y, x_2)dx_2 = \int_{-\infty}^{\infty} f_{X_1X_2}(y - x_2, x_2)dx_2$$

which is the probability density function of the new random variable $Y$.

If the random variables $X_1$ and $X_2$ are independent then

$$f_Y(y) = \int_{-\infty}^{\infty} f_{X_1}(y - x_2) f_{X_2}(x)dx_2.$$

Also, if the random variables $X_1$ and $X_2$ take only positive values such that their probability density functions are defined by

$$f_{X_1}(x_1) = 0 \qquad \text{for } x_1 < 0$$
$$f_{X_1}(x_2) = 0 \qquad \text{for } x_2 < 0$$

then the probability density function of the new random variable $Y$ takes the form

$$f_Y(y) = \int_0^y f_{X_1}(y - x_2) f_{X_2}(x_2)dx_2 \qquad \text{for } y > 0$$

$$= 0 \qquad \text{for } y < 0.$$

By way of illustration, consider the independent random variables $X_1, X_2$ described by the probability density functions

$$f_{X_1}(x_1) = ae^{-ax_1}, \qquad f_{X_2}(x_2) = be^{-bx_2}$$

where

$$x_1, x_2 > 0, \qquad a, b > 0.$$

From the above theory we see that the probability density function of the new random variable $Y = X_1 + X_2$ is given by

$$
\begin{aligned}
f_Y(y) &= \int_0^y abe^{-a(y-x_2)}e^{-bx_2}dx_2 \\
&= \frac{ab}{a-b}(e^{-by} - e^{-ay}) && a \neq b, \\
&= a^2 ye^{-ay} && a = b,
\end{aligned}
$$

and

$$f_Y(y) = 0 \qquad y < 0.$$

# 4 Characteristics of probability density functions

Sometimes, the probability density function $f(x)$ may be known. Thus on the basis of reasonable assumptions concerning dice-rolling or card-drawing it is possible to arrive at functions $f(x)$. More usually, however, we can only postulate that $f(x)$ exists, and the form it takes is very much in doubt. We must now consider this state of affairs.

If $f(x)$ is unknown, it is usually possible to measure, or to postulate on the basis of experience, certain of its features. These features have well understood physical interpretations in so much that if $f(x)$ *were* known and these features were not, we should probably wish to calculate the latter anyway.

## 4.1 Expected value

The 'expected value' or 'mean' of a continuous random variable $X$ is defined as

$$E[X] = \int_{-\infty}^{\infty} x f_X(x) dx$$

where it is assumed that the integral does exist, or in other words that $|x| f_X(x)$ may be integrated in the interval $(-\infty, \infty)$. For brevity the expected value of the random variable $X$ is sometimes denoted by $\mu_X$.

Notice two things about $E[X]$. Firstly the integral referred to is the first moment of the $f_X(x)$ curve about the value $x = 0$ and secondly it has a clear physical interpretation. For if $f_X(x)$ represents the variable mass per unit length of a rod of unit mass lying along the $x$-axis, then the expected value of the random variable $X$ is analogous to the position of the centre of mass of the rod.

It also follows for any constants $a$, $b$ that,

$$E[aX + b] = \int_{-\infty}^{\infty} (ax + b)f_X(x)dx$$

$$= a \int_{-\infty}^{\infty} xf_X(x)dx + b \int_{-\infty}^{\infty} f_X(x)dx$$

$$= a\,\mu_X + b.$$

Thus, adding a fixed amount to a random variable changes the mean of the random variable by the same amount, whilst any constant multiplying the random variable multiplies the mean of the random variable.

### 4.1.1 Mean square value

The above argument may be generalised. The expected value of $X^2$ is

$$E[X^2] = \int_{-\infty}^{\infty} x^2 f_X(x)dx.$$

This is referred to as the 'mean square value' of $X$ and again it has at least some physical 'feel'. We here have a quantity representing the second moment of the $f_X(x)$ curve, such that the mean square value can be interpreted as the moment of inertia about an axis perpendicular to the $x$-axis and passing through the origin in the rod. This mean square value represents a little more information about the function $f_X(x)$ and it happens to be a fact that, in general, if the moments

$$\int_{-\infty}^{\infty} x^n f_X(x)dx$$

of *all* orders $n$ were known, $f_X(x)$ could be deduced.

### 4.1.2 Variance and standard deviation

We may generalise still further. The expected value of *any* function $g(X)$ is given by

$$E[g(X)] = \int_{-\infty}^{\infty} g(x)f(x)dx$$

provided that the integral converges absolutely, i.e.

$$\int_{-\infty}^{\infty} |g(x)| f(x) dx < \infty .$$

Consider a function $g(X) = (X - \mu_X)^2$. The expected value of this is a measure of 'looseness' of the value of $X$ with respect to $\mu_X$; it is a measure of dispersion and it is known as the 'variance', being denoted by $\sigma_X^2$. That is to say

$$\sigma_X^2 = E[(X - \mu_X)^2] = \int_{-\infty}^{\infty} (x - \mu_X)^2 f_X(x) dx$$

$$= \int_{-\infty}^{\infty} x^2 f_X(x) dx - 2\mu_X \int_{-\infty}^{\infty} x f_X(x) dx$$

$$+ \mu_X^2 \int_{-\infty}^{\infty} f_X(x) dx$$

$$= E[X^2] - \mu_X^2 \geqslant 0.$$

Since $f_X(x)\,\delta x$ represents a number, the units of variance are those of $x^2$. This is not convenient for some purposes so that the measure of dispersion is taken as the postive square root of $\sigma_X^2$. It is referred to as the 'standard deviation', or 'root mean square value', (r.m.s.).

In terms of the mass analogy, the variance may be interpreted as the moment of inertia with respect to a perpendicular axis passing through the centre of mass of the rod, and the relationship

$$\sigma_X^2 = E[X^2] - \mu_X^2$$

is simply a statement of the 'parallel axis theorem'.

It also follows that the variance of the random variable $(aX + b)$ is given by

$$\sigma_{aX+b}^2 = E[\{(aX + b) - (a\mu_X + b)\}^2]$$

$$= E[a^2(X - \mu_X)^2] = a^2 \int_{-\infty}^{\infty} (x - \mu_X)^2 f_X(x) dx$$

$$= a^2 \sigma_X^2$$

and

$$\sigma_{aX+b} = |a| \sigma_X .$$

Adding a fixed amount to a random variable has no effect on the variance of the random variable, whilst multiplying the random variable by a constant multiplies the variance of the random variable by the square of the constant and the standard deviation by the absolute value of the constant.

### 4.1.3 Two random variables

The existence of two random variables does not require us to introduce any fresh concepts when determining the 'moments' of the variables. If $g(X)$ and $h(Y)$ are known functions then the expectations of their product may be found. The probability that the value of $X$ lies between $x$ and $x + \delta x$ *and* the value of $Y$ lies between $y$ and $y + \delta y$ is $f_{XY}(x, y)\delta x \delta y$. The expected value is thus

$$E[g(X)h(Y)] = \int_{-\infty}^{\infty}\int_{-\infty}^{\infty} g(x)h(y)f_{XY}(x, y)dx\, dy$$

In particular if $g(X) = X - \mu_X$ and $h(Y) = Y - \mu_Y$, then

$$\begin{aligned} E[(X - \mu_X)(Y - \mu_Y)] &= \int_{-\infty}^{\infty}\int_{-\infty}^{\infty} (x - \mu_X)(y - \mu_Y)f_{XY}(x, y)dx\, dy \\ &= E[XY] - E[X]E[Y] \end{aligned}$$

is defined as the 'covariance' $C_{XY}$ of the random variables $X$ and $Y$. When $X = Y$ the covariance $C_{XX}$ reduces to the variance $\sigma_X^2$ of the random variable $X$.

The 'correlation coefficient' of the random variables is defined by

$$\rho_{XY} = \frac{C_{XY}}{\sigma_X \sigma_Y}$$

provided that the covariance and standard deviations exist.

If we let $g(X) = b(X - \mu_X)$ and $h(Y) = (Y - \mu_Y)$ then

$$E[\{b(X - \mu_X) - (Y - \mu_Y)\}^2] = b^2\sigma_X^2 - 2bC_{XY} + \sigma_Y^2$$

gives a quadratic relationship in $b$ which by definition is non-negative for any $b$. Thus the discriminant $(4C_{XY}^2 - 4\sigma_X^2\sigma_Y^2)$ must be non-positive, yielding

$$C_{XY}^2 - \sigma_X^2\sigma_Y^2 \leqslant 0.$$

It follows that the correlation coefficient lies in the range

$$-1 \leqslant \rho_{XY} \leqslant 1.$$

When $\rho_{XY} = 0$, the random variables $X$ and $Y$ are said to be *uncorrelated.* If $X$ and $Y$ are independent random variables then they are also uncorrelated since

$$E[XY] = \int_{-\infty}^{\infty}\int_{-\infty}^{\infty} xyf_{XY}(x, y)dx\,dy$$

$$= \int_{-\infty}^{\infty} xf_X(x)dx \int_{-\infty}^{\infty} yf_Y(y)dy$$

$$= E[X]E[Y]$$

and

$$C_{XY} = E[XY] - E[X]E[Y] = 0 = \rho_{XY}.$$

Note, however, that uncorrelated random variables are not necessarily independent random variables.

It also follows that the expected value of the sum random variable $(aX + bY)$ is

$$E[(aX + bY)] = \int_{-\infty}^{\infty}\int_{-\infty}^{\infty} (ax + by)f_{XY}(x, y)dx\,dy$$

$$= a\int_{-\infty}^{\infty} x \int_{-\infty}^{\infty} f_{XY}(x, y)dy\,dx$$

$$+ b\int_{-\infty}^{\infty} y \int_{-\infty}^{\infty} f_{XY}(x, y)dx\,dy$$

$$= a\int_{-\infty}^{\infty} xf_X(x)dx + b\int_{-\infty}^{\infty} yf_Y(y)dy$$

$$= a\mu_X + b\mu_Y$$

and the variance is

$$\sigma^2_{aX+bY} = E[\{(aX + bY) - (a\mu_X + b\mu_Y)\}^2]$$

$$= E[\{a(X - \mu_X) + b(Y - \mu_Y)\}^2]$$

$$= a^2E[(X - \mu_X)^2] + 2abE[(X - \mu_X)(Y - \mu_Y)]$$

$$+ b^2E[(Y - \mu_Y)^2]$$

$$= a^2\sigma^2_X + 2abC_{XY} + b^2\sigma^2_Y.$$

If $X$ and $Y$ are independent random variables, then the covariance function $C_{XY} = 0$ and

$$\sigma^2_{aX+bY} = a^2\sigma^2_X + b^2\sigma^2_Y.$$

We may extend these results to the set of $n$ independent random variables $X_1, X_2, \ldots, X_n$ with individual expected values $\mu_i (i = 1, 2, \ldots, n)$ and variance $\sigma_i^2 (i = 1, 2, \ldots, n)$. It is found that the summed random variable

$$X = X_1 + X_2 + \ldots + X_n$$

has the expected value

$$\mu_X = \sum_{i=1}^{n} \mu_{X\,i}$$

and variance

$$\sigma_X^2 = \sum_{i=1}^{n} \sigma_i^2 .$$

## 4.2 Moment generating function

The expected value of a function $g(X) = e^{i\theta X}$ (where $i = \sqrt{-1}$) is given by

$$E[g(X)] = E[e^{i\theta X}] = M_X(\theta) = \int_{-\infty}^{\infty} e^{i\theta x} f_X(x)dx$$

provided that the integral exists, $\theta$ being an arbitrary dummy real variable. $M_X(\theta)$ is referred to as the 'moment generating function' of the random variable $X$ and, since

$$e^{i\theta x} = 1 + i\theta x + \frac{(i\theta x)^2}{2!} + \frac{(i\theta x)^3}{3!} + \ldots,$$

it follows that

$$E[e^{i\theta x}] = M_X(\theta) = 1 + i\theta E[X] + \frac{(i\theta)^2}{2!} E[X^2] + \frac{(i\theta)^3}{3!} E[X^3] + \ldots$$

By differentiating this expression with respect to $\theta$, we find that the $n$th moment is given by

$$E[X^n] = \frac{1}{i^n} \frac{d^n M_X(\theta)}{d\theta^n}$$

evaluated at $\theta = 0$. Thus if we form the function $M_X(\theta)$ for a given probability density function $f_X(x)$ we have a ready means of finding any moment of the latter. We shall see the usefulness of $M_X(\theta)$ later on when we come to examine the central limit theorem.

For the two jointly distributed random variables $X$ and $Y$, the moment generating function is obtained by letting the functions $g(X) = e^{i\theta x}$ and $h(Y) = e^{i\lambda y}$ so that

$$M_{XY}(\theta, \lambda) = E[e^{i(\theta X + \lambda Y)}] = \int_{-\infty}^{\infty} \int_{-\infty}^{\infty} e^{i(\theta x + \lambda y)} f_{XY}(x, y) dx\, dy.$$

If the exponential term is now expanded in a Taylor series, it is found that

$$M_{XY}(\theta, \lambda) = 1 + i\theta E[X] + i\lambda E[Y]$$

$$+ \frac{i^2}{2!} \{\theta^2 E[X^2] + 2\theta\lambda E[XY] + \lambda^2 E[Y^2]\} + \ldots$$

and the moments of the random variables $X$ and $Y$ are given by

$$E[X] = \frac{1}{i} \frac{\partial M_{XY}(\theta, \lambda)}{\partial \theta}, \qquad E[Y] = \frac{1}{i} \frac{\partial M_{XY}(\theta, \lambda)}{\partial \lambda}$$

etc., evaluated at $\theta = 0 = \lambda$.

The moment generating function of several jointly distributed random variables may be similarly obtained.

## 4.3 The 'Gaussian' or 'normal' probability density function

Consider the function

$$y = f_X(x) = \frac{1}{\sigma_X \sqrt{2\pi}} e^{-(x - \mu_X)^2 / 2\sigma_X^2}.$$

Its curve is shown in Figure 12a. Variation of $\mu_X$ merely shifts the peak along the $x$-axis. If $\sigma_X$ is altered, the sharpness of the peak is changed.

We shall now summarise certain properties of this function, and in the working which follows it will be found helpful to bear in mind the result

$$\int_{-\infty}^{\infty} e^{-x^2} dx = \sqrt{\pi}.$$

Our purpose is to show that the function $f_X(x)$ has certain features that make it useful as a probability density function.

The curve is non-negative and also

$$\int_{-\infty}^{\infty} y\, dx = 1.$$

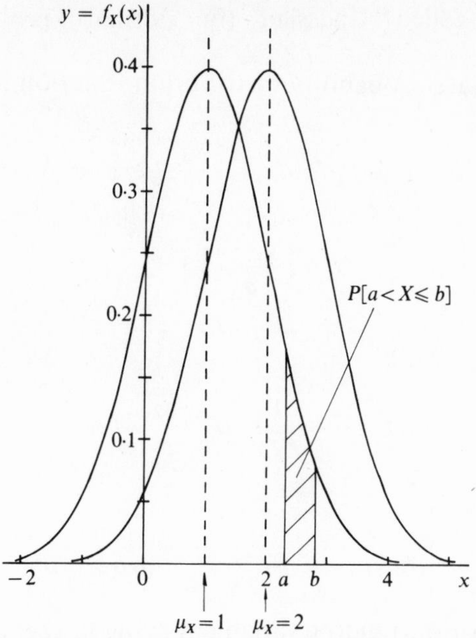


**Figure 12a.** The 'normal' (or 'Gaussian') probability density function with $\sigma_X = 1$, taking $\mu_X = 1$ and 2.

The abscissa of the centroid of the area under the curve is

$$\int_{-\infty}^{\infty} xy\,dx = \mu_X.$$

The mean value of the function therefore appears explicitly in the function. The integral

$$\int_{-\infty}^{\infty} (x - \mu_X)^2\, y\, dx = \sigma_X^2$$

so that not only the mean value, but also the variance, is immediately recognisable. To sum up,

$$f_X(x) = \frac{1}{\sigma_X\sqrt{2\pi}}\, e^{-(x-\mu_X)^2/2\sigma_X^2}$$

is the equation of a valid probability density function, for the random variable $X$, in which

$\mu_X$ = expected value,

$\sigma_X$ = standard deviation.

This is the so-called 'Gaussian' (or 'normal') probability density function.

The appropriate probability distribution function is

$$P[X \leqslant x] = F_X(x) = \int_{-\infty}^{x} f_X(z)dz = \frac{1}{\sigma_X\sqrt{2\pi}} \int_{-\infty}^{x} e^{-(z-\mu_X)^2/2\sigma_X^2}\, dz$$

$$= \tfrac{1}{2} + erf\left(\frac{x-\mu_X}{\sigma_X}\right)$$

where the error function

$$erf(x) = \frac{1}{\sqrt{2\pi}} \int_0^x e^{-(z^2/2)}\, dz$$

has the properties

$$erf(-x) = -erf(x), \qquad erf(\infty) = \tfrac{1}{2}.$$

The probability of the random variable $X$ lying in the range $(x_1, x_2)$ is

$$P[x_1 < X \leqslant x_2] = \int_{x_1}^{x_2} f_X(z)dz$$

$$= erf\,\frac{x_2-\mu_X}{\sigma_X} - erf\,\frac{x_1-\mu_X}{\sigma_X}\,.$$

Values of the error function are given in Table 1a.

### 4.3.1 Standardised Gaussian probability density function

Let us consider a continuous random variable $X$ with a mean value $\mu_X$ and variance $\sigma_X^2$. Suppose that it is described by the Gaussian probability density function

$$f_X(x) = \frac{1}{\sigma_X\sqrt{2\pi}}\, e^{-(x-\mu_X)^2/2\sigma_X^2}, \qquad -\infty < x < \infty.$$

The centre of symmetry of the density function is at $x = \mu_X$ as shown in Figure 12a. The total area bounded by the curve and the $x$-axis is unity, whilst the area between two ordinates $a < x \leqslant b$ represents the probability of the random variable $X$ lying between $a$ and $b$, i.e. $P[a < X \leqslant b]$.

Changes in the value of the standard deviation $\sigma_X$ affect the sharpness of the peak of the curve as shown in Figure 12b (for which $\mu_X = 0$). As $\sigma_X$ increases the curves become flatter and more

**Table 1a**

*Areas under the standardised Gaussian probability density curve*

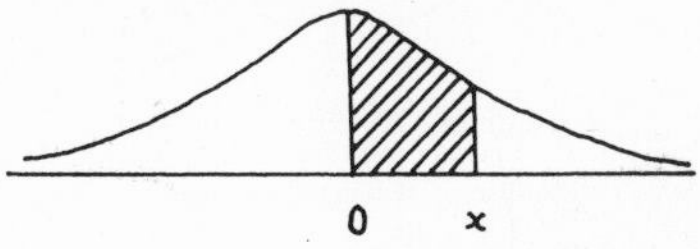


| *x* | 0·00 | 0·01 | 0·02 | 0·03 | 0·04 | 0·05 | 0·06 | 0·07 | 0·08 | 0·09 |
|---|---|---|---|---|---|---|---|---|---|---|
| 0·0 | 0·0000 | 0·0040 | 0·0080 | 0·0120 | 0·0160 | 0·0199 | 0·0239 | 0·0279 | 0·0319 | 0·0359 |
| 0·1 | 0·0398 | 0·0438 | 0·0478 | 0·0517 | 0·0557 | 0·0596 | 0·0636 | 0·0675 | 0·0714 | 0·0753 |
| 0·2 | 0·0793 | 0·0832 | 0·0871 | 0·0910 | 0·0948 | 0·0987 | 0·1026 | 0·1064 | 0·1103 | 0·1141 |
| 0·3 | 0·1179 | 0·1217 | 0·1255 | 0·1293 | 0·1331 | 0·1368 | 0·1406 | 0·1443 | 0·1480 | 0·1517 |
| 0·4 | 0·1554 | 0·1591 | 0·1628 | 0·1664 | 0·1700 | 0·1736 | 0·1772 | 0·1808 | 0·1344 | 0·1879 |
| 0·5 | 0·1915 | 0·1950 | 0·1985 | 0·2019 | 0·2054 | 0·2088 | 0·2123 | 0·2157 | 0·2190 | 0·2224 |
| 0·6 | 0·2257 | 0·2291 | 0·2324 | 0·2357 | 0·2389 | 0·2422 | 0·2454 | 0·2486 | 0·2517 | 0·2549 |
| 0·7 | 0·2580 | 0·2611 | 0·2642 | 0·2673 | 0·2704 | 0·2734 | 0·2764 | 0·2794 | 0·2823 | 0·2852 |
| 0·8 | 0·2881 | 0·2910 | 0·2939 | 0·2967 | 0·2995 | 0·3023 | 0·3051 | 0·3078 | 0·3106 | 0:3133 |
| 0·9 | 0·3159 | 0·3186 | 0·3212 | 0·3238 | 0·3264 | 0·3289 | 0·3315 | 0·3340 | 0·3365 | 0:3389 |
| 1·0 | 0·3413 | 0·3438 | 0·3461 | 0·3485 | 0·3508 | 0:3531 | 0·3554 | 0·3577 | 0·3599 | 0·3621 |
| 1·1 | 0·3643 | 0·3665 | 0·3686 | 0·3708 | 0·3729 | 0:3749 | 0·3770 | 0·3790 | 0·3810 | 0·3830 |
| 1·2 | 0·3849 | 0·3869 | 0·3688 | 0·3907 | 0·3925 | 0·3944 | 0·3962 | 0·3980 | 0·3997 | 0·4015 |
| 1·3 | 0·4032 | 0·4049 | 0·4066 | 0·4082 | 0·4099 | 0·4115 | 0·4131 | 0·4147 | 0·4162 | 0·4177 |
| 1·4 | 0·4192 | 0·4207 | 0·4222 | 0·4236 | 0·4251 | 0·4265 | 0·4279 | 0·4292 | 0·4306 | 0·4319 |
| 1·5 | 0·4332 | 0·4345 | 0·4387 | 0·4370 | 0·4382 | 0·4394 | 0·4406 | 0·4418 | 0·4429 | 0·4441 |
| 1·6 | 0·4452 | 0·4463 | 0·4474 | 0·4484 | 0·4495 | 0·4505 | 0·4515 | 0·4525 | 0·4535 | 0·4545 |
| 1·7 | 0·4554 | 0·4564 | 0·4573 | 0·4582 | 0·4591 | 0·4599 | 0·4608 | 0·4616 | 0·4625 | 0·4633 |
| 1·8 | 0·4641 | 0·4649 | 0·4656 | 0·4664 | 0·4671 | 0·4678 | 0·4686 | 0·4693 | 0·4699 | 0·4706 |
| 1·9 | 0·4713 | 0·4719 | 0·4726 | 0·4732 | 0·4738 | 0·4744 | 0·4750 | 0·4756 | 0·4761 | 0·4767 |
| 2·0 | 0·4772 | 0·4778 | 0·4783 | 0·4788 | 0·4793 | 0·4798 | 0·4803 | 0·4808 | 0·4812 | 0·4817 |
| 2·1 | 0·4821 | 0·4826 | 0·4830 | 0·4834 | 0·4838 | 0·4842 | 0·4846 | 0·4850 | 0·4854 | 0·4857 |
| 2·2 | 0·4861 | 0·4864 | 0·4868 | 0·4871 | 0·4875 | 0·4878 | 0·4881 | 0·4884 | 0·4887 | 0·4890 |
| 2·3 | 0·4893 | 0·4896 | 0·4898 | 0·4901 | 0·4904 | 0·4906 | 0·4909 | 0·4911 | 0·4913 | 0·4916 |
| 2·4 | 0·4918 | 0·4920 | 0·4922 | 0·4925 | 0·4927 | 0·4929 | 0·4931 | 0·4932 | 0·4934 | 0·4936 |
| 2·5 | 0·4938 | 0·4940 | 0·4941 | 0·4943 | 0·4945 | 0·4946 | 0·4948 | 0·4949 | 0·4251 | 0·4952 |
| 2·6 | 0·4953 | 0·4955 | 0·4956 | 0·4957 | 0·4959 | 0·4960 | 0·4961 | 0·4962 | 0·4963 | 0·4964 |
| 2·7 | 0·4965 | 0·4966 | 0·4967 | 0·4968 | 0·4969 | 0·4970 | 0·4971 | 0·4972 | 0·4973 | 0·4974 |
| 2·8 | 0·4974 | 0·4975 | 0·4976 | 0·4977 | 0·4977 | 0·4978 | 0·4979 | 0·4979 | 0·4980 | 0·4981 |
| 2·9 | 0·4981 | 0·4982 | 0·4982 | 0·4983 | 0·4984 | 0·4984 | 0·4985 | 0·4985 | 0·4986 | 0·4986 |
| 3·0 | 0·4987 | 0·4987 | 0·4987 | 0·4988 | 0·4988 | 0·4989 | 0·4989 | 0·4989 | 0·4990 | 0·4990 |
| 3·1 | 0·4990 | 0·4991 | 0·4991 | 0·4991 | 0·4992 | 0·4992 | 0·4992 | 0·4992 | 0·4993 | 0·4993 |
| 3·2 | 0·4993 | 0·4993 | 0·4994 | 0·4994 | 0·4994 | 0·4994 | 0·4994 | 0·4995 | 0·4995 | 0·4995 |
| 3·3 | 0·4995 | 0·4995 | 0·4995 | 0·4996 | 0·4996 | 0·4996 | 0·4996 | 0·4996 | 0·4996 | 0·4997 |
| 3·4 | 0·4997 | 0·4997 | 0·4997 | 0·4997 | 0·4997 | 0·4997 | 0·4997 | 0·4997 | 0·4997 | 0·4998 |

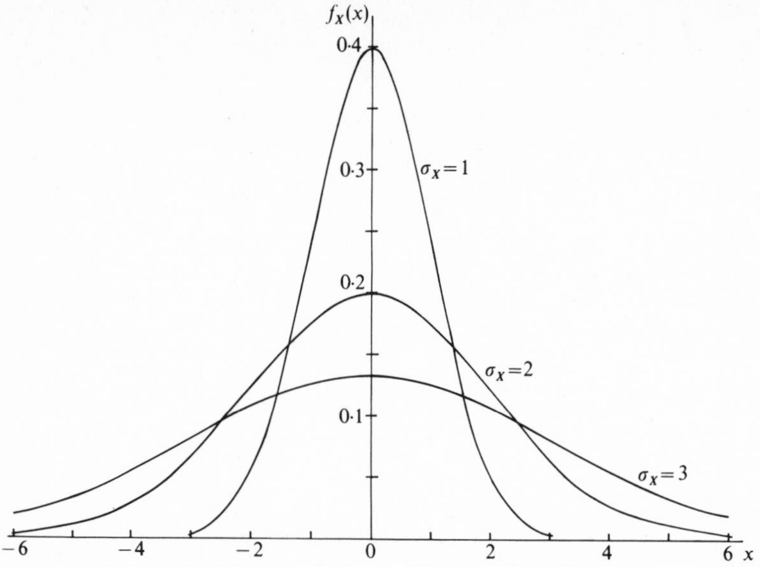


**Figure 12b.** $\sigma_X$ = 1, 2 and 3, taking $\mu_X = 0$.

spead out. When $x > 3\sigma_X$ the tails of the curves become very flat and nearly touch the $x$-axis. The maximum value of the density function is at $x = 0$ giving $f_X(0) = (\sigma_X\sqrt{2\pi})^{-1}$ and the points of inflection in the curve occur at $x = \pm\sigma_X$.

The 'standardised' Gaussian random variable is defined as

$$Z = \frac{X - \mu_X}{\sigma_X}$$

It has a mean value $\mu_Z = 0$ and standard deviation $\sigma_Z = 1$. By a change of variable, as described in section 3.3, the standardised Gaussian probability density function of the variable $Z$ is found to be

$$f_Z(z) = \frac{1}{\sqrt{2\pi}} e^{-z^2/2}, \qquad -\infty < z < \infty.$$

Figure 12c shows the percentage area between ordinates under the standardised Gaussian curve. Fuller details of the ordinate values of the standard normal curve are given in Table 1b.

It will be noticed that the Gaussian density function has retained its characteristic form under the linear transformation of variable. This is typical of any linear transformation of a Gaussian density function, and it is this property which enables us to determine the effects of random phenomena (at least to a first approximation) in relatively simple form.

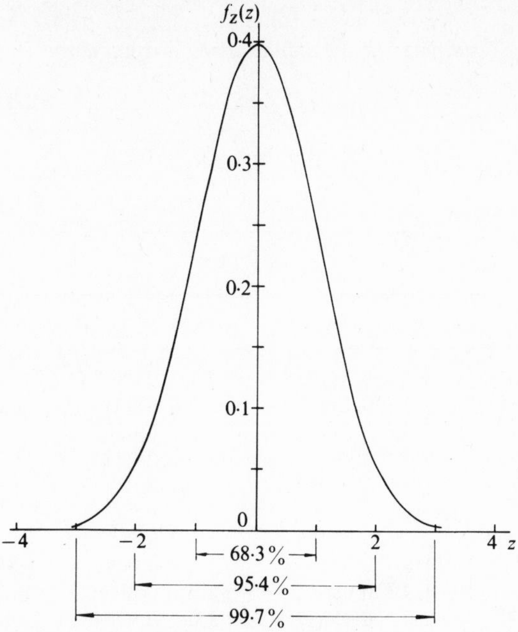


**Figure 12c.** Percentage areas under the standardised normal curve ($\mu_Z = 0$, $\sigma_Z = 1$) indicated.

### 4.3.2 Moments of standardised random normal variable

The moment generating function of the standardised Gaussian random variable $Z$ is given by

$$M_Z(\theta) = \int_{-\infty}^{\infty} e^{i\theta z} f_Z(z)\,dz$$

$$= e^{-\theta^2/2} \int_{-\infty}^{\infty} \frac{1}{\sqrt{2\pi}} e^{-(z-i\theta)^2/2}\,dz$$

$$= e^{-\theta^2/2}$$

or

$$\log_e M_Z(\theta) = -\theta^2/2.$$

The moments of the random variable are defined by

$$E[Z^n] = \frac{1}{i^n} \frac{d^n M_Z(\theta)}{d\theta}\bigg|_{\theta=0}$$

**Table 1b**

*Ordinates of the standardised normal curve*

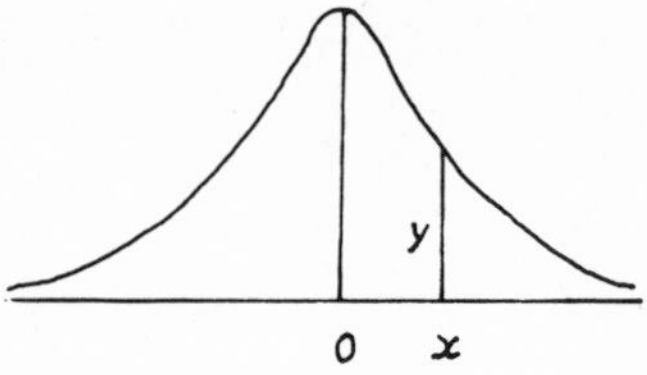


| $x$ | $y$ | $x$ | $y$ | $x$ | $y$ | $x$ | $y$ |
|---|---|---|---|---|---|---|---|
| 0·0 | 0·39894 | 1·1 | 0·21785+ | 2·1 | 0·04398 | 3·1 | 0·00327 |
| 0·1 | 0·39695+ | 1·2 | 0·19419 | 2·2 | 0·03547 | 3·2 | 0·00238 |
| 0·2 | 0·39104 | 1·3 | 0·17137 | 2·3 | 0·02833 | 3·3 | 0·00172 |
| 0·3 | 0·38139 | 1·4 | 0·14973 | 2·4 | 0·02239 | 2·4 | 0·00123 |
| 0·4 | 0·36827 | 1·5 | 0·12952 | 2·5 | 0·01753 | 3·5 | 0·00087 |
| 0·5 | 0·35207 | 1·6 | 0·11092 | 2·6 | 0·01358 | 3·6 | 0·00061 |
| 0·6 | 0·33322 | 1·7 | 0·09405− | 2·7 | 0·01042 | 3·7 | 0·00042 |
| 0·7 | 0·31225+ | 1·8 | 0·07895+ | 2·8 | 0·00792 | 3·8 | 0·00029 |
| 0·8 | 0·28969 | 1·9 | 0·06562 | 2·9 | 0·00595+ | 3·9 | 0·00020 |
| 0·9 | 0·26609 | 2·0 | 0·05399 | 3·0 | 0·00443 | 4·0 | 0·00013 |
| 1·0 | 0·24197 | | | | | | |

such that, for $n = 1, 2, 3, \ldots$ we have

$$E[Z] = 0, \qquad E[Z^2] = 1, \qquad E[Z^3] = 0, \qquad E[Z^4] = 1.3,$$
$$E[Z^5] = 0, \text{ etc.}$$

In general

$$E[Z^{n+1}] = 0$$

for $n$ even, and

$$E[Z^{n+1}] = n(n-2)\ldots 5.3.1 = \sqrt{\left(\frac{2^{n+1}}{\pi}\right)}\,\Gamma\left(\frac{n+2}{2}\right)$$

for $n$ odd, where $\Gamma$ denotes the Gamma function with the properties

$$\Gamma(n+1) = n\Gamma(n), \qquad \Gamma(½) = \sqrt{\pi}, \qquad \Gamma(1) = 1, \qquad \Gamma(\infty) = \infty.$$

For the random variable $X(= \sigma_X Z)$ with zero mean value we have

$$E[X^{n+1}] = 0$$

for $n$ even, and

$$E[X^{n+1}] = \sqrt{\left(\frac{2^{n+1}}{\pi}\right)}\,\sigma_X^{n+1}\,\Gamma\left(\frac{n+2}{2}\right)$$

for $n$ odd.

### 4.3.3 The central limit theorem

Let $X_1, X_2, \ldots X_n$ be $n$ statistically independent random variables whose individual probability distributions are not specified and may be different. The individual random variables $X_j (j = 1, 2, \ldots, n)$ have mean values $\mu_j$, variance $\sigma_j^2$ and the summed random variable

$$X = X_1 + X_2 + \ldots + X_n = \sum_{j=1}^{n} X_j.$$

From section 4.1.3 we learn that the expected value of the summed random variable is

$$E[X] = \mu_X = \sum_{j=1}^{n} \mu_j,$$

while the variance is

$$\sigma_X^2 = \sum_{j=1}^{n} \sigma_j^2.$$

The assumptions that we have made are common in the study of, for example, turbulent flow. An effect of any given localised vortex in the flow can be identified with a single random variable $X_j$.

The central limit theorem states that, under conditions that are fairly commonly met, the summed random variable $X$ will be *normally* distributed (i.e. Gaussian in form) as $n \to \infty$ with an expected value $\mu_X$ and variance $\sigma_X^2$ as previously defined.

In order to simplify our proof of this theorem, let us assume that all the random variables $X_j (j = 1, 2, \ldots, n)$ are identically distributed (though not necessarily Gaussian) with individual expected values zero, and all variances equal, so that

$$\mu_X = 0, \qquad \sigma_X^2 = n\sigma_j^2$$

or, here,

$$E[X^2] = nE[X_j^2], \qquad E[X^3] = nE[X_j^3], \qquad \text{etc.}$$

We shall show that, by assembling the moment generating functions for the individual random variables, we can form the moment generating function of the aggregate. This latter function shows the truth of the theorem.

The theory of section 4.2 shows that the standardised random variable

$$Z = \frac{X}{\sigma_X} = \frac{\sum_{j=1}^{n} X_j}{\sigma_X}$$

has a moment generating function

$$M_Z(\theta) = E[e^{iZ\theta}] = E\left[\exp\left(i\sum_{j=1}^{n} X_j\theta/\sigma_X\right)\right] = \prod_{j=1}^{n} M_{X_j}(\theta/\sigma_X)$$

and

$$\log M_Z(\theta) = \sum_{j=1}^{n} \log M_{X_j}(\theta/\sigma_X).$$

The series expansion of the $j$th component characteristic function $M_{X_j}(\theta/\sigma_X)$ is given by

$$\begin{aligned} M_{X_j}(\theta/\sigma_X) &= E[e^{iX_j\theta/\sigma_X}] \\ &= E\left[1 + iX_j\frac{\theta}{\sigma_X} - X_j^2\frac{\theta^2}{2!\sigma_X^2} - iX_j^3\frac{\theta^3}{3!\sigma_X^3} + \ldots\right] \\ &= 1 - \frac{\theta^2}{2!\sigma_X^2}E[X_j^2] - \frac{i\theta^3}{3!\sigma_X^3}E[X_j^3] + \ldots \\ &= 1 - \frac{\theta^2}{2!n} + \text{terms of order}\left(\frac{1}{n^{3/2}}\right). \end{aligned}$$

Thus for large values of $n$ we have

$$\log M_{X_j}(\theta/\sigma_X) = \frac{-\theta^2}{2!n} + O\left[\frac{1}{n^{3/2}}\right]$$

and in the limit as $n \to \infty$

$$\begin{aligned} \lim_{n\to\infty} [\log M_Z(\theta)] &= \lim_{n\to\infty}\left[\sum_{j=1}^{n} \log M_{X_j}(\theta/\sigma_X)\right] \\ &= -\frac{\theta^2}{2!}. \end{aligned}$$

Section 4.3.2 shows that $M_Z(\theta) = e^{-\theta^2/2}$ is precisely the moment generating function of a standardised *Gaussian* random variable $Z$. It follows that

$$\lim_{n\to\infty} f_Z(z) = \frac{1}{\sqrt{2\pi}}\, e^{-z^2/2}$$

Hence, by the theory of section 3.3, as $n \to \infty$

$$f_X(x) = \frac{1}{\sigma_X\sqrt{2\pi}}\, e^{-x^2/2\sigma_X^2}.$$

A more general proof of this theorem has been given by Cramér[1].

## 4.4 Gaussian probability density function for two random variables

If $\mu_X = 0$, the Gaussian probability density function for a single random variable is

$$f_X(x) = \frac{1}{\sigma_X\sqrt{2\pi}}\, e^{-x^2/2\sigma_X^2}.$$

We shall now briefly examine the properties of a surface which can be thought of as a generalisation of this curve. It can be shown that the surface may be used to represent a joint probability density function $f_{XY}(x, y)$ for which

$$E[X] = 0 = E[Y].$$

The surface is

$$f_{XY}(x, y) = \frac{1}{2\pi\sqrt{|\gamma|}}\, e^{-(\beta_{XX}x^2 + \beta_{XY}xy + \beta_{YX}yx + \beta_{YY}y^2)/2}$$

where $\gamma$ is the matrix

$$\gamma = \begin{bmatrix} \gamma_{XX} & \gamma_{XY} \\ \gamma_{YX} & \gamma_{YY} \end{bmatrix}$$

and where $\gamma_{XY} = \gamma_{YX}$. The quantities $\beta_{XX}$ etc. are the elements of the inverse matrix of $\gamma$, so that

$$\boldsymbol{\beta} = \gamma^{-1}$$

Before examining this function, it will be convenient to express it in a slightly different form. Let

$$\rho_{XY} = \frac{\gamma_{XY}}{\sqrt{\gamma_{XX}\gamma_{YY}}}$$

so that, when $\gamma$ is inverted to obtain $\boldsymbol{\beta}$, it is found that

$$\beta_{XX} = \frac{1}{(1 - \rho_{XY}^2)\,\gamma_{XX}}$$

$$\beta_{YY} = \frac{1}{(1 - \rho_{XY}^2)\,\gamma_{YY}}$$

$$\beta_{XY} = \beta_{YX} = \frac{\rho_{XY}}{\sqrt{(\gamma_{XX}\gamma_{YY})(1 - \rho_{XY}^2)}}$$

Then $f_{XY}(x, y)$ may be written in the alternative form

$$f_{XY}(x, y) = \frac{1}{2\pi\sqrt{\{\gamma_{XX}\,\gamma_{YY}\,(1-\rho_{XY}^2)\}}}\, e^{-H/2}.$$

The expression

$$H = \frac{1}{(1-\rho_{XY}^2)}\left\{\frac{x^2}{\gamma_{XX}} - \frac{2\,\rho_{XY}xy}{\sqrt{(\gamma_{XX}\,\gamma_{YY})}} + \frac{y^2}{\gamma_{YY}}\right\}$$

is a quadratic in the variables $x$ and $y$. Unless $x = 0 = y$, $H$ must be positive and non-negative separately in $x$ and $y$, since the integral of the function $f_{XY}(x, y)$ over the whole range of $x$ and $y$ may exceed unity. Thus, for $H$ to be positive definite the following conditions must be met:

when $x = 0$, $H = \dfrac{1}{(1-\rho_{XY}^2)\gamma_{YY}}\,y^2$, so that $\dfrac{1}{(1-\rho_{XY}^2)\gamma_{YY}} > 0$;

when $y = 0$, $H = \dfrac{1}{(1-\rho_{XY}^2)\gamma_{XX}}\,x^2$, so that $\dfrac{1}{(1-\rho_{XY}^2)\gamma_{XX}} > 0$.

Moreover, since

$$H = \frac{1}{(1-\rho_{XY}^2)}\left\{\left(\frac{x}{\sqrt{\gamma_{XX}}} - \frac{\rho_{XY}y}{\sqrt{\gamma_{YY}}}\right)^2 + \frac{(1-\rho_{XY}^2)y^2}{\gamma_{YY}}\right\}$$

it is evident that if

$$\left(\frac{x}{\sqrt{\gamma_{XX}}} - \frac{\rho_{XY}y}{\sqrt{\gamma_{YY}}}\right)$$

is zero then it is necessary that $1/\gamma_{YY} > 0$. Thus the conditions needed are:

$$\gamma_{XX}^{-1} > 0, \qquad \gamma_{YY}^{-1} > 0, \qquad |\rho_{XY}| < 1.$$

Turning now to the properties of this surface we note that it is always positive when these three conditions are satisfied, and that it has a hump at the origin $x = 0 = y$. It can be shown that the volume under the surface is

$$\int_{-\infty}^{\infty}\int_{-\infty}^{\infty} f_{XY}(x, y)dxdy = 1$$

so that $f_{XY}(x, y)$ is indeed suitable as a joint probability density function. If it is taken as such, it is found that

$$E[X^2] = \gamma_{XX};\ E[Y^2] = \gamma_{YY};$$

$$E[XY] = \rho_{XY}\sqrt{(\gamma_{XX}\gamma_{YY})} = \gamma_{XY} = \gamma_{YX};$$

so that these expectations appear explicitly in the expression for the joint probability density function.

The quantity $\gamma_{XX}$ is the variance $\sigma_X^2$ of $X$, $\gamma_{YY}$ is the variance $\sigma_Y^2$ of $Y$, $\gamma_{XY}$ is the covariance $C_{XY}$ of the product $XY$ and the quantity

$$\rho_{XY} = \frac{E[XY]}{\sigma_X\ \sigma_Y}$$

is the correlation coefficient for the random variables $X$ and $Y$, as we see from section 4.1.3. The correlation coefficient lies in the range $(-1, 1)$ and when it is zero the joint probability density function has the form

$$f_{XY}(x, y) = \left(\frac{1}{\sigma_X\sqrt{2\pi}}\, e^{-x^2/2\sigma_X^2}\right) \cdot \left(\frac{1}{\sigma_Y\sqrt{2\pi}}\, e^{-y^2/2\sigma_Y^2}\right)$$

$$= f_X(x) \,.\, f_Y(y)$$

where $f_X(x)$ and $f_Y(y)$ are Gaussian probability density functions for the single random variables $X$ and $Y$. That is to say, the condition $\rho_{XY} = 0$ corresponds to the statistical independence of $X$ and $Y$.

As one would expect from a joint probability density function $f_{XY}(x, y)$, it is possible to determine $f_X(x)$ and $f_Y(y)$ even when $\rho_{XY} \neq 0$. Thus

$$\int_{-\infty}^{\infty} f_{XY}(x, y)dy = f_X(x)$$

and similarly for $f_Y(y)$.

In our discussion of the 'two dimensional' Gaussian probability density function we have assumed all along that

$$E[X] = 0 = E[Y].$$

This is not an essential feature of the function, but the inclusion of explicit mean values $\mu_X$ and $\mu_Y$ introduces mathematical complications which we prefer to avoid.

In general, the $n$ random variables $X_1, X_2, \ldots, X_n$ are said to be jointly Gaussian distributed if their probability density function is given by

$$f_{X_1 X_2 \cdots X_n}(x_1, x_2, \ldots x_n)$$

$$= \frac{1}{(2\pi)^{n/2}|\mathbf{\Delta}|^{1/2}} \exp\left[\frac{-1}{2|\mathbf{\Delta}|}\sum_{j=1}^{n}\sum_{k=1}^{n} |\mathbf{\Delta}|_{jk}(x_j - E[X_j])(x_k - E[X_k])\right]$$

where $|\mathbf{\Delta}| \neq 0$ is the determinant of the matrix of covariances and variances, so that

$$\mathbf{\Delta} = \begin{bmatrix} \sigma_{X_1}^2 & C_{X_1X_2} & \cdots & C_{X_1X_2} \\ C_{X_2X_1} & \sigma_{X_2}^2 & \cdots & C_{X_2X_n} \\ \cdots & \cdots & \cdots & \cdots \\ C_{X_nX_1} & C_{X_nX_2} & \cdots & \sigma_{X_n}^2 \end{bmatrix}$$

where

$$C_{X_iX_j} = E[(X_i - E[X_i])(X_j - E[X_j])] = C_{X_jX_i}, \quad i \neq j,$$

and $|\mathbf{\Delta}|_{jk}$ is the cofactor of the element in the $j$th row and $k$th column of $\mathbf{\Delta}$.

In particular when $n = 3$, the joint probability density function of the zero mean valued random variables $X$, $Y$ and $Z$ is given by

$$f_{XYZ}(x, y, z) = \frac{1}{(2\pi)^{3/2}|\mathbf{\Delta}|^{1/2}} \times$$

$$\exp\left[\frac{-1}{2|\mathbf{\Delta}|}(\Delta_{xx}x^2 + \Delta_{yy}y^2 + \Delta_{zz}z^2 + 2\Delta_{xy}xy + 2\Delta_{xz}xz + 2\Delta_{yz}yz)\right],$$

$$\mathbf{\Delta} = \begin{bmatrix} \sigma_X^2 & C_{XY} & C_{XZ} \\ C_{XY} & \sigma_Y^2 & C_{YZ} \\ C_{XZ} & C_{YZ} & \sigma_Z^2 \end{bmatrix}, \qquad |\mathbf{\Delta}| = \det \mathbf{\Delta},$$

and

$$\Delta_{xx} = \sigma_Y^2\sigma_Z^2 - C_{YZ}^2; \qquad \Delta_{xy} = C_{XZ}C_{YZ} - C_{XY}\sigma_Z^2;$$

$$\Delta_{yy} = \sigma_X^2\sigma_Z^2 - C_{XZ}^2; \qquad \Delta_{xz} = C_{XY}C_{YZ} - C_{XZ}\sigma_Y^2;$$

$$\Delta_{zz} = \sigma_X^2\sigma_Y^2 - C_{XY}^2; \qquad \Delta_{yz} = C_{XY}C_{XZ} - C_{YZ}\sigma_X^2.$$

### 4.4.1 Moment generating function of Gaussian random variables

From section 4.2, the moment generating function for the random variable $X$ is seen to be

$$M_X(\theta) = \int_{-\infty}^{\infty} e^{i\theta x} f_X(x)dx.$$

For a Gaussian distributed variable with zero mean value the moment generating function becomes

$$M_X(\theta) = \int_{-\infty}^{\infty} e^{i\theta x} \frac{1}{\sqrt{2\pi}\,\sigma_X} e^{-x^2/2\sigma_X^2}\, dx = e^{-E[X^2]\theta^2/2}$$

whereas for the jointly Gaussian distributed variables $X_1, X_2$ with zero mean values, the moment generating function is

$$M_{X_1X_2}(\theta_1, \theta_2) = \int_{-\infty}^{\infty}\int_{-\infty}^{\infty} e^{i(\theta_1 x_1 + \theta_2 x_2)} f_{X_1X_2}(x_1, x_2)\, dx_1\, dx_2$$

which, after substitution of $f_{X_1X_2}(x_1, x_2)$ from the previous section and integration, gives

$$M_{X_1X_2}(\theta_1, \theta_2) = \exp\left[-\frac{1}{2}\sum_{j=1}^{n}\sum_{k=1}^{n} E[X_jX_k]\theta_j\theta_k\right].$$

It follows that, for $n$ jointly Gaussian distributed variables $X_1, X_2, \ldots, X_n$ with zero mean values, the moment generating function is given by

$$M_{X_1, \ldots, X_n}(\theta_1, \ldots, \theta_n) = M'_X(\theta) = \exp\left[-\frac{1}{2}\sum_{j=1}^{n}\sum_{k=1}^{n} E[X_jX_k]\theta_j\theta_k\right]$$

which may be written as $e^R$ where

$$R = -\tfrac{1}{2}\sum_{j=1}^{n}\sum_{k=1}^{n} E[X_jX_k]\theta_j\theta_k$$

and

$$\frac{\partial R}{\partial \theta_j} = \sum_{k=1}^{n} E[X_jX_k]\theta_k; \quad \frac{\partial^2 R}{\partial\theta_j\partial\theta_k} = -E[X_jX_k]; \quad \frac{\partial^3 R}{\partial\theta_j\partial\theta_k\partial\theta_l} = 0.$$

By successive differentiation of the moment generating function we find that

$$\frac{\partial M'_X(\theta)}{\partial\theta_j} = M'_X(\theta)\frac{\partial R}{\partial\theta_j}$$

$$\frac{\partial^2 M'_X(\theta)}{\partial\theta_j\partial\theta_k} = \frac{\partial M'_X(\theta)}{\partial\theta_k}\frac{\partial R}{\partial\theta_j} + M'_X(\theta)\frac{\partial^2 R}{\partial\theta_j\partial\theta_k}$$

$$\frac{\partial^3 M'_X(\theta)}{\partial\theta_j\partial\theta_k\partial\theta_l} = \frac{\partial^2 M'_X(\theta)}{\partial\theta_k\partial\theta_l}\frac{\partial R}{\partial\theta_j} + \frac{\partial M'_X(\theta)}{\partial\theta_k}\frac{\partial^2 R}{\partial\theta_j\partial\theta_l} + \frac{\partial M'_X(\theta)}{\partial\theta_l}\frac{\partial^2 R}{\partial\theta_j\partial\theta_k}$$

$$\frac{\partial^4 M'_X(\theta)}{\partial\theta_j\partial\theta_k\partial\theta_l\partial\theta_m} = \frac{\partial^3 M'_X(\theta)}{\partial\theta_k\partial\theta_l\partial\theta_m}\frac{\partial R}{\partial\theta_j} + \frac{\partial^2 M'_X(\theta)}{\partial\theta_k\partial\theta_l}\frac{\partial^2 R}{\partial\theta_j\partial\theta_m}$$

$$+ \frac{\partial^2 M'_X(\theta)}{\partial\theta_k\partial\theta_m}\frac{\partial^2 R}{\partial\theta_j\partial\theta_l} + \frac{\partial^2 M'_X(\theta)}{\partial\theta_l\partial\theta_m}\frac{\partial^2 R}{\partial\theta_j\partial\theta_k}$$

. . . . . . . . . . . . . . . . . . . . . . . . . . . . . . . . . . . . . . . . . . . . . . . . .

$$\frac{\partial^n M'_X(\theta)}{\partial\theta_1 \ldots \partial\theta_n} = \frac{\partial^{n-1} M'_X(\theta)}{\partial\theta_1 \ldots \partial\theta_{j-1}\partial\theta_{j+1} \ldots \partial\theta_n}\frac{\partial R}{\partial\theta_j}$$

$$+ \sum_{\substack{k=1\\k\neq j}}^{n} \frac{\partial^{n-2} M'_X(\theta)}{\partial\theta_{v_1} \ldots \partial\theta_{v_{n-2}}}\frac{\partial^2 R}{\partial\theta_j\partial\theta_k}$$

where the sequence $v_1 \ldots . v_{n-2}$ does not include the two numbers $k$ and $j$.

From the theory of section 4.2, we see that the moments of the random variable are obtained by letting

$$\theta_1 = 0 = \theta_2 = \theta_3 = \ldots = \theta_n$$

in the last equation. For these values of $\theta$, $\partial R/\partial\theta_j = 0$ and the moments of the $n$ random variables are

$$\frac{\partial^n M'_X(\theta)}{\partial\theta_1 \ldots \partial\theta_n} = i^n E[X_1 \ldots X_m]$$

$$= \sum_{\substack{k=1\\k\neq j}}^{n} -i^{n-2} E[X_{v_1} \ldots X_{v_{n-2}}]E[X_j X_k]$$

so that

$$E[X_1 \ldots X_n] = \sum_{\substack{k=1\\k\neq j}}^{n} E[X_{v_1} \ldots X_{v_{n-2}}]E[X_j X_k].$$

For $n = 3$ we have

$$E[X_1X_2X_3] = \sum_{k=2}^{3} E[X_{v_1}]E[X_1X_k] = 0$$

since we have considered variables with zero mean values. For $n = 4$ we have

$$E[X_1X_2X_3X_4] = \sum_{k=2}^{4} E[X_{v_1}X_{v_2}]E[X_1X_k]$$

$$= E[X_1X_2]E[X_3X_4] + E[X_2X_3]E[X_1X_4]$$

$$+ E[X_1X_3]E[X_2X_4].$$

In general,

$$E[X_1X_2 \dots X_{2n+1}] = 0$$

$$E[X_1X_2 \dots X_{2n}] = \sum E[X_jX_k]E[X_lX_m]$$

where the summation is taken over all the different ways in which $2n$ elements may be arranged in $n$ pairs, that is in $2n!/n!2^n$ ways.

In the particular situation where $n = 2$ and $X_1 = X_2 = X_3$, we have

$$E[X_1^3X_4] = 3E[X_1^2]E[X_1X_4]$$

whereas for $n = 3$ and $X_1 = X_2 = X_3 = X_4 = X_5$ we have

$$E[X_1^5X_6] = 5E[X_1^4]E[X_1X_6].$$

For $2n$ random variables where $X_1 = X_2 = X_3 = \dots = X_{2n-1}$, it follows that

$$E[X_1^{2n-1}X_{2n}] = (2n-1)E[X_1^{2n-1}]E[X_1X_{2n}].$$

## 4.5 Rayleigh probability density function

The probability density function of a random variable chi, $\chi$, is sometimes required, where $\chi$ is related to independent component random variables $X_1, X_2, \dots$ by a relationship of the type

$$\chi^2 = X_1^2 + X_2^2 + \dots$$

and where each of the individual random variables $X_1, X_2, \dots$ has a Gaussian distribution. We now derive $f_\chi(\chi)$, starting with a simple special case.

Let $X_1$ and $X_2$ be two independent random variables each of which has a Gaussian distribution with zero mean value and the same variance $\sigma^2$. Let the new random variable be defined as $\chi$, where

$$\chi^2 = X_1^2 + X_2^2.$$

If the transformations $Y_1 = X_1^2$ and $Y_2 = X_2^2$ are adopted, the probability density function of the random variable $Y_1$ is found to be

$$f_{Y_1}(y_1) = \begin{cases} \dfrac{f_{X_1}(\sqrt{y_1}) + f_{X_1}(-\sqrt{y_1})}{2\sqrt{y_1}} & \text{for } 0 < y_1 < \infty \\ 0 & \text{otherwise,} \end{cases}$$

as shown in section 3.3. A similar result may be found for $f_{Y_2}(y_2)$. Again if

$$Z = Y_1 + Y_2$$

for $0 < z < \infty$, where

$$Z = \chi^2,$$

then the probability density function for the random variable $Z$ is given in section 3.3 as,

$$f_Z(z) = \int_0^z f_{Y_1}(z - y_2) f_{Y_2}(y_2) dy_2.$$

When $f_{X_1}(x_1)$ and $f_{X_2}(x_2)$ are Gaussian distributed, such that

$$f_{X_1}(x_1) = \frac{1}{\sigma\sqrt{2\pi}} e^{-x_1^2/2\sigma^2}$$

for $-\infty < x < \infty$, it follows that

$$f_{X_1}(x) = \frac{f_{X_1}(\sqrt{y_1}) + f_{X_1}(-\sqrt{y_1})}{2\sqrt{y_1}} = \frac{1}{\sigma\sqrt{2\pi}} \frac{e^{-y_1/2\sigma^2}}{\sqrt{y_1}}$$

for $0 < y_1 < \infty$, with a similar result for $f_{Y_2}(y_2)$, Substituting in the above integral we find that

$$f_Z(z) = \frac{1}{2\pi\sigma^2} \int_0^z \frac{e^{-z/2\sigma^2}}{\sqrt{\{y_2(z - y_2)\}}} dy_2 = \frac{1}{2\sigma^2} e^{-z/2\sigma^2}$$

for $0 < z < \infty$. Since

$$Z = \chi^2 \quad \text{and} \quad f_\chi(\chi) = f_Z(z) / \left| \frac{d\chi}{dz} \right|,$$

the probability density function of the $\chi$ random variable reduces to

$$f_\chi(\chi) = \frac{\chi}{\sigma^2} e^{-\chi^2/2\sigma^2}$$

for $0 < \chi < \infty$ which is the Rayleigh probability density function. It is of the form shown in Figure 13.

These results may be generalised to cover $n$ summed independent random variables. If

$$\chi^2 = X_1^2 + X_2^2 + X_3^2 + \ldots + X_n^2$$

it can be shown[2] that the probability density function of the random variable $Z$ is given by

$$f_Z(z) = \frac{1}{2^{n/2} \sigma^n \Gamma(n/2)} z^{(n-2)/2} e^{-z/2\sigma^2}$$

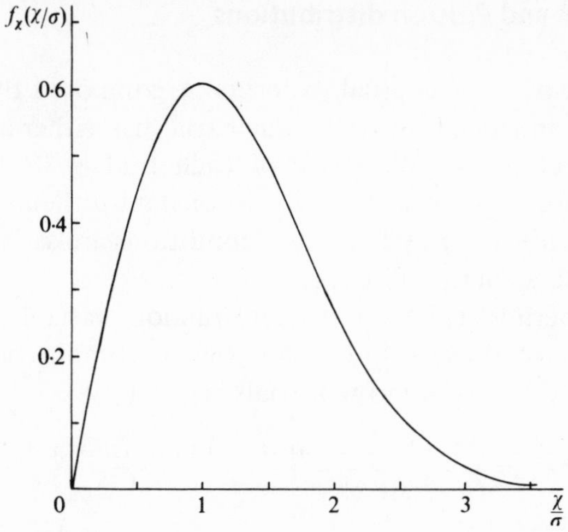


**Figure 13.** The Rayleigh probability density function of the random variable $\chi/\sigma$.

for $0 < z < \infty$, where the Gamma function $\Gamma(n + 1) = n\Gamma(n)$, $\Gamma(1) = 1$ and $\Gamma(\frac{1}{2}) = \sqrt{\pi}$. This is the 'Chi-square distribution with $n$ degrees of freedom'. Moreover the probability density of the random variable $\chi$ is given by

$$f_\chi(\chi) = \frac{2}{2^{n/2}\,\sigma^n\,\Gamma(n/2)}\,\chi^{n-1}\,e^{-\chi^2/2\sigma^2}$$

for $0 < \chi < \infty$. This latter is the 'Chi-distribution' and, when $n = 2$, it becomes a special case of the Rayleigh probability density function.

### 4.5.1 Moments of the Chi-distribution

The moments of the Chi-distribution are given by

$$E[\chi^m] = \frac{2}{2^{n/2}\,\sigma^n\,\Gamma(n/2)}\int_0^\infty \chi^{m+n-1}\,e^{-\chi^2/2\sigma^2}\,d\chi$$

$$= \frac{(2\sigma^2)^{(m+n)/2}}{2^{n/2}\,\sigma^n\,\Gamma(n/2)}\,\Gamma\left(\frac{m+n}{2}\right).$$

When $n = 2$ the Chi-distribution reduces to a Rayleigh distribution, and then it is seen that

$$E[\chi] = \sigma\sqrt{\frac{\pi}{2}},\quad E[\chi^2] = 2\sigma^2, \ldots \text{ etc.}$$

### 4.6 Binomial and Poisson distributions

Consider a real or conceptual experiment composed of a large finite number $n$ of independent trials. Each trial has either a success, S, or failure, F, and the sample space of each trial is {S, F}. If in each trial, the probability of a success is a constant $p$ then the probability of a failure is $q = 1 - p$. When these conditions are satisfied, the trials are referred to as Bernoulli trials.

In the experiment, let the discrete random variable $X$ denote the total number of successes. The number of combinations of subsets with exactly $x$ successes in the $n$ trials is given by

$$^nC_x = \frac{n!}{(n-x)!\,x!} = \frac{n(n-1)\ldots(n-x+1)}{x!}.$$

Before proceeding further, let us consider an experiment consisting of three independent trials. The sample space of the experiment is a combination of the sample space of each trial {S, F} including all possible outcomes of the experiment and given by

$$\mathscr{S} = \{\text{SSS, SSF, SFS, FSS, FFS, FSF, SFF, FFF}\}.$$

The probabilities associated with each element of the sample space are determined by the form of the element. For example, the probability of obtaining two successes in the three independent trials is

$$P[\text{SSF}] = P[\text{SFS}] = P[\text{FSS}] = p^2 q.$$

However, this has not accounted for the number of elements which have two successes and one failure in the experiment. This we see is given by

$$^3C_2 = \frac{3!}{1!2!} = 3$$

and the total probability of having two successes in three independent trials is $3p^2q$.

When the $n$ trials are independent, the probability of an element of the sample space having $x$ successes and $(n - x)$ failures in a single trial is $p^x q^{n-x}$. Thus the probability of the discrete random variable $X$ equalling $x$ successes is

$$\begin{aligned} P[X = x] &= f_X(x) = {}^nC_x p^x q^{n-x} \\ &= x^{\text{th}} \text{ term of the binomial expansion of } (q+p)^n \end{aligned}$$

where $f_X(x)$ denotes the binomial probability density function, and the binomial distribution function is given by

$$P[X \leqslant x] = F_X(x) = \sum_{r=0}^{x} {}^nC_r p^r q^{n-r}$$

for $x = 0, 1, 2 \ldots n$ successes.

If in the experiment the number of trials increases indefinitely such that the probability of a success is very small and of a failure is near unity then, with

$$\lim_{p \to 0} (1 - p)^x = 1 \quad \text{and} \quad np = m,$$

we have

$$\lim_{p \to 0} \{P[X = x]\} = \lim_{p \to 0} [m(m - p) \ldots \{m - (x - 1)p\}] = m^x$$

and

$$\lim_{n \to \infty} (1 - p)^n = \lim_{n \to \infty} \left(1 - \frac{m}{n}\right)^n = e^{-m}.$$

Thus in the limit as $n \to \infty$, $p \to 0$ and $q \to 1$, the binomial probability density function reduces to

$$P[X = x] = f_X(x) = e^{-m}\left(\frac{m^x}{x!}\right)$$

for $x = 0, 1, 2, \ldots, \infty$ successes, and is called the 'Poisson probability density function'. The corresponding distribution function is given by

$$P[X \leqslant x] = F_X(x) = \sum_{r=0}^{x} e^{-m}\left(\frac{m^r}{r!}\right)$$

for $x = 0, 1, 2, \ldots$ successes. Since these Poisson functions contain only one constant they are easier to handle than their two constant binomial counterparts.

The moment generating function for the discrete random variable $X$ is,

$$M_X(\theta) = \sum_{x=0}^{n} e^{i\theta x} f_X(x) = \sum_{x=0}^{n} {}^nC_x (pe^{i\theta})^x q^{n-x} = (q + pe^{i\theta})^n$$

which after expanding or differentiating gives the mean and variance as

$$\mu_X = np, \qquad \sigma_X^2 = npq$$

for a binomially distributed variable and

$$\mu_X = m = \sigma_X^2$$

for a Poisson distributed variable, since $q \doteqdot 1$.

These distributions are sometimes referred to as the counting distributions and are used in problems dealing with calls on a telephone, the number of customers arriving at a petrol station, electron emission, etc.

### 4.6.1 The time dependent Poisson distribution

Although in the previous section we derived the Poisson density function as a limiting case of a binomial density function it may also be obtained by considering the arrivals, counts or successes of a particular event distributed randomly over a time or space domain. In effect, we wish to determine the probability of the discrete random variable $X$ having $x(\geqslant 0)$ successful events or arrivals occurring in the time interval $(0, t)$; this is denoted by $P[X = x; t] = f_X(x; t)$. It is assumed that the arrivals satisfy the following conditions:

(a) The arrivals in the time interval under consideration are independent of other arrivals in the past or in the future.
(b) The probability of exactly one arrival in the interval $(t, t + \delta t)$ is $\lambda \delta t$ where $\lambda$ is a constant. This probability is the same as the probability of exactly one arrival in the time interval $(t + T, t + T + \delta t)$.
(c) The probability of simultaneous arrivals in the interval $(t, t + \delta t)$ is assumed negligible compared with the probability of one arrival in the same time interval.

By using these assumptions, it is seen that the occurrence of $x$ arrivals in the interval $(0, t + \delta t)$ is the union of the two mutually exclusive events. Either

$$x \text{ arrivals occur in } (0, t) \text{ and zero in } (t, t + \delta t)$$

or

$$(x - 1) \text{ arrivals occur in } (0, t) \text{ and } 1 \text{ in } (t, t + \delta t).$$

Since the arrivals are independent, the probability of $x$ arrivals in $(0, t + \delta t)$ is given by

$$P[X = x; t + \delta t] = P[X = x; t]P[X = 0; \delta t] + P[X = x - 1; t]P[X = 1; \delta t].$$

However, in the interval $(t, t + \delta t)$, we have from condition (b)

$$P[X = 1; \delta t] = \lambda \delta t \text{ and } P[X = 0; \delta t] = 1 - \lambda \delta t$$

with the result that,

$$P[X = x; t + \delta t] - P[X = x; t] = \lambda \delta t \{P[X = x - 1; t] - P[X = x; t]\}.$$

In the limit as $\delta t \to 0$ we have the following differential equation for $P[X = x; t]$:

$$\frac{dP[X = x; t]}{dt} = \lambda\{P[X = x - 1; t] - P[X = x; t]\}.$$

A solution for $P[X = x; t]$ may be obtained by solving the separate equations for $x = 0, 1, 2, \ldots,$ etc. Thus at $x = 0$, $P[X = -1, t] = 0$ and the differential equation reduces to

$$\frac{dP[X = 0; t]}{dt} = -\lambda P[X = 0; t]$$

which has a solution

$$P[X = 0; t] = e^{-\lambda t}$$

where the constant of integration is determined by the assumption that $P[X = 0; 0] = 1$. This is the value of $P[X = 0; \delta t] = 1 - \lambda \delta t$ at $\delta t = 0$.

At $x = 1$, the differential equation becomes

$$\frac{dP[X = 1; t]}{dt} = \lambda\{P[X = 0; t] - P[X = 1; t]\}$$

$$= \lambda e^{-\lambda t} - \lambda P[X = 1; t]$$

which, under the assumption that $P[X = 1; 0] = 0$, has a solution

$$P[X = 1; t] = (\lambda t)e^{-\lambda t}.$$

By repeating the procedure to $X = x$ it is found that

$$P[X = x; t] = f_X(x; t) = \frac{(\lambda t)^x}{x!} e^{-\lambda t}$$

for $x = 0, 1, 2, \ldots, \infty$ successes, which is the probability of the discrete random variable $X$ having $x$ arrivals in the continuous time interval $(0, t)$. By comparison with the previous section it is seen that $f_X(x; t)$ has the form of a Poisson density function with mean value $\lambda t$ where the constant $\lambda$ equals the expected arrival rate. However, if the $\lambda$ is itself a function of time, then the Poisson density function becomes

$$f_X(x; t) = \frac{\{\int_0^t \lambda(\tau) d\tau\}^x}{x!} e^{-\int_0^t \lambda(\tau) d\tau}$$

where the mean value is $\int_0^t \lambda(\tau) d\tau$ and reduces to $\lambda t$ when $\lambda(\tau) = \lambda$, a constant.

### 4.6.2 The exponential and Weibull distributions

The previous section shows that the probability of zero arrivals in the interval (0, $t$) is given by $e^{-\int_0^t \lambda(\tau)d\tau}$. If we assume that this implies a 'success' then the probability of failure in the interval (0, $t$) is

$$P[L \leqslant t] = 1 - e^{-\int_0^t \lambda(\tau)d\tau} = F_L(t)$$

where the random variable $L$ implies failure.

For $\lambda(\tau) = \lambda$, a constant,

$$F_L(t) = 1 - e^{-\lambda t}$$

is referred to as the exponential distribution function. It is used for example in the reliability analysis of electronic components in electrical systems.

In naval architecture, a modified form of this distribution is used in a probabilistic approach to wave-induced bending moment[3]. If the random variable $X$ denotes the amplitude of the wave bending moment, then, by applying the above result we have

$$P[X \leqslant x] = F_X(x) = 1 - e^{-\int_0^x \lambda(\tau)\,d\tau}.$$

When, for example,

$$\int_0^x \lambda(\tau)d\tau = \left(\frac{x}{k}\right)^r$$

then

$$F_X(x) = 1 - e^{-(x/k)^r}$$

and

$$f_X(x) = \frac{r}{k}\left(\frac{x}{k}\right)^{r-1} e^{-(x/k)^r}$$

for $x > 0$ are referred to as the 'Weibull' probability distribution function and density function respectively. The constant parameters $k$ and $r$ are determined either from data obtained from full scale experiment or from theoretical investigations. In particular, if $r = 2$ and $k = \sqrt{2}\sigma_X$ then the Weibull probability density function reduces to

$$f_X(x) = \frac{x}{\sigma_X^2} e^{-x^2/2\sigma_X^2}$$

which is the Rayleigh probability density function.

## References

[1] CRAMÉR, H., 1946, *Mathematical Methods of Statistics*, Princeton University Press, New Jersey.

[2] PAPOULIS, A., 1965, *Probability, Random Variables and Stochastic Processes*, McGraw-Hill, New York.

[3] MANSOUR, A., 1972, 'Methods of computing and probability of failure under extreme values of bending moment'. *J. Ship Res.*, **16**, 113–123.

# 5 Random processes

So far we have restricted attention to experiments in which discrete samples are taken, or else a discrete measurement is made. There is no reason why this should necessarily be so and the random variable may in fact depend on time. We now focus attention on measurements of time dependent quantities (such as pressure at some point in a turbulent flow). Let the measured quantity be as shown in curve (a) of Figure 14. At any instant $t$, the value is $X^{(1)}(t)$.

Now since the measured quantity is a random variable, $X^{(1)}(t)$ is just one realisation of an infinite number of possibilities. The variation measured could as well have been some other, $X^{(2)}(t)$ say. An 'ensemble' of all the possible individual deterministic realisations is sometimes referred to, and it is associated with a 'random process' $X(t)$. Notice that it is not possible to observe a random process – only one realisation of it.

The concept of a random process may be widened in two ways that are of interest to us. In the first place, a given phenomenon may be associated with two or more random processes. Thus the heave and pitch of a ship in a seaway are distinct, measurable quantities and at any instant the realisations of both, $X^{(1)}(t)$ and $Y^{(1)}(t)$ respectively, may be recorded. We should then have random processes $X(t)$ and $Y(t)$ (and we should not doubt expect them to be related to one another in some way). In fact any one experiment may provide us with one realisation each of as many random processes as we care to define.

Secondly, we may introduce more independent variables than the one (i.e. time) into our discussion. In the analysis of ocean waves, for instance, the height of the surface above some suitable datum varies with location as well as time. If the position of an observation point

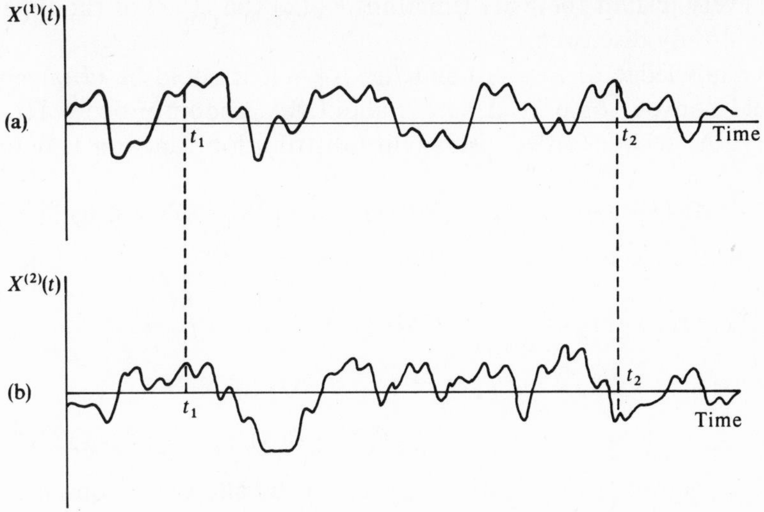


**Figure 14.** Realisations '1' and '2' of a random process $X(t)$.

with respect to a chosen origin is denoted by a vector **r**, the height of the surface in the measured realisation would be $X^{(1)}(\mathbf{r}, t)$. The random process is $X(\mathbf{r}, t)$.

From Figure 14 it can be seen that there exist two possible ways of defining the statistical properties of the random process $X(t)$. In one, we can consider the statistical properties taken 'across the ensemble' at fixed values of time $t = t_1$, $t = t_2$ etc., or we may consider the properties of the random process taken 'along the ensemble' where $t_1$, $t_2$ etc. are assumed to vary.

## 5.1 Probability functions in relation to a random process

A random process is determined by the probabilities associated with it. Consider, for instance, the random process $X(t)$ with realisations $X^{(i)}(t)$, $(i = 1, 2, \ldots, \infty)$ at the time $t = t_1$. At this specified instant the random process $X(t)$ becomes the continuous random variable $X(t_1)$ with a first order probability distribution defined by

$$F_X(x; t_1) = P[X(t_1) \leqslant x].$$

The corresponding density function is

$$f_X(x; t_1) = \frac{\partial F_X(x; t_1)}{\partial x}$$

which satisfies the relationship

$$f_X(x; t_1)\delta x = P[x < X(t_1) \leqslant x + \delta x].$$

For every instant there are functions $F_X(x)$ and $f_X(x)$ of the type we have already discussed.

A knowledge of $F(x; t_1)$ and $f(x; t_1)$ – if it could be obtained – would convey some information about the random process $X(t)$, at $t = t_1$. A 'second order' probability distribution function is defined as

$$F_{XX}(x_1; t_1 : x_2; t_2) = P[\{X(t_1) \leqslant x_1\} \cap \{X(t_2) \leqslant x_2\}]$$

and the second order density function is

$$f_{XX}(x_1; t_1 : x_2; t_2) = \frac{\partial^2}{\partial x_1 \partial x_2} F_{XX}(x_1; t_1 : x_2; t_2)$$

$$f_{XX}(x_1; t_1 : x_2; t_2)\delta x_1 \delta x_2 = P[\{x_1 < X(t_1) \leqslant x_1 + \delta x_1\} \cap \{x_2 < X(t_2) \leqslant x_2 + \delta x_2\}].$$

Again, the unwieldy notation should not be allowed to obscure the theory that has already been outlined.

In just the same way one could, in theory at least, build up distribution and density functions of higher order. We may also build up distribution and density functions for two or more random processes. Thus (to take the simplest case)

$$F_{XY}(x; t_1 : y; t_2) = P[\{X(t_1) \leqslant x\} \cap \{Y(t_2) \leqslant y\}]$$

and

$$f_{XY}(x; t_1 : y; t_2) = \frac{\partial^2}{\partial x \, \partial y} F_{XY}(x; t_1 : y; t_2)$$

such that

$$f_{XY}(x; t_1 : y; t_2)\delta x \delta y = P[\{x < X(t_1) \leqslant x + \delta x\} \cap \{y < Y(t_2) \leqslant y + \delta y\}].$$

If we consider the random process $X(\mathbf{r}, t)$ at a specified position $\mathbf{r} = \mathbf{r}_1$ and time $t = t_1$, information of the continuous random variable $X(\mathbf{r}_1, t_1)$ is given by the first order probability distribution function

$$F_X(x; \mathbf{r}_1, t_1) = P[X(\mathbf{r}_1, t_1) \leqslant x]$$

and second order probability distribution function

$$F_{XX}(x_1; \mathbf{r}_1, t_1 : x_2; \mathbf{r}_2, t_2) = P[\{X(\mathbf{r}_1, t_1) \leqslant x_1\} \cap \{X(\mathbf{r}_2, t_2) \leqslant x_2\}].$$

Even though the notation is more complicated, the variables $\{\mathbf{r}_1, \mathbf{r}_2\}$ indicate the position and $\{t_1, t_2\}$ the times where measurements of the random process $X(\mathbf{r}, t)$ are made.

A random process, or a set of random processes, may be thought of as being completely defined by its probability distributions or densities *of all orders*. Needless to say, it would be quite impossible to specify any one with much precision for any practical experiment, let alone an infinite number of them. We have therefore to make do with a limited amount of information on just a few.

## 5.2 Practical measures of a random process

### 5.2.1 Mean and mean square values

The simplest measure of a random process $X(t)$ at time $t = t_1$ is its mean or expected value

$$\mu_X(t_1) = E[X(t_1)] = \int_{-\infty}^{\infty} x\, f_X(x; t_1)\, dx.$$

As the notation indicates, $\mu_X(t_1)$ varies with $t_1$.

A little more information about the one dimensional probability density function is given by the mean square value

$$E[X^2(t_1)] = \int_{-\infty}^{\infty} x^2 f_X(x; t_1)\, dx.$$

### 5.2.2 Auto-correlation and auto-covariance

If we are prepared to consider the more complicated density function $f_{XX}(x_1; t_1 : x_2; t_2)$ then the simplest statistic of value is the 'auto-correlation' function defined as

$$R_{XX}(t_1, t_2) = E[X(t_1)X(t_2)].$$

This function may also be written as $R(t_1, t_2)$ and is the mean of the product of the random process at times $t_1$ and $t_2$ on the same realisation. In effect, the auto-correlation function helps establish the influence on the random process at time $t_2$ ($>t_1$) of values from the earlier time $t_1$. In terms of the density function,

$$R_{XX}(t_1, t_2) = \int_{-\infty}^{\infty} \int_{-\infty}^{\infty} x_1 x_2 f_{XX}(x_1; t_1 : x_2; t_2)\, dx_1 dx_2.$$

Notice that $E[X^2(t)]$ is the special case in which $t_2 = t_1 = t$.

The 'auto-covariance' of the random process $X(t)$ averages the product of the deviations from the mean at times $t_1$ and $t_2$ on the same realisation. The auto-covariance of $X(t)$ is thus the covariance of $X(t_1)$ and $X(t_2)$, given by

$$C_{XX}(t_1, t_2) = E[\{X(t_1) - \mu_X(t_1)\}\{X(t_2) - \mu_X(t_2)\}]$$

so that

$$C_{XX}(t_1, t_2) = \int_{-\infty}^{\infty}\int_{-\infty}^{\infty} \{x_1 - \mu_X(t_1)\}\{x_2 - \mu_X(t_2)\} \times$$

$$f_{XX}(x_1; t_1 : x_2; t_2)\, dx_1 dx_2$$

$$= R_{XX}(t_1, t_2) - \mu_X(t_1)\mu_X(t_2).$$

Notice that when $t_1 = t_2 = t$, the auto-covariance function reduces to $E[\{X(t) - \mu_X(t)\}^2]$, the variance of $X(t)$ given by

$$\sigma_X^2(t) = C_{XX}(t) = R_{XX}(t) - \mu_X^2(t).$$

Notice also that the auto-covariance function can be standardised by dividing it by the product of the standard deviation of $X(t_1)$ and of $X(t_2)$.

### 5.2.3 Cross-correlation and cross-covariance

The cross-correlation function of the random processes $X(t)$ at time $t_1$ and $Y(t)$ at time $t_2$ is defined by

$$R_{XY}(t_1, t_2) = E[X(t_1)Y(t_2)]$$

whilst the cross-covariance function is given by

$$C_{XY}(t_1, t_2) = E[\{X(t_1) - \mu_X(t_1)\}\{Y(t_2) - \mu_Y(t_2)\}]$$

$$= R_{XY}(t_1, t_2) - \mu_X(t_1)\mu_Y(t_2).$$

These functions describe the dependence of one random variable on the other.

The random processes are uncorrelated if, for any time, we have

$$C_{XY}(t_1, t_2) = 0,$$

which implies that

$$R_{XY}(t_1, t_2) = \mu_X(t_1)\mu_Y(t_2).$$

### 5.2.4 Position dependence

When the random process is a function of position $\mathbf{r}$ as well as time $t$ its mean value at a specified location $\mathbf{r}_1$ and time $t_1$ is given by

$$\mu_X(\mathbf{r}_1, t_1) = E[X(\mathbf{r}_1, t_1)] = \int_{-\infty}^{\infty} x f_X(x; \mathbf{r}_1, t_1)\, dx$$

and varies with both $\mathbf{r}_1$ and $t_1$. Though we have introduced the additional variable $\mathbf{r}$ the basic definitions of the statistical averages are unchanged. The auto-correlation function is defined as

$$R_{XX}(\mathbf{r}_1, t_1; \mathbf{r}_2, t_2) = E[X(\mathbf{r}_1, t_1)X(\mathbf{r}_2, t_2)]$$

$$= \int_{-\infty}^{\infty} \int_{-\infty}^{\infty} x_1 x_2 f_{XX}(x_1; \mathbf{r}_1, t_1 : x_2; \mathbf{r}_2, t_2)\, dx_1\, dx_2,$$

and the remaining averages are similarly transformed.

## 5.3 Special types of random process

Description of a random process $X(t)$ or $X(\mathbf{r}, t)$ by means of its associated probability density functions would require an enormous amount of information. Fortunately it is usually possible to assume that the process is of a special form. It may be 'stationary', 'homogeneous' or possibly 'ergodic'. We come now to discuss these concepts.

### 5.3.1 Stationary process

Stationarity of a process can be pictured intuitively as the absence of any drift in the ensemble of realisations as time proceeds. Mathematically this means that the probability distribution and density functions are unchanged by a shift of the time or position scale. They are applicable now and will remain so for all time. Thus the random processes $X(\mathbf{r}, t)$ and $X(\mathbf{r}, t + T)$ are *stationary in time* when they have the same statistics for any time $T$.

For the random process $X(\mathbf{r}, t)$ to be stationary in time, then,

$$f_X(x; \mathbf{r}, t) = f_X(x; \mathbf{r}, t + T)$$

for any value of $T$. It therefore follows that

$$f_X(x; \mathbf{r}, t) = f_X(x; \mathbf{r}).$$

The second order density function is such that

$$f_{XX}(x_1; \mathbf{r}_1, t_1 : x_2; \mathbf{r}_2, t_2) = f_{XX}(x_1; \mathbf{r}_1, t_1 + T : x_2; \mathbf{r}_2, t_2 + T)$$

for any $T$. This does not imply that time may be omitted altogether from the density function however, since it is *not* suggested that time differences are irrelevant. Let $t_2 - t_1 = \tau$; then

$$f_{XX}(x_1; \mathbf{r}_1, t_1 : x_2; \mathbf{r}_2, t_2) = f_{XX}(x_1; \mathbf{r}_1, 0 : x_2; \mathbf{r}_2, \tau)$$

$$= f_{XX}(x_1; \mathbf{r}_1, -\tau : x_2; \mathbf{r}_2, 0).$$

Similarly the $n$th order density functions of a stationary (in time) random process $X(\mathbf{r}, t)$ must be such that

$$f_{X_1X_2\ldots X_n}(x_1;\mathbf{r}_1, t_1:\ldots:x_n;\mathbf{r}_n, t_n)$$
$$= f_{X_1X_2\ldots X_n}(x_1;\mathbf{r}_1, t_1+T:\ldots:x_n;\mathbf{r}_n, t_n+T)$$

for any $T$. For a random process that is stationary in time, therefore,

$$E[X(\mathbf{r}, t)] = \int_{-\infty}^{\infty} xf_X(x;\mathbf{r}, t)\,dx = \int_{-\infty}^{\infty} xf_X(x;\mathbf{r})\,dx = \mu_X(\mathbf{r}).$$

And if the process happened to be a function of time only then $E[X(t)]$ would be constant.

Since the second order density function of the random process $X(\mathbf{r}, t)$ is a function of the time difference ($\tau = t_2 - t_1$) we conclude that the auto-correlation function $R_{XX}(\mathbf{r}_1, t_1 : \mathbf{r}_2, t_2)$ is dependent on $\tau$ and may be written as

$$R_{XX}(\mathbf{r}_1 : \mathbf{r}_2, \tau) = E[X(\mathbf{r}_1, t_1)X(\mathbf{r}_2, t_2)] = R_{XX}(\mathbf{r}_1, -\tau : \mathbf{r}_2).$$

We note that, if the random process is a function of time only, this reduces to

$$E[X(t_1)X(t_2)] = R_{XX}(\tau) = R_{XX}(-\tau)$$

indicating that the auto-correlation function is an even function of $\tau$. Moreover, if $t_1 = t_2 = t$,

$$E[X^2(t)] = R_{XX}(0)$$

so that the value of the auto-correlation function when $\tau = 0$ is the mean square value of $X(t)$.

The auto-covariance of a stationary random process is

$$E[\{X(\mathbf{r}_1, t_1) - \mu_X(\mathbf{r}_1)\}\{X(\mathbf{r}_2, t_2) - \mu_X(\mathbf{r}_2)\}] = C_{XX}(\mathbf{r}_1, -\tau : \mathbf{r}_2)$$
$$= C_{XX}(\mathbf{r}_1 : \mathbf{r}_2, \tau).$$

If the process is a function of time alone, this gives

$$E[\{X(t_1) - \mu_X\}\{X(t_2) - \mu_X\}] = C_{XX}(-\tau) = C_{XX}(\tau),$$

which reduces to

$$E[\{X(t) - \mu_X\}^2] = C_{XX}(0)$$

when $t_1 = t_2 = t$.

In the same way, the random processes $X(\mathbf{r}, t)$ and $Y(\mathbf{r}, t)$ are jointly stationary in time when the joint statistics of $X(\mathbf{r}, t)$, $Y(\mathbf{r}, t)$ are the same as the joint statistics of $X(\mathbf{r}, t + T)$, $Y(\mathbf{r}, t + T)$ for any time $T$. When this is so, the joint density function $f_{XY}$ is a function of the time difference $\tau$ such that the cross-correlation function is

$$R_{XY}(\mathbf{r}_1 : \mathbf{r}_2, \tau) = E[X(\mathbf{r}_1, t_1)Y(\mathbf{r}_2, t_2)] = R_{YX}(\mathbf{r}_2 : \mathbf{r}_1, -\tau)$$

and the cross covariance is given by

$$C_{XY}(\mathbf{r}_1 : \mathbf{r}_2, \tau) = E[\{X(\mathbf{r}_1, t_1) - \mu_X(\mathbf{r}_1)\}\{Y(\mathbf{r}_2, t_2) - \mu_Y(\mathbf{r}_2)\}]$$
$$= C_{YX}(\mathbf{r}_2 : \mathbf{r}_1, -\tau).$$

When the random processes are functions of time alone, the last two results reduce to

$$R_{XY}(\tau) = E[X(t_1)Y(t_2)] = R_{YX}(-\tau)$$

and

$$C_{XY}(\tau) = E[\{X(t_1) - \mu_X\}\{Y(t_1) - \mu_Y\}] = C_{YX}(-\tau)$$

where $\tau = t_2 - t_1$.

### 5.3.2 Homogeneous process [1]

It is commonly assumed in analysis of the surface of the sea that the random process $X(\mathbf{r}, t)$ is not only stationary in time but also 'homogeneous', or stationary in space. This means that no change would be made in the process (i.e. in its probability and density functions) if the origin from which $\mathbf{r}$ is measured were shifted to a new location.

Since the fundamental principles of stationarity in time and homogeneity (or stationarity in space) convey the same idea, we shall in future, refer to all processes as just stationary random processes unless we are discussing the individual concepts of stationarity in space or time.

For a stationary random process $X(\mathbf{r}, t)$ then,

$$f_X(x; \mathbf{r}, t) = f_X(x; \mathbf{r} + \mathbf{s}, t + T)$$

for all s and all $T$. That is to say

$$f_X(x; \mathbf{r}, t) = f_X(x),$$

independent of $\mathbf{r}$ and $t$.

Turning next to the second order density function, and writing $\mathbf{r}_2 - \mathbf{r}_1 = \boldsymbol{\rho}$, we find

$$f_{XX}(x_1; \mathbf{r}_1, t_1 : x_2; \mathbf{r}_2, t_2) = f_{XX}(x_1; \mathbf{r}_1 + \mathbf{s}, t_1 + T : x_2; \mathbf{r}_2 + \mathbf{s}, t_2 + T)$$

for all s and $T$. Hence,

$$f_{XX}(x_1; \mathbf{r}_1, t_1 : x_2; \mathbf{r}_2, t_2) = f_{XX}(x_1; 0, 0 : x_2; \mathbf{r}_2 - \mathbf{r}_1, t_2 - t_1)$$
$$= f_{XX}(x_1 : x_2, \boldsymbol{\rho}, \tau)$$

and

$$f_{XX}(x_1; \mathbf{r}_1, t_1 : x_2; \mathbf{r}_2, t_2) = f_{XX}(x_1; \mathbf{r}_1 - \mathbf{r}_2, t_1 - t_2 : x_2; 0, 0)$$
$$= f_{XX}(x_1; -\rho, -\tau : x_2).$$

So one might go on for density functions of higher order.

It follows from these simplifications that the practical measures of a stationary random process $X(\mathbf{r}, t)$ are also simplified. Thus

$$E[X(\mathbf{r}, t)] = \int_{-\infty}^{\infty} x f_X(x)\, dx = \mu_X,$$

a constant. Again

$$E[X(\mathbf{r}_1, t_1)X(\mathbf{r}_2, t_2)] = \int_{-\infty}^{\infty}\int_{-\infty}^{\infty} x_1 x_2 f_{XX}(x_1 : x_2, \rho, \tau)dx_1\, dx_2$$
$$= R_{XX}(\rho, \tau)$$

and

$$E[\{X(\mathbf{r}_1, t_1) - \mu_X\}\{X(\mathbf{r}_2, t_2) - \mu_X\}]$$
$$= \int_{-\infty}^{\infty}\int_{-\infty}^{\infty} (x_1 - \mu_X)(x_2 - \mu_X) f_{XX}(x_1 : x_2, \rho, \tau)dx_1\, dx_2$$
$$= C_{XX}(\rho, \tau).$$

We have therefore achieved some substantial simplifications.

Notice particularly that for a stationary random process, the statistical measures are independent of position $\mathbf{r}$ and time $t$ but they usually depend on the shift $\rho$ and time interval $\tau$.

### 5.3.3 Ergodic process

Strictly speaking, only a single realisation $X^{(1)}(t)$ of a random process $X(t)$ can be recorded. Having obtained it, however, we note that every observable feature of it may be identified *axiomatically* with appropriate probability distribution and density functions. The question of how these functions may be found is another matter – and a very difficult one. Generally speaking these functions are never known with any precision and recourse is had to the crude information imparted by the expectations mentioned in section 5.2. How these expectations are found is a matter of guessing, experience or of measurement by necessarily approximate means.

It will be helpful at this point to digress and to point out a common alternative approach.

We have referred to the ensemble of realisations $X^{(1)}(t)$, $X^{(2)}(t), \ldots$ which, together, determine a random process $X(t)$. The idea of an average 'across the ensemble' is commonly used to define the expectations. Thus the expected value of $X(t)$ at some instant $t_1$ is taken as

$$E[X(t_1)] = \lim_{N \to \infty} \frac{1}{N} \sum_{i=1}^{N} X^{(i)}(t_1).$$

This approach is based upon the traditional theory of probability mentioned in section 1. It is just an alternative way of looking at the theory.

Suppose we have a realisation $X^{(1)}(t)$ of the random process $X(t)$ and that its duration is $T$. By some means it is required to obtain the measures of the expectations of the process (c.f. section 5.2). Generally it will not be possible to postulate these *a priori* and so it becomes necessary to enquire if $X^{(1)}(t)$ will provide them.

Notice that if the experiment in which $X^{(1)}(t)$ is measured could be repeated a number of times at *exactly* the same location, over the same period, with identical equipment, then at least approximations to $X^{(2)}(t)$, $X^{(3)}(t), \ldots$ could be obtained. From these realisations it might be possible to assemble something like the ensemble and so form approximate ensemble averages. But this approach is of questionable validity in theory and is likely to be inconclusive in practice.

One approach that may be adopted with a known realisation is to obtain *temporal* averages along it. Thus the temporal mean is

$$\langle X^{(1)}(t) \rangle = \lim_{T \to \infty} \frac{1}{T} \int_{-T/2}^{T/2} X^{(1)}(t) dt.$$

The temporal mean square is

$$\langle X^{(1)2}(t) \rangle = \lim_{T \to \infty} \frac{1}{T} \int_{-T/2}^{T/2} X^{(1)2}(t) dt$$

and so forth. The brackets $\langle \; \rangle$ will henceforth be taken to indicate a temporal mean.

A random process $X(t)$ is said to be 'ergodic' if it is stationary and the expectations are equal to the corresponding temporal averages taken along the single realisation. Thus

$$E[X(t)] = \mu_X = \langle X^{(1)}(t) \rangle$$

and

$$E[X(t)X(t+\tau)] = R_{XX}(\tau) = \langle X^{(1)}(t)X^{(1)}(t+\tau)\rangle$$

$$= \lim_{T\to\infty} \frac{1}{T} \int_{-T/2}^{T/2} X^{(1)}(t)X^{(1)}(t+\tau)dt.$$

Intuitively it may be said that the realisation $X^{(1)}(t)$ must be 'typical' of all the possible realisations if the process is to be ergodic. An ergodic random process $X(t)$ must be stationary, but a stationary random process need not be ergodic.

If a random process is space dependent the ergodic hypothesis implies that averaging must take place over space as well as time, if that is relevant too. This requires that the process must be homogeneous, as well as stationary in time. For example, if $X^{(1)}(\mathbf{r}, t)$ is a realisation of an ergodic process $X(\mathbf{r}, t)$ measured over a space $S$ and time $T$, then

$$R_{XX}(\rho, \tau) = \langle X^{(1)}(\mathbf{r}, t)X^{(1)}(\mathbf{r}+\rho, t+\tau)\rangle$$

$$= \lim_{\substack{S\to\infty \\ T\to\infty}} \frac{1}{ST} \int_T \int_S X^{(1)}(\mathbf{r}, t)X^{(1)}(\mathbf{r}+\rho, t+\tau)d\mathbf{r}dt.$$

An *estimate* of the auto-correlation function $R_{XX}(\rho, \tau)$ is found if the limit is not taken; that is

$$R_{XX}(\rho, \tau) \doteqdot \frac{1}{ST} \int_T \int_S X^{(1)}(\mathbf{r}, t)X^{(1)}(\mathbf{r}+\rho, t+\tau)d\mathbf{r}dt.$$

The idea of an 'ergodic process' implies that all the statistical measures of the process are ergodic although it it usual to define only the ergodicity of particular statistical measures.

## 5.4 The correlation functions of an ergodic process

The auto-correlation function of an ergodic process $X(t)$ is

$$R_{XX}(\tau) = \langle X^{(1)}(t)X^{(1)}(t+\tau)\rangle = \langle X^{(1)}(t-\tau)X^{(1)}(t)\rangle$$

$$= R_{XX}(-\tau)$$

so that $R_{XX}(\tau)$ is a real even function of $\tau$. The value of the function corresponding to $\tau = 0$ is

$$R_{XX}(0) = \langle X^{(1)2}(t)\rangle$$

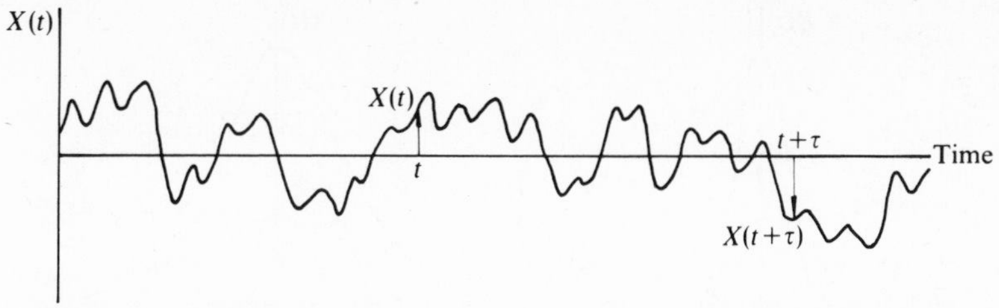


**Figure 15.** A single realisation of a random process $X(t)$ in which it is assumed that $E[X(t)X(t+\tau)] = \langle X(t)X(t+\tau)\rangle$ for all realisations. The realisation shown is therefore not identified (e.g. as $X^{(1)}(t)$) and the process is said to be 'ergodic'.

which is the mean square value of the random process. Notice that the ratio $R_{XX}(\tau)/R_{XX}(0)$ is sometimes used as a convenient non-dimensional form of $R_{XX}(\tau)$.

It will be seen on inspection of Figure 15 that $R_{XX}(\tau)$ is a measure of the connection between the value of the random variable $X(t)$ at instants that are separated by a time interval $\tau$. It is the same for each realisation and is therefore a characteristic of the random process. As $\tau$ increases, the connection becomes less apparent and eventually the values are practically unrelated. Consequently, provided that there are no periodic components present in the random process, $R_{XX}(\tau)$ tends to zero since products will be equally positive and negative. That is $R_{XX}(\tau) \to 0$ as $\tau \to \infty$.

Consideration of the inequality

$$E[\{X(t) \pm X(t+\tau)\}^2] \geqslant 0$$

shows

$$2R_{XX}(0) \pm 2R_{XX}(\tau) \geqslant 0.$$

It follows that

$$R_{XX}(0) \geqslant |R_{XX}(\tau)|$$

indicating that the auto-correlation function has a maximum at $\tau = 0$. Typical forms of $R(\tau)$ are shown in Figure 16.

It will be shown later that the derivatives of $R_{XX}(\tau)$ are of interest, particularly at $\tau = 0$. If an over dot signifies differentiation with respect to $\tau$ we have

$$\dot{R}_{XX}(\tau) = \langle X^{(1)}(t)\dot{X}^{(1)}(t+\tau)\rangle = R_{X\dot{X}}(\tau).$$

But equally

$$\dot{R}_{XX}(\tau) = \langle -\dot{X}^{(1)}(t-\tau)X^{(1)}(t)\rangle$$

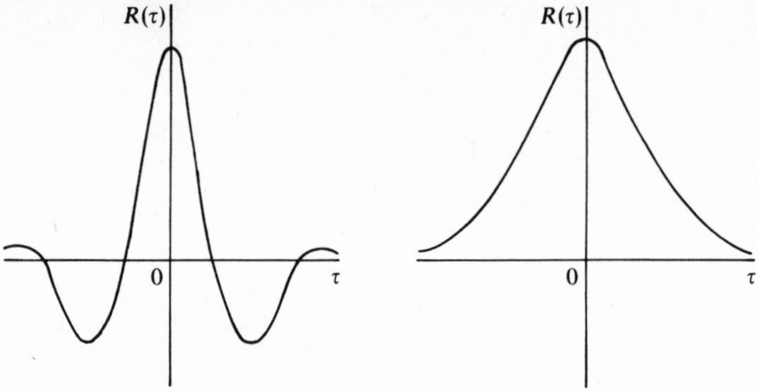


**Figure 16.** Possible forms of the auto-correlation function.

so that, at $\tau = 0$,

$$\dot{R}_{XX}(0) = \langle X^{(1)}(t)\dot{X}^{(1)}(t)\rangle = -\langle \dot{X}^{(1)}(t)X^{(1)}(t)\rangle.$$

It follows that

$$\dot{R}_{XX}(0) = 0 = R_{X\dot{X}}(0) = R_{\dot{X}X}(0).$$

In the same way,

$$\ddot{R}_{XX}(\tau) = \frac{\partial}{\partial\tau}\langle X^{(1)}(t)\dot{X}^{(1)}(t+\tau)\rangle$$

$$= \frac{\partial}{\partial\tau}\langle X^{(1)}(t-\tau)\dot{X}^{(1)}(t)\rangle$$

$$= -\langle \dot{X}^{(1)}(t-\tau)\dot{X}^{(1)}(t)\rangle$$

so that

$$\ddot{R}_{XX}(0) = -\langle \dot{X}^{(1)2}(t)\rangle.$$

This process of differentiation may be repeated and it is found in fact that

$$\dddot{R}_{XX}(0) = 0,$$

$$\ddddot{R}_{XX}(0) = \langle \ddot{X}^{(1)2}(t)\rangle.$$

Thus the mean square value of any derivative of $X^{(1)}(t)$ – and, since the random process is ergodic, of $X(t)$ – can be determined if $R_{XX}(\tau)$ is known.

If we consider the random process $X(t)$ as the sum of two ergodic random processes $u(t)$ and $v(t)$, i.e. if

$$X(t) = u(t) + v(t),$$

then the auto-correlation function

$$R_{XX}(\tau) = \lim_{T\to\infty} \frac{1}{T} \int_{-T/2}^{T/2} X(t)X(t+\tau)dt$$

$$= \lim_{T\to\infty} \frac{1}{T} \int_{-T/2}^{T/2} \{u(t)+v(t)\}\{u(t+\tau)+v(t+\tau)\}dt$$

$$= R_{uu}(\tau) + R_{uv}(\tau) + R_{vu}(\tau) + R_{vv}(\tau).$$

The first and last terms are the auto-correlation functions of the random processes $u(t)$ and $v(t)$ respectively. The remaining terms are cross-correlation functions measuring the dependence between the random processes and defined by

$$R_{uv}(\tau) = \lim_{T\to\infty} \frac{1}{T} \int_{-T/2}^{T/2} u(t)v(t+\tau)dt$$

$$= \lim_{T\to\infty} \frac{1}{T} \int_{-T/2}^{T/2} u(t-\tau)v(t)dt$$

$$= R_{vu}(-\tau).$$

Unlike the auto-correlation function the cross correlation function, in general, is neither even nor odd and does not necessarily have a maximum at $\tau = 0$. It is a real valued function which may be either positive or negative and by consideration of the inequality

$$E[\{u(t) \pm v(t+\tau)\}^2] \geqslant 0$$

we see that

$$R_{uu}(0) + R_{vv}(0) \pm 2R_{uv}(\tau) \geqslant 0$$

and

$$R_{uv}(\tau) = |R_{uv}(\tau)| \leqslant \tfrac{1}{2}\{R_{uu}(0) + R_{vv}(0)\}.$$

Further for any real constant $\alpha$ the expectation

$$E[\{\alpha u(t) \pm v(t+\tau)\}^2] \geqslant 0$$

so that

$$\alpha^2 R_{uu}(0) \pm 2\alpha R_{uv}(\tau) + R_{vv}(0) \geqslant 0.$$

This inequality is always true provided that the discriminant of this quadratic equation in $\alpha$ is non-positive. That is,

$$4R_{uv}^2(\tau) - 4R_{uu}(0)R_{vv}(0) \leqslant 0,$$

or

$$R_{uv}^2(\tau) = |R_{uv}(\tau)|^2 \leqslant R_{uu}(0)R_{vv}(0).$$

## Reference

[1] CRAMÉR, H. and LEADBETTER, M. R., 1967, *Stationary and Related Stochastic Processes: Sample Function Properties and Their Applications.* Wiley, New York.

# 6 Frequency analysis

There are two distinct methods of specifying a random process $X(t)$:

(a) through its probability density functions

(b) through certain convenient mean values.

Of these, the first is impractical. The second reveals the course of the variation of $X(t)$ through the auto-correlation function and the covariance function, these functions taking on simple forms $R_{XX}(\tau)$ and $C_{XX}(\tau)$ respectively when the process is stationary and even being measurable if the process is known to be ergodic.

It is usually more convenient to deal with time variation as measured by frequency. (This is obviously so with periodic phenomena and it will be shown later still to be true with random processes.) We shall therefore turn our attention next to the Fourier techniques that permit us to deal with frequency rather than with time.

## 6.1 Fourier series

A quantity

$$g(t) = A \cos \omega t + B \sin \omega t$$

could be represented by the pair of curves shown in Figure 17. If we were not concerned about phase with respect to the instant $t = 0$, we might choose instead to use the simpler representation of Figure 18. Yet again we might represent the sinusoidal fluctuation by means of complex exponentials, since

$$g(t) = \frac{A - iB}{2} e^{i\omega t} + \frac{A + iB}{2} e^{-i\omega t}$$

$$= Ce^{i\omega t} + C^* e^{-i\omega t},$$

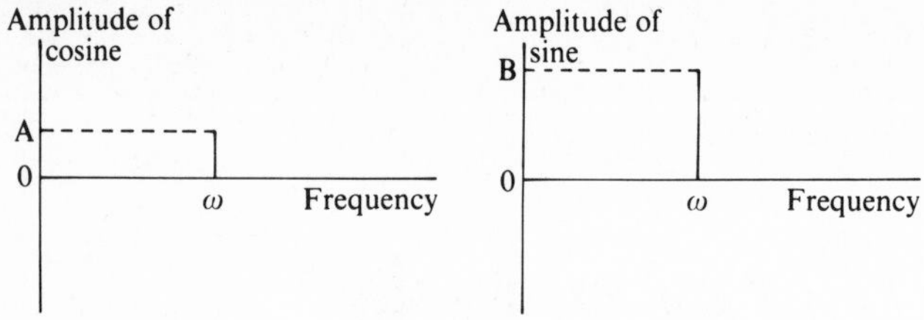


**Figure 17.** Cosine and sine frequency spectra for the function $g(t) = A \cos \omega t + B \sin \omega t$.

$C^*$ being the complex conjugate of $C$ (so that $|C^*| = |C|$). If we are prepared to disregard certain information about the fluctuating quantity $g(t)$, we can again use a graphical representation (as in Figure 19). The information that is thrown away is $\arg C = -\arg C^*$.

The curves of Figures 17, 18, 19 are all rudimentary 'frequency spectra'. The idea of such spectra is less trivial when the fluctuating quantity is periodic but not simple harmonic. Consider for instance the periodic function $g(t)$ with period $T$ which can be expressed in the form of a Fourier series given by,

$$g(t) = \tfrac{1}{2}A_0 + \sum_{n=1}^{\infty} (A_n \cos n\,\omega_0 t + B_n \sin n\,\omega_0 t)$$

where the fundamental frequency $\omega_0 = 2\pi/T$. The coefficients $A_n$ and $B_n$ are given by the following integral relationships:

$$A_n = \frac{2}{T}\int_{-T/2}^{T/2} g(t) \cos n\,\omega_0 t\, dt \qquad \text{for } n = 0, 1, 2 \ldots$$

$$B_n = \frac{2}{T}\int_{-T/2}^{T/2} g(t) \sin n\,\omega_0 t\, dt \qquad \text{for } n = 1, 2, \ldots$$

where it is assumed that

$$\int_{-T/2}^{T/2} |g(t)|dt < \infty.$$

The periodic function $g(t)$ can be written as a series in the complex form since

$$A_n \cos n\,\omega_0 t + B_n \sin n\,\omega_0 t = g(n)e^{in\omega_0 t} + g(-n)e^{-in\omega_0 t}$$

where

$$g(n) = \tfrac{1}{2}(A_n - iB_n) = \tfrac{1}{2}\sqrt{(A_n^2 + B_n^2)}\,e^{-i(\tan^{-1} B_n/A_n)}$$

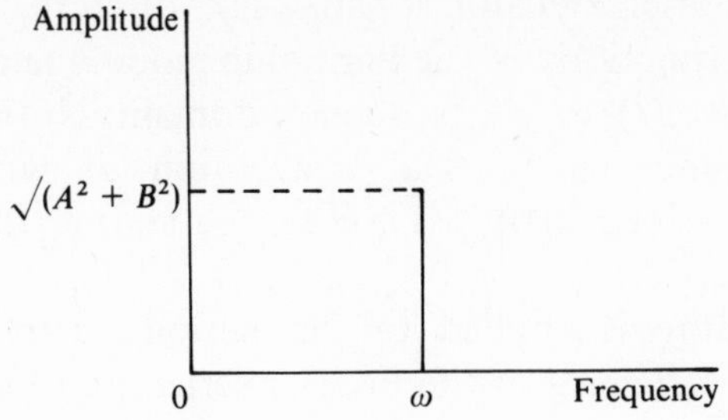


**Figure 18.** Rudimentary frequency spectrum for the function $g(t)$ of Figure 17.

and

$$g(-n) = \tfrac{1}{2}(A_n + iB_n) = g^*(n)$$

is the complex conjugate of $g(n)$. Thus, the periodic function may be written as

$$\begin{aligned} g(t) &= \tfrac{1}{2}A_0 + \sum_{n=1}^{\infty} (A_n \cos n\,\omega_0 t + B_n \sin n\,\omega_0 t) \\ &= \sum_{n=-\infty}^{\infty} g(n)\, e^{in\,\omega_0 t} \end{aligned}$$

where the coefficients are expressed in the integral form

$$g(n) = \frac{1}{T} \int_{-T/2}^{T/2} g(t) e^{-in\,\omega_0 t} dt \qquad n = 0, \pm 1, \pm 2, \ldots$$

which is referred to as the 'Fourier transform' of the periodic function. The function $g(t)$ and the quantity $g(n)$ are said to

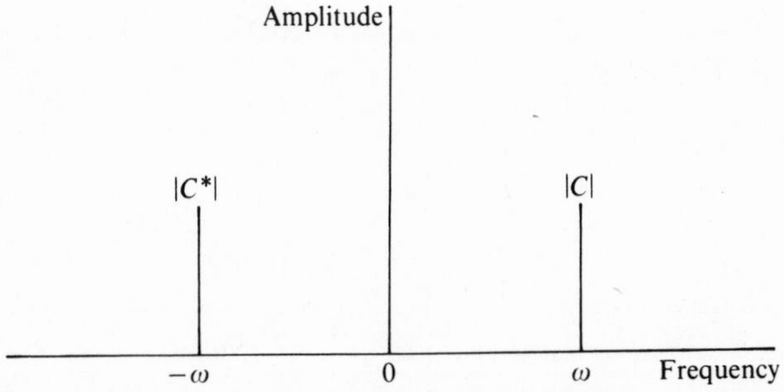


**Figure 19.** Rudimentary frequency spectrum for the function $g(t)$ of Figure 17, using the frequency range $-\infty < \omega < \infty$.

constitute a 'Fourier transform pair'. The discrete, complex coefficients $g(n)$ are functions of the harmonic mode $n$ and represent the periodic function $g(t)$ in the frequency domain. Notice that we are adopting the somewhat unusual convention of writing constant coefficients in the form $g(0)$, $g(1)$, $g(2)$, . . . (rather than $g_0, g_1, g_2$, . . . for example).

If this procedure is applied to the periodic curve as shown in Figure 20, the coefficients of the Fourier series are found to be

$$A_0 = 0$$

$$A_n = \frac{2}{T}\int_{-T/2}^{T/2} g(t)\cos n\,\omega_0 t\,dt = \frac{4A}{n\pi}\sin\frac{n\pi}{2} \qquad (n \neq 0)$$

$$B_n = \frac{2}{T}\int_{-T/2}^{T/2} g(t)\sin n\,\omega_0 t\,dt = 0.$$

The spectra are thus as shown in Figure 21. In this particular case, the $B_n$ are all zero so that the spectrum of $A_n$ would simply be rectified if the form of representation of Figure 18 were employed, and this is illustrated in Figure 22.

Since the Fourier transform

$$g(n) = \frac{1}{T}\int_{-T/2}^{T/2} g(t)\,e^{-in\omega_0 t}dt \qquad n = 0, \pm 1, \pm 2, \ldots,$$

it follows that, for this particular periodic function, the coefficients are given by

$$g(0) = 0; \qquad g(n) = \frac{2A}{n\pi}\sin\frac{n\pi}{2} \qquad (n \neq 0)$$

and the frequency spectrum analogous to that of Figure 19 is now that shown in Figure 23.

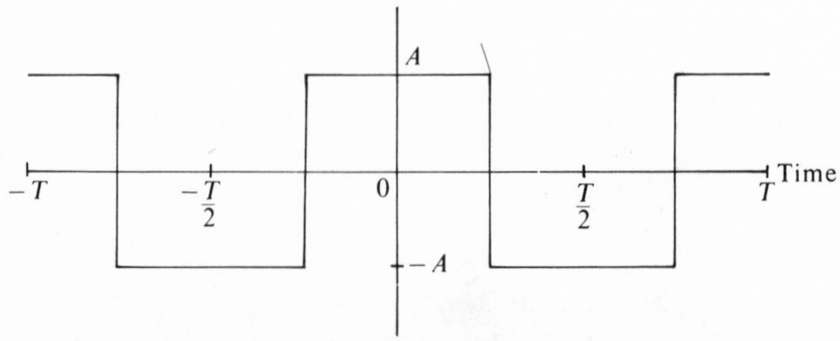


**Figure 20.** A typical periodic function.

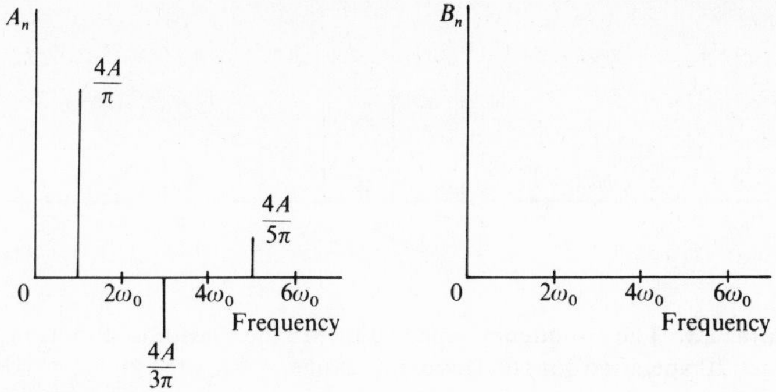


**Figure 21.** Cosine and sine frequency spectra of the periodic function represented in Figure 20.

In our subsequent discussion we shall employ the complex form of Fourier analysis. It is therefore as well to review one or two features of spectra such as that of Figure 23:

1. If $g(t)$ is a real function, the coefficients of the series will be such that $g(-n) = g^*(n)$.
2. If $g(t)$ is a real and even function, so that $g(t) = g(-t)$, then $g(n)$ and $g(-n)$ are both real and they are equal.
3. If $g(t)$ is a real and odd function, so that $g(t) = -g(-t)$, then $g(n) = -g(-n)$ and both are imaginary.

### 6.1.1 Spectral density

The time averages of the arbitrary periodic function $g(t)$ may be defined in terms of the coefficients in the Fourier series or the Fourier transform coefficients. The mean value of $g(t)$ is given by

$$\langle g(t) \rangle = \frac{1}{T} \int_{-T/2}^{T/2} g(t)\, dt = \tfrac{1}{2} A_0 = g(0)$$

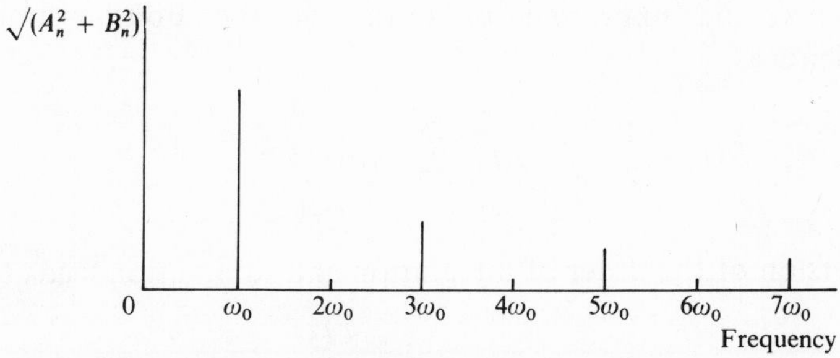


**Figure 22.** Frequency spectrum obtained from Figure 21, discarding information on phase differences.

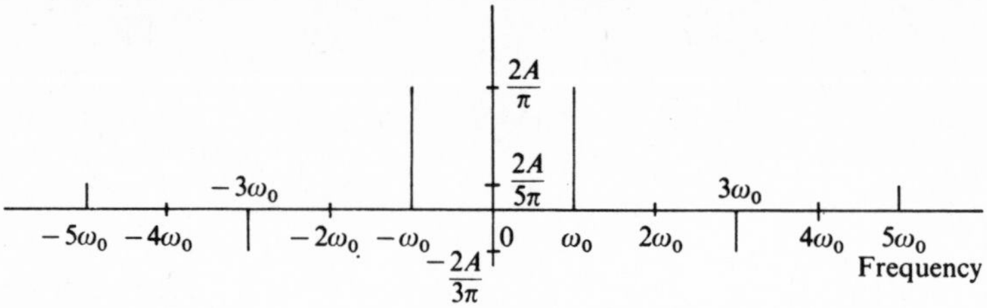


**Figure 23.** The frequency spectrum of the periodic function of Figure 20 specified for the frequency range $-\infty < \omega < \infty$.

and the mean square value is expressed as

$$\langle g^2(t)\rangle = \frac{1}{T}\int_{-T/2}^{T/2} g^2(t)\,dt$$

$$= \frac{A_0^2}{4} + \tfrac{1}{2}\sum_{n=1}^{\infty}(A_n^2 + B_n^2) = \sum_{n=-\infty}^{\infty}|g(n)|^2$$

where $|g(n)|$ is the modulus of the coefficient and

$$|g(n)|^2 = |g(-n)|^2 = \tfrac{1}{4}(A_n^2 + B_n^2)\,, \qquad n = 0, 1, 2, \ldots$$

This relationship

$$\frac{1}{T}\int_{-T/2}^{T/2} g^2(t)\,dt = \sum_{n=-\infty}^{\infty}|g(n)|^2$$

is known as Parseval's theorem for periodic functions.

The auto-correlation function of the periodic function is given by

$$R(\tau) = \frac{1}{T}\int_{-T/2}^{T/2} g(t)g(t+\tau)\,dt$$

which may be expressed in terms of the Fourier transform coefficients as

$$R(\tau) = \frac{1}{T}\int_{-T/2}^{T/2} g(t)\sum_{n=-\infty}^{\infty} g(n)e^{in\omega_0(t+\tau)}\,dt.$$

An inversion of the order of integration and summation leads to

$$R(\tau) = \sum_{n=-\infty}^{\infty} g(n)e^{in\omega_0\tau}\,\frac{1}{T}\int_{-T/2}^{T/2} g(t)e^{in\omega_0 t}\,dt$$

where the integral term is recognised as the complex conjugate of $g(n)$. Thus we have.

$$R(\tau) = \sum_{n=-\infty}^{\infty} |g(n)|^2 e^{in\omega_0\tau}$$

$$= \sum_{n=-\infty}^{\infty} S(n)e^{in\omega_0\tau}$$

where $S(n) = |g(n)|^2$ is called the two sided spectral density function and is related to the harmonic coefficients $A_n$ and $B_n$ by

$$S(n) = |g(n)|^2 = \tfrac{1}{4}(A_n^2 + B_n^2) \qquad n = 0, \pm 1, \pm 2, \ldots$$

Thus the spectral density is proportional to the sum of the squares of the individual sinusoidal component amplitudes. When the time interval $\tau = 0$, we have the form of Parseval's theorem for periodic functions, i.e.

$$R(0) = \sum_{n=-\infty}^{\infty} |g(n)|^2 = \sum_{n=-\infty}^{\infty} S(n) = \frac{1}{T}\int_{-T/2}^{T/2} g^2(t)\, dt.$$

The spectral density function may be expressed in the form

$$S(n) = \frac{1}{T}\int_{-T/2}^{T/2} R(\tau)e^{-in\omega_0 t}\, d\tau$$

so that the auto-correlation function $R(\tau)$ and spectral density function $S(n)$ form a Fourier transform pair. Since the auto-correlation function is a real and even function it follows that

$$S(n) = \frac{1}{T}\int_{-T/2}^{T/2} R(\tau) \cos n\,\omega_0\tau\, d\tau$$

and

$$R(\tau) = \sum_{n=-\infty}^{\infty} S(n) \cos n\,\omega_0\tau$$

$$= S(0) + 2\sum_{n=1}^{\infty} S(n) \cos n\,\omega_0\tau$$

$$= \tfrac{1}{4}A_0{}^2 + \tfrac{1}{2}\sum_{n=1}^{\infty} (A_n^2 + B_n^2) \cos n\,\omega_0\tau$$

$$= \sum_{n=0}^{\infty} \Phi(n) \cos n\,\omega_0\tau$$

where the physically realisable one-sided spectral density function is defined by

$$\Phi(0) = \tfrac{1}{4}A_0^2$$

$$\Phi(n) = \tfrac{1}{2}(A_n^2 + B_n^2) \qquad n = 1, 2, 3, \ldots$$

$$= 0 \qquad n < 0.$$

For the periodic function discussed in the previous section, the mean square value is given by

$$\langle g^2(t)\rangle = A^2 = \sum_{n=-\infty}^{\infty} |g(n)|^2 = \sum_{n=-\infty}^{\infty} S(n)$$

and Figure 24 illustrates the form of $S(n)$ and $\Phi(n)$.

## 6.2 The Fourier integral

An aperiodic, non-periodic or transient function can be only represented as a Fourier series if the period of the series tends to infinity. When this is allowed the fundamental frequency $\omega_0$ becomes the

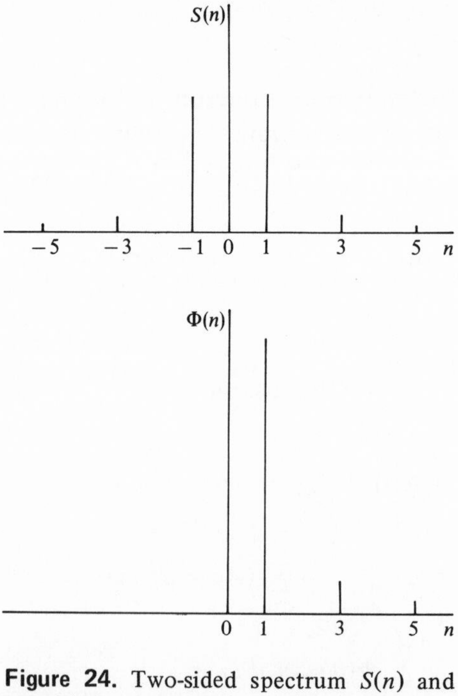


**Figure 24.** Two-sided spectrum $S(n)$ and one-sided physically realisable spectrum $\Phi(n)$ for the periodic curve of Figure 20.

infinitesimal angular frequency $\delta\omega_0 = \lim_{T\to\infty} 2\pi/T$ such that the periodic function $g(t)$ reduces to

$$g(t) = \frac{1}{2\pi} \sum_{n=-\infty}^{\infty} e^{in\delta\omega_0 t}\, \delta\omega_0 \int_{-T/2}^{T/2} g(t) e^{-in\omega_0 t}\, dt.$$

As the period $T$ approaches infinity, the infinitesimal angular frequency $\delta\omega_0$ becomes a differential of the angular frequency, $d\omega$, and the $n$th infinitesimal harmonic angular frequency $n\delta\omega_0$ becomes the continuous angular frequency $\omega$. The summation now includes a continuous range of frequencies so that the summation becomes an integration over the entire frequency range $(-\infty, \infty)$. Thus when there is no periodicity, the function $g(t)$ can be expressed as

$$g(t) = \frac{1}{2\pi} \int_{-\infty}^{\infty} e^{i\omega t}\, d\omega \lim_{T\to\infty} \int_{-\infty}^{\infty} g(t)\, e^{-i\omega t}\, dt$$

$$= \int_{-\infty}^{\infty} G(\omega)\, e^{i\omega t}\, d\omega$$

where

$$G(\omega) = \frac{1}{2\pi} \int_{-\infty}^{\infty} g(t)\, e^{-i\omega t}\, dt.$$

It is, of course, assumed that $g(t)$ is the 'given' function and two questions now arise:

(a) What conditions must be placed on $g(t)$ to ensure that

$$\int_{-\infty}^{\infty} g(t) e^{-i\omega t}\, dt.$$

exists (so that $G(\omega)$ may be found)?

(b) What additional conditions on $g(t)$ ensure that

$$\int_{-\infty}^{\infty} G(\omega) e^{i\omega t} d\omega$$

exists and is in fact equal to $g(t)$?

The answers to these two questions are:

(a) It is sufficient that $\int_{-\infty}^{\infty} |g(t)|\, dt$ exists.

(b) It is sufficient that $g(t)$ is continuous and

$$\int_{-\infty}^{\infty} \{g(t)\}^2\, dt$$

is finite.

The functions $g(t)$ and $G(\omega)$ display a sort of duality and are known as a 'Fourier transform pair'. Since the factor $(1/2\pi)$ may be interchanged between $g(t)$ and $G(\omega)$ there is no universally accepted convention governing the definition of a Fourier transform. In this book we shall adopt the form previously defined.

Provided $g(t)$ satisfies the requirements just mentioned, a function $G(\omega)$ can be found to correspond to it, and if $g(t)$ is real and even, $G(\omega)$ is real and even. Thus if $g(t)$ is a rectangular pulse as shown in Figure 25a,

$$G(\omega) = \frac{1}{2\pi} \int_{-\infty}^{\infty} g(t) e^{-i\omega t} dt = \frac{AB}{2\pi} \left\{ \frac{\sin(\omega B/2)}{\omega B/2} \right\}$$

which is the even function shown sketched in Figure 25b.

More generally, if $g(t)$ is merely a real function (and not real and even), then $G(\omega) = G^*(-\omega)$ where $G^*$ is the complex conjugate of $G$. The problem illustrated in Figure 25 is a special case of this. If the

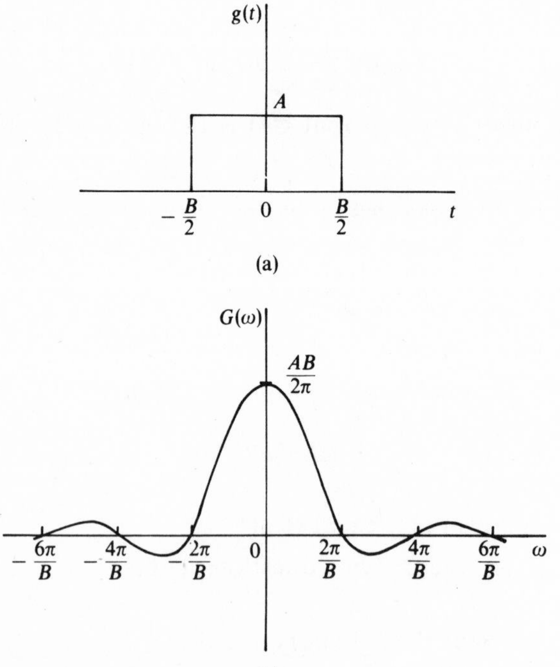


**Figure 25.** A typical function and its Fourier transform:
(a) rectangular pulse $g(t)$, and
(b) the Fourier transform $G(\omega)$ of $g(t)$.

area $AB$ of the rectangular pulse is kept equal to unity as the breadth of the pulse approaches zero and the height becomes infinite, then a single impulse is obtained at the origin. As $B \to 0$, we have,

$$G(\omega) = \lim_{B \to 0} \frac{AB}{2\pi} \frac{\sin\left(\frac{\omega B}{2}\right)}{\left(\frac{\omega B}{2}\right)} = \frac{1}{2\pi}$$

since $AB = 1$. The impulse – or Dirac – function[1] is defined by

$$\delta(t) = \int_{-\infty}^{\infty} G(\omega) e^{i\omega t} d\omega$$

$$= \frac{1}{2\pi} \int_{-\infty}^{\infty} e^{i\omega t} d\omega.$$

The properties of such a function are as follows:

i) $\delta(t) = 0 \quad \text{if } t \neq 0.$

ii) $\int_{-\infty}^{\infty} \delta(t) dt = 1.$

iii) If $q(t)$ is continuous at $t_0$ and $t_1$, $t_2$ are constants, with $t_2 > t_1$, then

$$\int_{t_1}^{t_2} q(t)\delta(t - t_0) dt = \begin{cases} q(t_0) & \text{if } t_1 < t_0 \leqslant t_2, \\ 0 & \text{otherwise.} \end{cases}$$

iv) The unit step function $U(t)$ is defined by

$$U(t) = \begin{cases} 1 & t \geqslant 0 \\ 0 & t < 0, \end{cases}$$

and is such that

$$\delta(t) = \frac{dU(t)}{dt}.$$

### 6.2.1 An important special case

The function

$$g(t) = A \cos \Omega t$$

does not satisfy the conditions for the existence of $G(\omega)$. But we can approach it as an interesting special case. Suppose

$$g(t) = \begin{cases} Ae^{at} \cos \Omega t & t \leqslant 0 \\ Ae^{-at} \cos \Omega t & t \geqslant 0 \end{cases}$$

where $a > 0$. We now have

$$G(\omega) = \frac{A}{2\pi} \left\{ \int_{-\infty}^{0} e^{(a-i\omega)t} \cos \Omega t \, dt + \int_{0}^{\infty} e^{-(a+i\omega)t} \cos \Omega t \, dt \right\}.$$

Consultation of integral tables leads to the conclusion that

$$G(\omega) = \frac{Aa}{2\pi} \left\{ \frac{1}{a^2 + (\Omega + \omega)^2} + \frac{1}{a^2 + (\Omega - \omega)^2} \right\},$$

which is a real and even function of $\omega$.

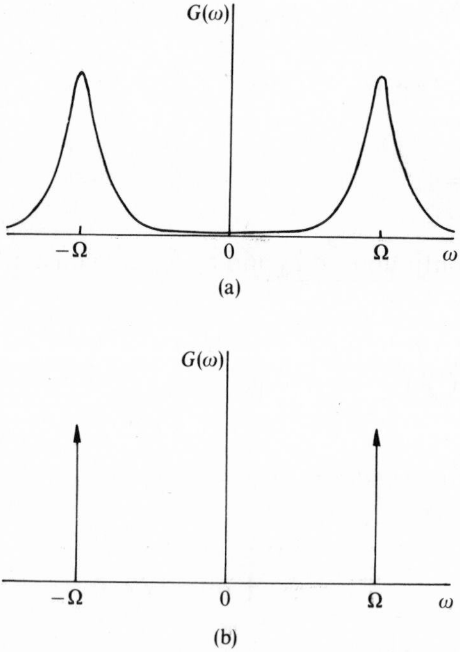


**Figure 26.** The Fourier transform of a sinusoidal function:
(a) Fourier transform of the sinusoidal function when it is modified so that it decays, and
(b) the limiting form of the Fourier transform of the modified sinusoidal function when the rate of decay is zero.

If $a$ is made small, the curve of $G(\omega)$ displays pronounced humps at $\omega = \pm\Omega$ as shown in Figure 26a. As $a \to 0$, these humps become sharper and, in the limit, the curve becomes a pair of delta functions as shown in Figure 26b. That is, if $g(t)$ is sinusoidal with frequency $\Omega$, the Fourier transform curve becomes a couple of lines at the appropriate frequency.

### 6.3 Application of the Fourier transform to random processes

We shall show that it is useful to consider a random process in terms of frequency. Now the idea of breaking up a random process into frequency components suggests that a realisation of the process should be so analysed. From the quantity $X^{(1)}(t)$ one might obtain a curve $G^{(1)}(\omega)$. It might then be profitable to assign probabilities to this new function (for it too is only a realisation of a random process). Alternatively it might be worthwhile to estimate 'parametric averages' for this latter realisation, comparable with the time averages $\langle\ \rangle$ we have already discussed.

The difficulty with this approach is that the function $X^{(1)}(t)$ will not in general possess a Fourier transform since it does not meet the conditions we mentioned previously. That is

$$\int_{-\infty}^{\infty} |X^{(1)}(t)|\, dt$$

is not finite for a stationary process. This drawback can be overcome however, by suitably truncating $X^{(1)}(t)$. Let

$$X_T(t) = \begin{cases} X^{(1)}(t) & -T/2 < t < T/2 \\ 0 & \text{otherwise} \end{cases}$$

so that $X_T(t)$ does meet the requirements and so does possess a Fourier transform, which is the complex function

$$G_T(\omega) = \frac{1}{2\pi}\int_{-\infty}^{\infty} X_T(t)e^{-i\omega t}dt.$$

Thus,

$$\langle X_T(t)\rangle = \frac{1}{T}\int_{-T/2}^{T/2} X_T(t)dt = \frac{1}{T}\int_{-\infty}^{\infty} X_T(t)dt = \frac{2\pi}{T}\, G_T(0)$$

whence

$$\langle X^{(1)}(t)\rangle = \lim_{T\to\infty}\left\{\frac{2\pi}{T}\,G_T(0)\right\}.$$

The mean square of the realisation over the range $T$ is

$$\langle X_T^2(t)\rangle = \frac{1}{T}\int_{-T/2}^{T/2} X_T^2(t)dt = \frac{1}{T}\int_{-\infty}^{\infty} X_T^2(t)dt$$

which can be written

$$\langle X_T^2(t)\rangle = \frac{1}{T}\int_{-\infty}^{\infty} X_T(t)\left\{\int_{-\infty}^{\infty} G_T(\omega)e^{i\omega t}d\omega\right\}dt$$

$$= \frac{1}{T}\int_{-\infty}^{\infty} G_T(\omega)\left\{\int_{-\infty}^{\infty} X_T(t)e^{i\omega t}dt\right\}d\omega$$

$$= \frac{2\pi}{T}\int_{-\infty}^{\infty} G_T(\omega)G_T^*(\omega)d\omega$$

where $G_T^*(\omega)$ is the complex conjugate of $G_T(\omega)$. Thus

$$\langle X^{(1)2}(t)\rangle = \int_{-\infty}^{\infty} \lim_{T\to\infty}\left\{\frac{2\pi}{T}|G_T(\omega)|^2\right\}d\omega = \int_{-\infty}^{\infty} S_{XX}(\omega)d\omega$$

where the quantity

$$S_{XX}(\omega) = \lim_{T\to\infty}\left\{\frac{2\pi}{T}|G_T(\omega)|^2\right\}$$

is defined as the 'mean square spectral density' and is a real and even function of $\omega$. The mean square spectral density describes the harmonic content of the random process in the frequency range.

Further, we may define the auto-correlation function $R_T(\tau)$ of the truncated random process to be

$$R_T(\tau) = \langle X_T(t)X_T(t+\tau)\rangle = \frac{1}{T}\int_{-T/2}^{T/2} X_T(t)X_T(t+\tau)dt.$$

By the procedure identical to the mean square value analysis we find that

$$R_T(\tau) = \frac{2\pi}{T}\int_{-\infty}^{\infty} G_T(\omega)G_T^*(\omega)e^{i\omega\tau}d\omega$$

and

$$R_{XX}(\tau) = \lim_{T\to\infty} R_T(\tau) = \int_{-\infty}^{\infty} \lim_{T\to\infty} \left\{\frac{2\pi}{T}|G_T(\omega)|^2\right\} e^{i\omega\tau} d\omega$$

$$= \int_{-\infty}^{\infty} S_{XX}(\omega) e^{i\omega\tau} d\omega.$$

Since the auto-correlation and spectral density functions are real and even, we may write

$$R_{XX}(\tau) = \int_0^{\infty} 2S_{XX}(\omega) \cos \omega\tau \, d\omega = \int_0^{\infty} \Phi_{XX}(\omega) \cos \omega\tau \, d\omega$$

where

$$\Phi_{XX}(\omega) = \begin{cases} 2S_{XX}(\omega) & \text{for } \omega \geqslant 0 \\ 0 & \text{otherwise,} \end{cases}$$

is the 'physically realisable one sided spectral density function'. It is this function which may be measured experimentally. A possible form of $S_{XX}(\omega)$ or $\Phi_{XX}(\omega)$ is shown in Figure 27. Note that $\Phi_{XX}(\omega)$ is physically identifiable in the range $0 \leqslant \omega \leqslant \infty$, whereas *for mathematical convenience* $S_{XX}(\omega)$ is used in the range $-\infty < \omega < \infty$.

### 6.3.1 The mean square spectral density

The foregoing definition of the mean square spectral density refers to a time average taken over a realisation. Provided the random process is ergodic the quantity $S_{XX}(\omega)$ will be equal to the spectral density that we shall now define in another way.

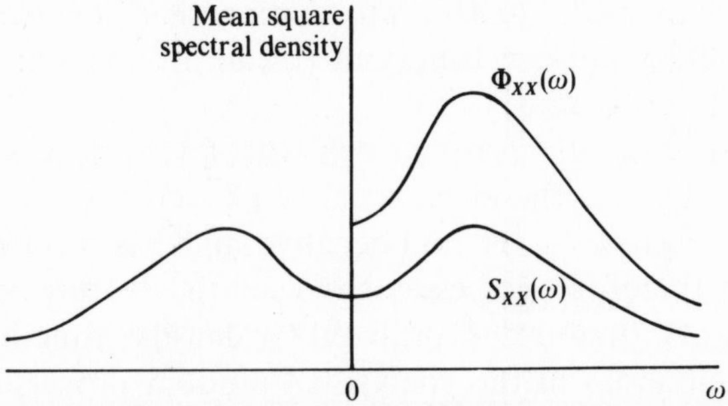


**Figure 27.** One- and two-sided mean square spectral density functions of a random process $X(t)$.

Consider a stationary random process $X(t)$ whose auto-correlation function is $R_{XX}(\tau)$. We have seen that $R_{XX}(\tau)$ is real and even; moreover unless $X(t)$ has sinusoidal components $R_{XX}(\tau) \to 0$ as $\tau \to \pm\infty$. Unlike a realisation, then, $R_{XX}(\tau)$ can be expected to possess a Fourier transform which will be a real and even function of $\omega$. (If $X(t)$ does contain sinusoidal components, then they will be exposed in $R_{XX}(\tau)$, whose transform will grow appropriate pairs of 'horns', as shown in section 6.2.1.)

The Fourier transform of $R_{XX}(\tau)$ and its inverse for a stationary random process $X(t)$ are respectively

$$S_{XX}(\omega) = \frac{1}{2\pi} \int_{-\infty}^{\infty} R_{XX}(\tau) e^{-i\omega\tau} d\tau$$

$$R_{XX}(\tau) = \int_{-\infty}^{\infty} S_{XX}(\omega) e^{i\omega\tau} d\omega.$$

These two equations are called the 'Wiener – Khintchine Relations'.

The justification for using the symbol $S_{XX}(\omega)$ again may be seen from the fact that

$$R_{XX}(0) = E[X^2(t)] = \int_{-\infty}^{\infty} S_{XX}(\omega) d\omega.$$

That is to say, the area under the curve of $S_{XX}(\omega)$ is equal to the mean square value of the random process. In other words $S_{XX}(\omega)\delta\omega$ is the contribution to the mean square of components having frequencies lying between $\omega$ and $\omega + \delta\omega$. $S_{XX}(\omega)$ is therefore the 'mean square spectral density' (although the name 'power spectral density' is sometimes used).

It is not uncommon to specify a random process $X(t)$ approximately by $\langle X^{(1)}(t) \rangle$ and $S_{XX}(\omega)$ under the assumption that it is ergodic. This represents far less information than is contained in the various probability density functions (of all orders) which, between them, define $X(t)$ strictly.

Since $S_{XX}(\omega)$ is the spectral density of the mean square, it is non-negative. Suppose the mean square $E[X^2(t)] = C$, say. The curve of $S_{XX}(\omega)/C$ against $\omega$ is non-negative and has an area of unity beneath it; it therefore possesses the essential features of a (rather special form of) first order probability density function. Certain theoretical techniques in the analysis of random processes are based on this fact, though we shall not follow the matter up here.

In summarising we see that the mean square spectral density

function of an ergodic random process $X(t)$ has the following properties:

1. The spectral density function is a real, even, non-negative function defined in the range of frequency $(-\infty, \infty)$. These limits are adopted for mathematical convenience.
2. The auto-correlation function and spectral density function form a Fourier transform pair provided that $|R_{XX}(\tau)|$ is integrable over the range $(-\infty, \infty)$.
3. If the auto-correlation function is periodic, the spectral density function is discrete and indicates the distribution of the harmonic content of the waveform over the frequency range.
4. The area under the continuous spectral density function curve represents the mean square value of the random process and the dimensions of $S_{XX}(\omega)$ depend on those of the random process.
5. The derived spectral density function $S_{XX}(\omega)$ is applicable to a large number of different waveforms. This results from a similar property discussed previously for the auto-correlation function.

We shall work usually in the frequency domain rather than the time domain since use of the spectral density function is a more concise way of describing a random process than use of the corresponding auto-correlation function.

## 6.4 Cross-correlation and cross-spectral density functions

The ergodic random processes $X(t)$ and $Y(t)$ may be represented by the truncated processes

$$X_T(t) = \begin{cases} X(t) & \text{for } -T/2 < t < T/2 \\ 0 & \text{otherwise,} \end{cases}$$

$$Y_T(t) = \begin{cases} Y(t) & \text{for } -T/2 < t < T/2 \\ 0 & \text{otherwise,} \end{cases}$$

with Fourier transforms $X_T(\omega)$ and $Y_T(\omega)$ respectively.

The cross-correlation function of the truncated random processes is defined by

$$R_{X_T Y_T}(\tau) = \frac{1}{T} \int_{-T/2}^{T/2} X_T(t) Y_T(t+\tau) dt$$

so that, by proceeding as previously described, we may determine the cross correlation function of the ergodic random processes $X(t)$ and $Y(t)$; it is given by

$$R_{XY}(\tau) = \lim_{T\to\infty} R_{X_T Y_T}(\tau) = \lim_{T\to\infty} \frac{1}{T} \int_{-T/2}^{T/2} X_T(t) Y_T(t+\tau)dt$$

$$= \int_{-\infty}^{\infty} \lim_{T\to\infty} \left\{ \frac{2\pi}{T} X_T^*(\omega) Y_T(\omega) \right\} e^{i\omega\tau} d\omega$$

$$= \int_{-\infty}^{\infty} S_{XY}(\omega) e^{i\omega\tau} d\omega$$

where $X_T^*(\omega)$ is the complex conjugate and where

$$S_{XY}(\omega) = \lim_{T\to\infty} \left\{ \frac{2\pi}{T} X_T^*(\omega) Y_T(\omega) \right\}$$

is defined as the 'cross-spectral density function' of the ergodic random processes $X(t)$ and $Y(t)$. The inverse Fourier transform is

$$S_{XY}(\omega) = \frac{1}{2\pi} \int_{-\infty}^{\infty} R_{XY}(\tau) e^{-i\omega\tau} d\tau.$$

It also follows that the cross-spectral density function of the ergodic random processes $Y(t)$ and $X(t)$ is defined by

$$S_{YX}(\omega) = \lim_{T\to\infty} \left\{ \frac{2\pi}{T} Y_T^*(\omega) X_T(\omega) \right\}$$

so that

$$S_{XY}(\omega) = S_{YX}^*(\omega) = S_{YX}(-\omega)$$

form a complex conjugate pair and the respective cross-correlation functions satisfy the relationship

$$R_{XY}(\tau) = R_{YX}(-\tau).$$

When the cross-spectral density function is written in terms of its real and imaginary parts, we have

$$S_{XY}(\omega) = C_{XY}(\omega) - iQ_{XY}(\omega)$$

$$= \frac{1}{2\pi} \int_{-\infty}^{\infty} R_{XY}(\tau)(\cos\omega\tau - i\sin\omega\tau)d\tau$$

where the even function of $\omega$

$$C_{XY}(\omega) = \frac{1}{2\pi}\int_{-\infty}^{\infty} R_{XY}(\tau)\cos\omega\tau d\tau$$

$$= \frac{1}{2\pi}\int_{0}^{\infty}\left\{R_{XY}(\tau) + R_{YX}(\tau)\right\}\cos\omega\tau d\tau = C_{XY}(-\omega)$$

is called the 'co-spectrum' and the odd function of $\omega$

$$Q_{XY}(\omega) = \frac{1}{2\pi}\int_{-\infty}^{\infty} R_{XY}(\tau)\sin\omega\tau d\tau$$

$$= \frac{1}{2\pi}\int_{0}^{\infty}\left\{R_{XY}(\tau) - R_{YX}(\tau)\right\}\sin\omega\tau d\tau = -\,Q_{XY}(-\omega)$$

is called the 'quadrature-spectrum'. Whereas the spectral density represents a contribution to the mean square value of the random process, the co-spectrum is a contribution to the product of the random processes provided by components that are in phase with one another. The quadrature spectral density is the contribution of components that are out of phase with one another by 90°. Note that in contrast to the mean square spectral density, the phase relationship is an inherent part of the cross-spectral density function. This is because we may write the cross-spectral density in the form

$$S_{XY}(\omega) = \sqrt{\{C_{XY}^2(\omega) + Q_{XY}^2(\omega)\}}\, e^{-i\delta_{XY}(\omega)}$$

where the phase relationship is given by

$$\delta_{XY}(\omega) = \tan^{-1}\frac{Q_{XY}(\omega)}{C_{XY}(\omega)}.$$

The previous results show that expressions of the form $\{C_{XY}(\omega)\sin\omega\tau - Q_{XY}(\omega)\cos\omega\tau\}$ and $\{C_{XY}(\omega)\cos\omega\tau + Q_{XY}(\omega)\sin\omega\tau\}$ are respectively odd and even functions of $\omega$. It now follows that the cross-correlation function

$$R_{XY}(\tau) = \int_{-\infty}^{\infty}\{C_{XY}(\omega)\cos\omega\tau + Q_{XY}(\omega)\sin\omega\tau\}\,d\omega$$

$$= \int_{0}^{\infty}\{\Phi_{XY}^R(\omega)\cos\omega\tau + \Phi_{XY}^I(\omega)\sin\omega\tau\}\,d\omega,$$

where

$$\Phi_{XY}^{R}(\omega) - i\Phi_{XY}^{I}(\omega) = \Phi_{XY}(\omega) = 2S_{XY}(\omega)$$
$$= 2C_{XY}(\omega) - 2iQ_{XY}(\omega).$$

In effect, we have defined the physically realisable one-sided cross-spectral density function with the property

$$\Phi_{XY}(\omega) = \begin{cases} 2S_{XY}(\omega) & 0 \leqslant \omega \leqslant \infty \\ 0 & \text{otherwise} \end{cases}$$

where the co-spectrum and quadrature spectrum are now respectively defined as

$$4C_{XY}(\omega) = \Phi_{XY}(\omega) + \Phi_{YX}(\omega)$$

$$4Q_{XY}(\omega) = i\,\{\Phi_{XY}(\omega) - \Phi_{YX}(\omega)\}$$

for $0 \leqslant \omega \leqslant \infty$ and zero otherwise.

Thus the joint properties of the two ergodic random processes $X(t)$ and $Y(t)$ have been described by the cross-correlation and cross-spectral density functions. It is unfortunate that these functions do not possess the symmetry properties of the auto-correlation and mean square spectral density functions. But in their way they do indicate the necessary relationships between the random processes.

When dealing with cross-spectral density functions, it is sometimes convenient to consider the real function $\gamma_{XY}^2(\omega)$ defined by

$$0 \leqslant \gamma_{XY}^2(\omega) = \frac{|S_{XY}(\omega)|^2}{S_{XX}(\omega)S_{YY}(\omega)} = \frac{|\Phi_{XY}(\omega)|^2}{\Phi_{XX}(\omega)\Phi_{YY}(\omega)} \leqslant 1.$$

It is called the coherence function (note the similarity between this function and the correlation coefficient described in section 4.1.3). When $\gamma_{XY}^2(\omega) = 0$ for all frequencies, the random processes $X(t)$ and $Y(t)$ are 'statistically independent' while if $\gamma_{XY}^2(\omega) = 0$ at a particular frequency, the random processes are 'incoherent' at that frequency. Finally, if $\gamma_{XY}^2(\omega) = 1$ at all frequencies, the random processes $X(t)$ and $Y(t)$ are said to be fully 'coherent'.

## Reference

[1] CHISHOLM, J. S. R. and MORRIS, R. M., 1964, *Mathematical Methods in Physics*, North-Holland, Amsterdam.

# Part 2 Waves

# 7 Travelling waves

In Part II of this book we shall examine the problem of specifying the elevation of the sea surface. There are two approaches:

(a) Treat the motion as the aggregate of motions associated with a number of travelling waves.
(b) Specify suitable statistics of the elevation without reference to component travelling waves.

In this chapter we examine possibility (a).

## 7.1 General mathematical theory

In this section wave theory is introduced, albeit rather superficially. The interested reader who requires more information should refer to books on the subject[1, 2, 3, 4].

The mathematical description of the motion of fluid particles contained in a body of incompressible liquid that is disturbed by the passage of waves over its free surface is based on the assumption that the motion is irrotational. In this case the velocity field in the liquid may be defined in terms of a single-valued velocity potential $\phi(x, y, z, t)$. The equation of continuity (conservation of mass) will be satisfied provided that the potential function $\phi$ is a solution of the Laplace equation

$$\nabla^2 \phi = 0$$

for all time. Rectangular coordinate axes $OXYZ$ are fixed so that the plane $OXY$ coincides with the undisturbed, horizontal, free surface of the liquid and $OZ$ is vertically upwards. The momentum equations governing possible motions may be integrated to give

$$\frac{\partial \phi}{\partial t} - \tfrac{1}{2}(u^2 + v^2 + w^2) - \frac{p}{\rho} - gz = 0$$

which is known as Bernoulli's equation. If the potential $\phi$ satisfies the Laplace equation, the components of the particle velocity are given by

$$u = -\frac{\partial \phi}{\partial x}, \quad v = -\frac{\partial \phi}{\partial y}, \quad w = -\frac{\partial \phi}{\partial z}$$

and the pressure $p$ may be determined from Bernoulli's equation.

The solutions of the differential equations are made to conform to boundary and initial conditions, and the boundaries usually considered are either free or rigid. It is assumed that on a free surface the pressure $p$ is prescribed, but the geometric elevation of the surface is an unknown function $\zeta(x, y, z, t)$. The kinematic condition of the surface is described by

$$u\frac{\partial \zeta}{\partial x} + v\frac{\partial \zeta}{\partial y} - w + \frac{\partial \zeta}{\partial t} = 0$$

on the surface

$$z = \zeta(x, y),$$

whilst Bernoulli's equation leads to the dynamic condition

$$\frac{\partial \phi}{\partial t} - \tfrac{1}{2}(u^2 + v^2 + w^2) - \frac{p}{\rho} - g\zeta = 0$$

on

$$z = \zeta(x, y).$$

On the free boundary the velocity potential $\phi$ must satisfy the non-linear conditions defined in the last two equations.

By contrast, at a rigid boundary such as the bed of the sea, at a distance $d$ below the surface $\phi$ is only required to satisfy the condition that there is no flow across the boundary, i.e.

$$v_n = \frac{\partial \phi}{\partial n} = 0$$

on the bottom

$$z = -d(x, y)$$

where the subscript $n$ denotes a directional normal to the solid surface at $z = -d(x, y)$.

The equations governing the wave propagation are completed by the addition to the above of equations describing the initial, undisturbed, equilibrium position of the liquid. These are

$$\zeta = 0 = u = v = w$$

for the instant

$$t = 0,$$

and an equation describing the disturbance which generates the waves.

It is generally necessary to simplify the general theory of wave propagation by incorporating limitations suggested by, and appropriate to, the physical nature of the particular problem of interest. Two assumptions have been of especial interest:

1. Wave amplitude is small compared with wave length.
2. Depth of water is small compared with wave length.

These assumptions may be made separately or together. Here we are concerned with assumption (1) alone, which leads to a linear theory of wave propagation and to the well known type of boundary value problem encountered in potential theory. (For (2) see ref.[4].)

If the velocity potential $\phi$ and surface elevation $\zeta$ are expressed as power series expansions and all but the linear terms are ignored, then it is implied that the velocities of the water particles, the free surface elevation, and their derivatives are all small quantities. The linearised equations on the free boundary reduce to

$$\frac{\partial \phi}{\partial t} - g\zeta = 0$$

and

$$\frac{\partial^2 \phi}{\partial t^2} + g\frac{\partial \phi}{\partial z} = 0$$

on the surface $z = 0$. Together with Laplace's equation, the rigid boundary criterion and the initial conditions at zero time, these equations represent a complete mathematical description of irrotational waves of amplitude small compared with the wave length.

### 7.1.1 Simple harmonic progressive waves

Consider a simple, two dimensional wave train progressing with velocity $c$ in the $OX$ direction over the free surface of water of uniform depth $d$ and resulting from a two dimensional disturbance of suitable form. The instantaneous wave profile is illustrated in Figure 28.

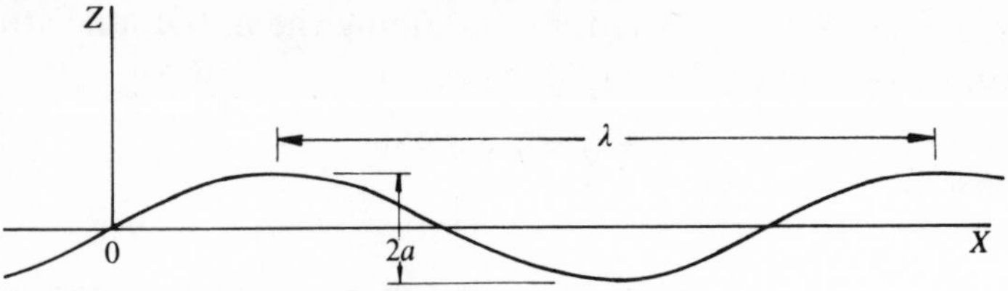


**Figure 28.** A sinusoidal wave showing the wavelength λ, amplitude $a$ and wave height (= $2a$).

If the velocity potential is assumed to be

$$\phi(x, z, t) = A \cosh k(z + d) \sin (kx - \omega t + \alpha),$$

then it may readily be shown that the Laplace equation and the rigid boundary condition at $z = -d(x, y)$ derived in the previous section are satisfied. Further, by substituting the assumed velocity potential function into the free boundary condition,

$$\frac{\partial^2 \phi}{\partial t^2} + g \frac{\partial \phi}{\partial z} = 0$$

on $z = 0$, we find the relationship

$$\omega^2 = gk \tanh (kd).$$

Substitution for $\phi$ in the other boundary condition leads to

$$\zeta(x, t) = a \cos(kx - \omega t + \alpha)$$

where $a \left(= -\frac{\omega A}{g} \cosh kd\right)$ is the amplitude of oscillation. The velocity potential is given by

$$\phi(x, z, t) = -\frac{ag}{\omega} \frac{\cosh k(z + d)}{\cosh kd} \sin (kx - \omega t + \alpha).$$

The parameters which define such a wave system are evidently,

'wavelength', λ, the horizontal distance between successive crests or troughs,

'wave period', $T$, the time taken for two successive crests to pass a fixed point on the axis OX or the time taken by a crest to travel a distance equal to one wavelength,

'wave or phase velocity', $c = \lambda/T$ is thus the velocity with which the wave crest moves, 'wave height', $h$, is the vertical distance between crest and trough,

'wave amplitude', $a = h/2$,
'wave number', $k = 2\pi/\lambda$,
'wave frequency', $\omega = 2\pi/T$,
'phase angle', $\alpha$ which is arbitrary.

The 'wave slope' of a profile with zero phase angle investigated at $t = 0$ is defined by

$$\frac{d\zeta}{dx} = -\, ak \sin kx$$

and if $kx = \pi/2$ this has a maximum value

$$\left|\frac{d\zeta}{dx}\right|_{\max} = ak = \pi h/\lambda,$$

occurring midway between a trough and a crest.

More generally, the elevation of a wave travelling, as shown in Figure 29, at an angle $\mu$ with respect to the axis $OX$ is

$$\zeta(x, y, t) = a \cos(k \cos \mu \,.\, x + k \sin \mu \,.\, y - \omega t + \alpha).$$

It will be convenient to shorten this expression by introducing the vectors

$$\mathbf{k} = k \cos \mu \,.\, \mathbf{i} + k \sin \mu \,.\, \mathbf{j}$$

$$\mathbf{r} = x\mathbf{i} + y\mathbf{j}$$

whence

$$\zeta(\mathbf{r}, t) = a \cos(\mathbf{k} \,.\, \mathbf{r} - \omega t + \alpha).$$

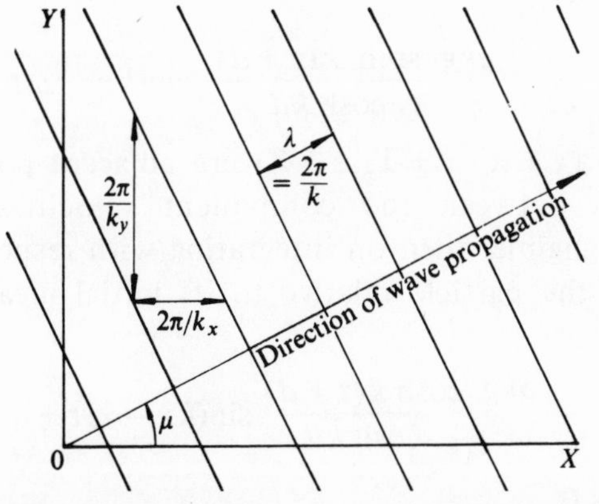


**Figure 29.** Sinusoidal waves travelling at an angle $\mu$ with respect to the fixed horizontal OX axis, illustrating the meanings of the 'wave numbers' $k_x$ and $k_y$.

The wave of phase velocity $c$ is given by

$$c = \frac{\lambda}{T} = \frac{\omega}{k} = \sqrt{\left\{\frac{g\lambda}{2\pi} \tanh\left(\frac{2\pi d}{\lambda}\right)\right\}}.$$

It is readily shown that when $d/\lambda \to 0$ or $kd \to 0$, the wave velocity

$$c = \sqrt{gd}$$

is independent of the wave-length when the depth of water is small compared to the wavelength.

Alternatively when $d/\lambda \to \infty$, or $kd \to \infty$, so that there exists an infinite depth of water, then

$$c = \sqrt{\left(\frac{g\lambda}{2\pi}\right)}$$

and

$$\omega^2 = gk$$

which is known as the 'dispersion relationship' for waves in deep water. Little error is incurred if the dispersion relationship is assumed to be valid when $d/\lambda > 0{\cdot}5$ which defines a criterion for 'deep water' waves. We also recall that the analysis is restricted to waves of small amplitude compared with the wavelength, so that $h/\lambda \leqslant 0{\cdot}05$ say.

From section 7.1 the component velocities of the fluid particles are seen to be

$$u = -\frac{\partial \phi}{\partial x} = \frac{akg}{\omega} \frac{\cosh k(z+d)}{\cosh kd} \cos(kx - \omega t + \alpha)$$

$$v = -\frac{\partial \phi}{\partial y} = 0$$

$$w = -\frac{\partial \phi}{\partial z} = \frac{akg}{\omega} \frac{\sinh k(z+d)}{\cosh kd} \sin(kx - \omega t + \alpha).$$

If $(x, y, z)$ and $(x + X,\ y + Y,\ z + Z)$ are adjacent points, such that the differences between the component velocities of the fluid particles are negligible, then on integrating with respect to time, the coordinates of the particle relative to its initial mean position are found to be

$$X = -\frac{akg}{\omega^2} \frac{\cosh k(z+d)}{\cosh kd} \sin(kx - \omega t + \alpha)$$

$$Y = 0$$

$$Z = \frac{akg}{\omega^2} \frac{\sinh k(z+d)}{\cosh kd} \cos(kx - \omega t + \alpha).$$

Thus

$$\frac{X^2}{\beta^2} + \frac{Z^2}{\gamma^2} = 1$$

where

$$\beta = \frac{akg}{\omega^2}\,\frac{\cosh k(z+d)}{\cosh kd}\,; \quad \gamma = \frac{akg}{\omega^2}\,\frac{\sinh k(z+d)}{\cosh kd}$$

and so

$$\beta^2 - \gamma^2 = \frac{a^2}{\cosh^2 kd}\left(\frac{kg}{\omega^2}\right)^2$$

indicating that the path of the particle is an ellipse with semi-major axis $\beta$ and semi-minor axis $\gamma$. In shallow water when $\omega^2/gk = \tanh(kd)$ and $d/\lambda < 0{\cdot}5$, the distance between the foci $(\beta^2 - \gamma^2)$ of all the ellipses is a constant but the length of the axes decrease with increase of the depth variable $z$. At $z = -d(x, y)$, the amplitude of the vertical motion is zero, $\gamma = 0$ and the fluid particle is moving to and fro along a horizontal path.

In deep water, where the dispersion relation $\omega^2 = kg$ is valid, and $kd \to \infty$, the coefficients

$$\beta = \gamma = ae^{+kz}$$

and the path of the particle is circular, satisfying the relationship

$$X^2 + Z^2 = a^2 e^{+2kz}$$

The maximum displacement of the particles decreases exponentially with increasing depth so that at $z = -d(x, y)$, the particle is stationary. In practice, particle movements at values of depth $|z| > 0{\cdot}5\lambda$ are negligible. For this reason, a vessel submerged below the surface at a distance $|z| > 0{\cdot}5\lambda$ is unaffected by the wave disturbance.

Since the motion of the water particles due to the passage of the wave train is oscillatory, there can be no net translation of fluid particles and therefore no net transport of mass. However, the fluid particles possess energy by virtue of their velocity and their elevation. The kinetic energy of the progressive wave may be shown[3] to be $a^2\rho g\lambda/4$, whilst the potential energy due to the elevated water in a wave length is $a^2\rho g\lambda/4$. Hence the total average energy per wave length is $a^2\rho g/2$.

### 7.1.2 Pressure variation

The pressure variation in a fluid is determined by the Bernoulli equation:

$$p = -\rho gz + \rho\frac{\partial\phi}{\partial t} - \tfrac{1}{2}\rho(u^2 + v^2 + w^2)$$

which is a combination of a hydrostatic term, $\rho gz$, a term due to the passage of the wave profile, $\rho(\partial\varphi/\partial t)$, and one due to the kinetic energy of the wave motion. Since the hydrostatic influence is always present, the pressure variation due solely to the wave is given by

$$\frac{\Delta p}{\rho} = \frac{\partial \phi}{\partial t} - \tfrac{1}{2}(u^2 + v^2 + w^2)$$

Under the assumption that the kinetic energy term can be neglected, in a linear theory we have

$$\Delta p = \rho \frac{\partial \phi}{\partial t} = ag\rho \frac{\cosh k(z+d)}{\cosh kd} \cos(kx - \omega t + \alpha)$$

$$= \rho g \frac{\cosh k(z+d)}{\cosh kd} \zeta(x, t).$$

In shallow water, the pressure variation is

$$\Delta p = \rho g \zeta(x, t)$$

whilst in deep water we have

$$\Delta p = \rho g e^{+kz} \zeta(x, t).$$

The pressure variation in deep water decreases exponentially such that the ratio of the surface pressure variation at $z = 0$ and that at $z = -\lambda/2$ is given by

$$\frac{\Delta p|_{z=-\lambda/2}}{\Delta p|_{z=0}} = e^{-\pi} \doteqdot 0{\cdot}04.$$

The pressure variation at $z = -\lambda/2$ has considerably decreased compared with the surface value, with the result that for $|z| > \lambda/2$ the total pressure is composed mainly of the hydrostatic influence. This exponential variation with depth is referred to as the 'Smith effect'. The inclusion of such an exponential term was found desirable when determining the effect of water pressure on a ship's hull in bending moment calculations.

### 7.1.3 Dispersion and wave groups

Suppose that, due to a disturbance, some area of water contains a range of deep water waves of various wavelengths. Since

$$c = \frac{\lambda}{T} = \frac{\omega}{k} = \sqrt{\frac{g}{k}} = \sqrt{\frac{g\lambda}{2\pi}},$$

the waves will be travelling at various velocities. The longer waves will possess higher speeds and will therefore be transmitted more rapidly away from the area. In this way the waves will become dispersed according to their wavelength and at some distance only waves of similar characteristics, will be observed.

A main feature of groups of waves may be illustrated by considering two elementary sinusoidal waves of equal amplitudes and nearly equal frequencies and wavelengths:

$$\zeta_1(x, t) = a \cos(k_1 x - \omega_1 t)$$

$$\zeta_2(x, t) = a \cos(k_2 x - \omega_2 t).$$

The total disturbance due to these waves is

$$\zeta(x, t) = \zeta_1 + \zeta_2 = 2a \cos\{½(k_1 + k_2)x - ½(\omega_1 + \omega_2)t\} \times \cos\{½(k_1 - k_2)x - ½(\omega_1 - \omega_2)t\}.$$

The first cosine term represents a wave very similar to the original waves, whose frequency and wavelength are an average of the two initial values, and which moves with velocity $(\omega_1 + \omega_2)/(k_1 + k_2)$. Since the initial wave lengths and frequencies are nearly equal the velocity of the resultant wave is practically the same as that of the original waves.

The second cosine term, which changes more slowly both with respect to position and time, may be regarded as a varying amplitude. Thus the resultant of the two elementary waves can be regarded as a wave of approximately the same wavelength and frequency but with an amplitude that changes both with position and time. That is

$$\zeta(x, t) = a(x, t) \cos\{½(k_1 + k_2)x - ½(\omega_1 + \omega_2)t\}$$

as is shown in Figure 30.

The resultant wave profile moves with a velocity $(\omega_1 - \omega_2)/(k_1 - k_2)$, which is different from the more rapidly oscillating part whose velocity is $(\omega_1 + \omega_2)/(k_1 + k_2)$. Hence, the individual waves

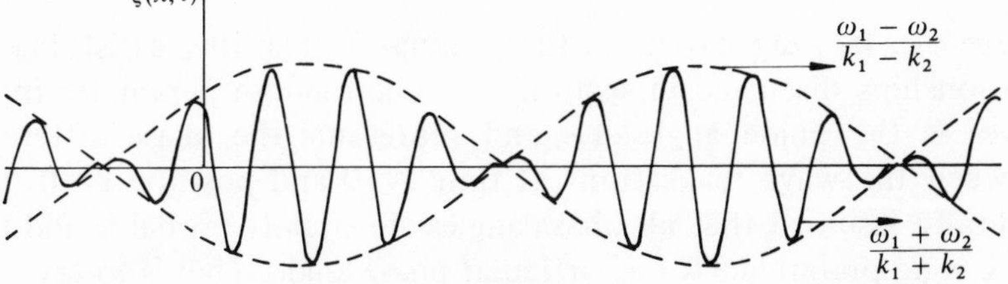


**Figure 30.** Summation of two sinusoidal waves of equal amplitude and nearly equal frequencies.

advance through the profile, gradually increasing and then decreasing their amplitude as they give place to other succeeding waves. As we have seen, this phenomenon occurs whenever the wave velocity, $c$, is not constant but depends on the frequency. The velocity of the wave envelope of profile curve is referred to as the 'group velocity' $U_e$, where

$$U_e = \frac{\omega_1 - \omega_2}{k_1 - k_2}.$$

If the two component waves are regarded as merely two contributions to a continuous distribution, this gives

$$\frac{d\omega}{dk} = \frac{d(ck)}{dk} = c + k\frac{dc}{dk} = c - \lambda\frac{dc}{d\lambda}$$

since

$$c = \frac{\omega}{k} = \sqrt{\left\{\frac{g}{k}\tanh(kd)\right\}}$$

and therefore the group velocity becomes

$$\frac{d(ck)}{dk} = U_e = \frac{c}{2}\left(1 + \frac{2kd}{\sinh 2kd}\right).$$

Thus the ratio $(U_e/c)$ reduces to unity for waves in shallow water $(d/\lambda \to 0)$ and to one half for waves in deep water $(d/\lambda \to \infty)$. It follows that in shallow water the resultant wave transmits its energy at the components' wave velocity $c$, whilst in deep water the transmission occurs at velocity $c/2$.

## 7.2 The probabilistic approach.

The previous specifications of the wave elevation $\zeta(x, y, t)$ are, of course, deterministic. The wave can be thought of in probabilistic terms, however, by thinking of it as just one realisation of a random process.

$$X(\mathbf{r}, t) = a_X \cos(\mathbf{k}_X \cdot \mathbf{r} - \omega_X t + A_X)$$

where $a_X$, $\mathbf{k}_X$, $\omega_X$ are given (determinate) quantities satisfying the relationships discussed in section 7.1. The random parameter in the waves is the phase angle $A_X$ and represents the phase difference between the wave realisations at time $t = 0$ and position $\mathbf{r} = \mathbf{0}$. It is generally assumed that all phase angles are equally probable and that there is no preference for a particular phase angle. That is to say

$$P[\alpha_X < A_X \leqslant \alpha_X + \delta\alpha_X] = f_A(\alpha_X)\delta\alpha_X$$

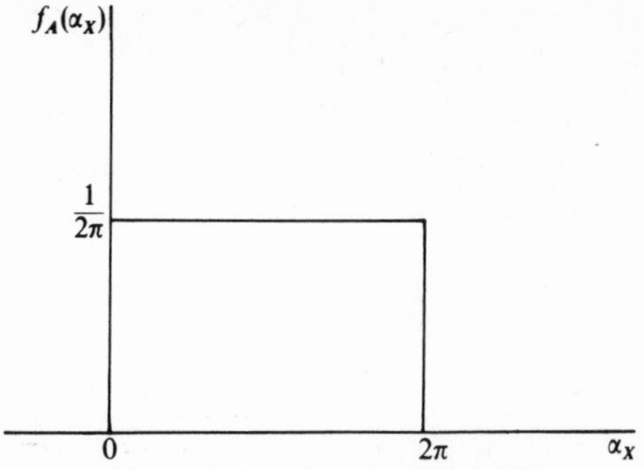


**Figure 31.** Probability density function of the random variable of phase.

where the probability density function $f_A(\alpha_X)$ is as shown in Figure 31.

As we shall see, a random process defined in this way has practical uses. It is therefore worthwhile to determine certain of its properties. In particular we shall require the probability density function of $X(\mathbf{r}, t)$, rather than of $A_X$.

Consider the new random variable at an arbitrary position $\mathbf{r}_1$ and time $t_1$

$$V = \mathbf{k}_X \cdot \mathbf{r}_1 - \omega_X t_1 + A_X$$

Now for each value $x$ of $X$ in the range $-a_X < X < a_X$ there correspond two values of $A_X$ in the range $0 < \alpha_X < 2\pi$, and hence two values of $V$, $v_A$ and $v_B$ say. Since the event $X$ occurring in $(x, x + \delta x)$ requires that either $V$ lies in $(v_A, v_A + \delta v)$ *or* in $(v_B, v_B + \delta v)$ and since these two events of $V$ do not intersect, it follows from the probability axioms that

$$\begin{aligned} P[x < X(\mathbf{r}_1, t_1) \leqslant x + \delta x] &= P[\{v_A < V(\mathbf{r}_1, t_1) \leqslant v_A + \delta v\} \\ &\qquad \cup \{v_B < V(\mathbf{r}_1, t_1) \leqslant v_B + \delta v\}] \\ &= P[v_A < V(\mathbf{r}_1, t_1) \leqslant v_A + \delta v] \\ &\qquad + P[v_B < V(\mathbf{r}_1, t_1) \leqslant v_B + \delta v] \end{aligned}$$

so that

$$f_X(x; \mathbf{r}_1, t_1)\delta x = \{f_V(v_A; \mathbf{r}_1, t_1) + f_V(v_B; \mathbf{r}_1, t_1)\}\,\delta v.$$

But for *any* value $v$

$$f_V(v; \mathbf{r}_1, t_1)\,\delta v = f_A(\alpha_X)\,\delta\alpha_X = \frac{1}{2\pi}\,\delta\alpha_X$$

since all values of $A_X$ between 0 and $2\pi$ are equally probable. Hence

$$f_X(x; \mathbf{r}_1, t_1)\delta x = \frac{1}{\pi}\,\delta\alpha_X.$$

For any particular realisation the values $x$ and $\alpha_X$ of $X(\mathbf{r}_1, t_1)$ and $A_X$ respectively are such that

$$x = a_X \cos(\mathbf{k}_X \,.\, \mathbf{r}_1 - \omega_X t_1 + \alpha_X).$$

It follows that

$$\left|\frac{dx}{d\alpha_X}\right| = (a_X^2 - x^2)^{1/2}$$

and so

$$f_X(x; \mathbf{r}_1, t_1) = \frac{1}{\pi(a_X^2 - x^2)^{1/2}}$$

for $-a_X < x < a_X$, since $f_X(x; \mathbf{r}_1, t_1) \geqslant 0$.

Having found the probability density function, we are in a position to find some practical measures of the random process $X(\mathbf{r}, t)$ at an arbitrary position $\mathbf{r}$ and time $t$. Thus,

$$E[X(\mathbf{r}, t)] = \int_{-a_X}^{a_X} x f_X(x; \mathbf{r}, t)dx = 0$$

$$E[X^2(\mathbf{r}, t)] = \int_{-a_X}^{a_X} x^2 f_X(x; \mathbf{r}, t)dx = a_X^2/2.$$

These results are important and it is worthwhile to dwell on them briefly. They can in fact be written down on the basis of semi-empirical arguments. Suppose that a location $\mathbf{r}$ and time $t$ are selected. The corresponding elevation $X(\mathbf{r}, t)$ can have any value such that $-a_X \leqslant x \leqslant a_X$ depending on the actual value $\alpha_X$ of the random variable $A_X$. A curve of $x$ plotted against $\alpha_X$ would have the form shown in Figure 32a and since all values of $\alpha_X$ between 0 and $2\pi$ are equally likely to occur, negative values of $x$ are as likely as positive. The expected value of $X(\mathbf{r}, t)$ is the likely value of $x$ and is, therefore, nil. By the same argument, the expected value of $X^2(\mathbf{r}, t)$ is the likely value of $x^2$ which is $a_X^2/2$ as indicated in Figure 32b.

In section 5.2 we defined the auto-correlation function of a

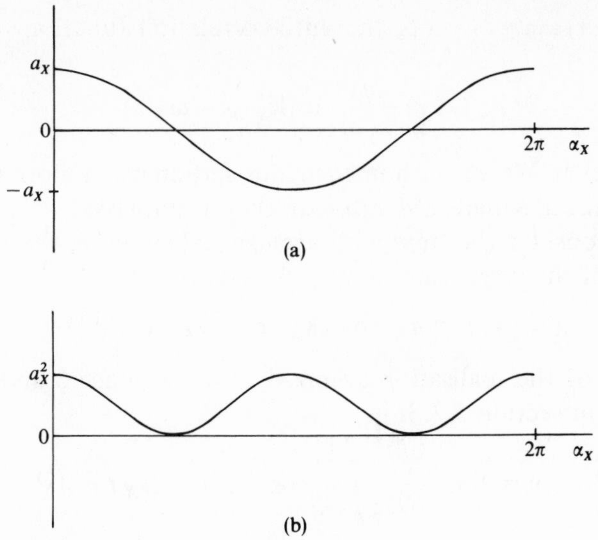


**Figure 32.** Illustration showing that:
(a) the mean value of a sinusoidal random process is nil, and
(b) the mean square value is $a_X^2/2$.

random variable $X(\mathbf{r}, t)$ to be

$$R_{XX}(\mathbf{r}_1, t_1; \mathbf{r}_2, t_2)$$

$$= E\,[X(\mathbf{r}_1, t_1)X(\mathbf{r}_2, t_2)]$$

$$= E[a_X^2 \cos(\mathbf{k}_X \,.\, \mathbf{r}_1 - \omega_X t_1 + A_X)\cos(\mathbf{k}_X \,.\, \mathbf{r}_2 - \omega_X t_2 + A_X)]$$

$$= \frac{a_X}{2}\, E[a_X \cos\{\mathbf{k}_X.(\mathbf{r}_2 + \mathbf{r}_1) - \omega_X(t_2 + t_1) + 2A_X\}$$
$$+ a_X \cos\{\mathbf{k}_X.(\mathbf{r}_2 - \mathbf{r}_1) - \omega_X(t_2 - t_1)\}]$$

$$= \frac{a_X}{2}\, E[X(\mathbf{r}_2 + \mathbf{r}_1, t_2 + t_1)]$$
$$+ \frac{a_X^2}{2}\, E[\cos\,\{\mathbf{k}_X.(\mathbf{r}_2 - \mathbf{r}_1) - \omega_X(t_2 - t_1)\}\,].$$

The first term we have shown to be zero and the second term is a constant independent of the initial random variable $A_X$. Hence,

$$R_{XX}(\mathbf{r}_1, t_1; \mathbf{r}_2, t_2) = \frac{a_X^2}{2} \cos\,\{\mathbf{k}_X.(\mathbf{r}_2 - \mathbf{r}_2) - \omega_X(t_2 - t_1)\}\,.$$

If $\rho = \mathbf{r}_2 - \mathbf{r}_1$, $\tau = t_2 - t_1$, the auto-correlation function reduces to

$$R_{XX}(\rho, \tau) = \frac{a_X^2}{2}\cos(\mathbf{k}_X \cdot \rho - \omega_X \tau)$$

indicating that $X(\mathbf{r}, t)$ is a homogeneous stationary random process in the mean, mean square and auto-correlation statistics.

Let us consider the 'temporal average' taken along the realisation $X^{(1)}(\mathbf{r}, t)$ of the given random process $X(\mathbf{r}, t)$, where

$$X^{(1)}(\mathbf{r}, t) = a_X \cos\{\mathbf{k}_X \cdot \mathbf{r} - \omega_X t + A_X^{(1)}\}\ .$$

The mean of the realisation averaged over the space $S$ and period $T$, as defined in section 5.3.3, is

$$\begin{aligned}\langle X^{(1)}(\mathbf{r}, t)\rangle &= \lim_{\substack{S\to\infty\\ T\to\infty}} \frac{1}{ST}\iint_{S\,T} a_X \cos\{\mathbf{k}_X . \mathbf{r} - \omega_X t + A_X^{(1)}\}\, d\mathbf{r}dt\\ &= 0\end{aligned}$$

The mean square value of the realisation is

$$\begin{aligned}\langle X^{(1)2}(\mathbf{r}, t)\rangle &= \lim_{\substack{S\to\infty\\ T\to\infty}} \frac{1}{ST}\iint_{S\,T} a_X{}^2 \cos^2\{\mathbf{k}_X . \mathbf{r} - \omega_X t + A_X^{(1)}\}\, d\mathbf{r}dt\\ &= a_X^2/2\end{aligned}$$

and the auto-correlation function is given by

$$\begin{aligned}&\langle X^{(1)}(\mathbf{r}, t)X^{(1)}(\mathbf{r}+\rho, t+\tau)\rangle\\ &= \langle a_X^2 \cos\{\mathbf{k}_X . \mathbf{r} - \omega_X t + A_X^{(1)}\}\cos\{\mathbf{k}_X . (\mathbf{r}+\rho) - \omega_X(t+\tau) + A_X^{(1)}\}\rangle\\ &= \tfrac{1}{2}a_X^2\,\langle\cos\{\mathbf{k}_X . (2\mathbf{r}+\rho) - \omega_X(2t+\tau) + 2A_X^{(1)}\} + \cos\{\mathbf{k}_X . \rho - \omega_X\tau\}\rangle.\end{aligned}$$

Since the brackets $\langle\,\rangle$ express an integration with respect to the space variable $\mathbf{r}$ and time $t$ over the range $S$ and $T$ respectively, it follows that

$$\langle X^{(1)}(\mathbf{r}, t)X^{(1)}(\mathbf{r}+\rho, t+\tau)\rangle = \tfrac{1}{2}\, a_X^2 \cos(\mathbf{k}_X \cdot \rho - \omega_X\tau).$$

On comparing the expectation statistical and temporal averages we see that

$$\begin{aligned}E[X(\mathbf{r}, t)] &= 0 = \langle X^{(1)}(\mathbf{r}, t)\rangle\\ E[X^2(\mathbf{r}, t)] &= a_X^2/2 = \langle X^{(1)}(\mathbf{r}, t)\rangle\end{aligned}$$

and

$$R_{XX}(\rho, \tau) = E[X(\mathbf{r}_1, t_1)X(\mathbf{r}_2, t_2)]$$
$$= \tfrac{1}{2}a_X^2 \cos(\mathbf{k}_X \cdot \rho - \omega_X \tau)$$
$$= \langle X^{(1)}(\mathbf{r}, t)X^{(1)}(\mathbf{r} + \rho, t + \tau)\rangle.$$

Under the proviso of the random phase variable $A_X$ being uniformly distributed, it follows that the random process $X(\mathbf{r}, t)$ is an *ergodic* random process in the mean, mean square and auto-correlation statistics.

### 7.2.1 Superposition of waves

Consider a second wave of the type already discussed, whose elevation is

$$Y(\mathbf{r}, t) = a_Y \cos(\mathbf{k}_Y \cdot \mathbf{r} - \omega_Y t + A_Y).$$

The parameters $a_Y$, $\mathbf{k}_Y$ and $\omega_Y$ are again to be thought of as determinate, while the phase angle $A_Y$ is a random variable which may take any value in the range $0 < \alpha_Y < 2\pi$. We know that

$$E[Y(\mathbf{r}, t)] = 0$$
$$E[Y^2(\mathbf{r}, t)] = a_Y^2/2$$

while the auto-correlation function is

$$R_{YY}(\mathbf{r}_1, t_1; \mathbf{r}_2, t_2) = \tfrac{1}{2} a_Y^2 \cos(\mathbf{k}_Y \cdot \rho - \omega_Y \tau).$$

The random processes $X(\mathbf{r}, t)$ and $Y(\mathbf{r}, t)$ are assumed to coexist and to be 'independent' of each other. Let a realisation of $Z(\mathbf{r}, t)$ be the sum of the realisations $X(\mathbf{r}, t)$ and $Y(\mathbf{r}, t)$ and the combined processes be referred to as the random process

$$Z(\mathbf{r}, t) = X(\mathbf{r}, t) + Y(\mathbf{r}, t).$$

Since the two component processes are stationary and homogeneous, so is the random process defined by their sum. Therefore we may define the auto-correlation function

$$R_{ZZ}(\mathbf{r}_1, t_1; \mathbf{r}_2, t_2) = E[Z(\mathbf{r}, t)\, Z(\mathbf{r} + \rho, t + \tau)]$$
$$= E[\{X(\mathbf{r}, t) + Y(\mathbf{r}, t)\}\{X(\mathbf{r} + \rho, t + \tau) + Y(\mathbf{r} + \rho, t + \tau\}]$$
$$= R_{XX}(\rho, \tau) + R_{XY}(\rho, \tau) + R_{YX}(\rho, \tau) + R_{YY}(\rho, \tau).$$

The auto-correlation function of the random process $Z(\mathbf{r}, t)$ therefore depends on the auto-correlation functions of the individual processes $R_{XX}(\rho, \tau)$ and $R_{YY}(\rho, \tau)$ and also upon the cross-correlation functions $R_{XY}(\rho, \tau)$ and $R_{YX}(\rho, \tau)$. However, from section 5.2 we see that these latter functions are both zero when the random processes $X(\mathbf{r}, t)$, $Y(\mathbf{r}, t)$ have zero mean value. In other words, the auto-correlation of the process $Z(\mathbf{r}, t)$ is

$$R_{ZZ}(\rho, \tau) = R_{XX}(\rho, \tau) + R_{YY}(\rho, \tau).$$

When the 'temporal average' is taken along the realisation

$$Z^{(1)}(\mathbf{r}, t) = X^{(1)}(\mathbf{r}, t) + Y^{(1)}(\mathbf{r}, t)$$

the temporal mean

$$\langle Z^{(1)}(\mathbf{r}, t)\rangle = \langle X^{(1)}(\mathbf{r}, t)\rangle + \langle Y^{(1)}(\mathbf{r}, t)\rangle = 0.$$

and the temporal mean square value along the realisation is given by

$$\langle Z^{(1)2}(\mathbf{r}, t)\rangle = \langle X^{(1)2}(\mathbf{r}, t)\rangle + 2\langle X^{(1)}(\mathbf{r}, t)Y^{(1)}(\mathbf{r}, t)\rangle + \langle Y^{(1)2}(\mathbf{r}, t)\rangle.$$

For the particular realisation $X^{(1)}(\mathbf{r}, t)$ and $Y^{(1)}(\mathbf{r}, t)$ we have

$$\langle X^{(1)}(\mathbf{r}, t)Y^{(1)}(\mathbf{r}, t)\rangle$$

$$= \tfrac{1}{2}a_X a_Y [\langle\cos\{(\mathbf{k}_X + \mathbf{k}_Y).\mathbf{r} - (\omega_X + \omega_Y)t + A_X^{(1)} + A_Y^{(1)}\}\rangle$$

$$+ \langle\cos\{(\mathbf{k}_X - \mathbf{k}_Y).\mathbf{r} - (\omega_X - \omega_Y)t + A_X^{(1)} - A_Y^{(1)}\}\rangle].$$

The brackets $\langle\,\rangle$ express an integration with respect to the variables $\mathbf{r}$ and $t$ over the range $S$ and period $T$ respectively. Thus, provided that the deterministic quantities $\omega_X - \omega_Y$ together with the components of $\mathbf{k}_X + \mathbf{k}_Y$ and $\mathbf{k}_X - \mathbf{k}_Y$ are all non zero,

$$\langle X^{(1)}(\mathbf{r}, t)Y^{(1)}(\mathbf{r}, t)\rangle = 0$$

and

$$\langle Z^{(1)2}(\mathbf{r}, t)\rangle = \langle X^{(1)2}(\mathbf{r}, t)\rangle + \langle Y^{(1)2}(\mathbf{r}, t)\rangle$$

$$= \tfrac{1}{2}(a_X^2 + a_Y^2).$$

Under these same deterministic conditions, the auto-correlation function of the realisation $Z^{(1)}(\mathbf{r}, t)$ is given by

$$\langle Z^{(1)}(\mathbf{r}, t)Z^{(1)}(\mathbf{r} + \rho, t + \tau)\rangle$$

$$= \langle X^{(1)}(\mathbf{r}, t)X^{(1)}(\mathbf{r} + \rho, t + \tau)\rangle + \langle Y^{(1)}(\mathbf{r}, t)Y^{(1)}(\mathbf{r} + \rho, t + \tau)\rangle$$

$$= \tfrac{1}{2}a_X^2 \cos(\mathbf{k}_X . \rho - \omega_X\tau) + \tfrac{1}{2}a_Y^2 \cos(\mathbf{k}_Y . \rho - \omega_Y\tau).$$

The statistical and temporal averages of the mean, mean square and auto-correlation functions are equal, indicating that the process $Z(\mathbf{r}, t)$ is an ergodic random process in these statistics.

### 7.2.2 Representation of an irregular seaway

It has been suggested[5] that an irregular sea surface can usefully be represented as the sum of a large number of regular waves, each component having a particular frequency, amplitude, direction, randomly distributed phase angle and being a solution of the linearised hydrodynamic equations for water waves (section 7.1). The deterministic characteristics of the component waves satisfy the relationships expressed in section 7.1.1. Thus it is assumed that the irregular sea surface is

$$\zeta(\mathbf{r}, t) = \sum_{i=1}^{n} X_i(\mathbf{r}, t)$$

where

$$X_i = a_i \cos(\mathbf{k}_i \,.\, \mathbf{r} - \omega_i t + A_i),$$

$A_i$ being a random variable.

The mean value of the elevation is

$$E[\zeta(\mathbf{r}, t)] = E\left[\sum_{i=1}^{n} X_i(\mathbf{r}, t)\right] = \sum_{i=1}^{n} E[X_i(\mathbf{r}, t)]$$

$$= 0.$$

Since the random processes $X_i(\mathbf{r}, t)$, $i = 1, 2, \ldots, n$ representing the individual waves are independent with zero mean values, terms of the type $E[X_i(\mathbf{r}_1, t_1) \,.\, X_j(\mathbf{r}_2, t_2)]$ are all zero for $i \neq j$ and are only non-zero when $i = j$, so that the mean square wave height is given by

$$E[\zeta^2(\mathbf{r}, t)] = E\left[\sum_{i=1}^{n} X_i^2(\mathbf{r}, t)\right] = \sum_{i=1}^{n} E[X_i^2(\mathbf{r}, t)] = \tfrac{1}{2} \sum_{i=1}^{n} a_i^2$$

and the auto-correlation function reduces to

$$R_{\zeta\zeta}(\mathbf{r}_1, t_1; \mathbf{r}_2, t_2) = E[\zeta(\mathbf{r}_1, t_1)\, \zeta(\mathbf{r}_2, t_2)]$$

$$= \sum_{i=1}^{n} \sum_{j=1}^{n} E[X_i(\mathbf{r}_1, t) X_j(\mathbf{r}_2, t)]$$

$$= \sum_{i=1}^{n} R_{ii}(\rho, \tau).$$

That is to say, the random process $\zeta(\mathbf{r}, t)$ is homogeneous and stationary, with an auto-correlation function of the form $R_{\zeta\zeta}(\rho, \tau)$.

By a procedure similar to that described in the previous sections, the process $\zeta(\mathbf{r}, t)$ may be shown to be an ergodic random process in the mean, mean square and auto-correlation function statistics.

In the one-dimensional analysis of chapter 6, we noted that the auto-correlation function $R_{XX}(\tau)$ of the random process $X(t)$ has a Fourier transform given by

$$R_{XX}(\tau) = \int_{-\infty}^{\infty} S_{XX}(\omega) e^{i\omega\tau} d\omega.$$

We are not yet in the position to see how this may be generalised for the three dimensional auto-correlation function $R_{\zeta\zeta}(\boldsymbol{\rho}, \tau)$ of the wave elevation random process $\zeta(\mathbf{r}, t)$, the three dimensions being the three scalar quantities $\mathbf{k} \cdot \mathbf{i}$, $\mathbf{k} \cdot \mathbf{j}$, $\omega$. We merely quote the relevant results for the moment:

$$R_{\zeta\zeta}(\boldsymbol{\rho}, \tau) = \int_{-\infty}^{\infty} \int_{-\infty}^{\infty} \int_{-\infty}^{\infty} S_{\zeta\zeta}(\mathbf{k}, \omega) e^{i(\mathbf{k} \cdot \boldsymbol{\rho} - \omega\tau)} d\mathbf{k} d\omega$$

where $d\mathbf{k} = dk_x \, dk_y$, $k_x$ being the quantity $\mathbf{k} \cdot \mathbf{i}$ and $k_y$ being $\mathbf{k} \cdot \mathbf{j}$. $S_{\zeta\zeta}(\mathbf{k}, \omega)$ is a three dimensional spectral density which represents $R_{\zeta\zeta}(\boldsymbol{\rho}, \tau)$ in the wave-number and frequency domains. At $\boldsymbol{\rho} = 0$, $\tau = 0$ we have

$$R_{\zeta\zeta}(0, 0) = E[\zeta^2(\mathbf{r}, t)] = \sum_{i=1}^{n} a_i^2/2 = \int_{-\infty}^{\infty} \int_{-\infty}^{\infty} \int_{-\infty}^{\infty} S_{\zeta\zeta}(\mathbf{k}, \omega) d\mathbf{k} d\omega.$$

It appears that the compounded frequencies of the sinusoidal wave components account for the varying amplitude and period of irregular waves, whilst the mixture of directions incorporated in the wave number $\mathbf{k}$ variable leads to a more irregular picture of the sea surface.

## References

[1] LAMB, H., 1945, *Hydrodynamics*, C.U.P., Cambridge.
[2] MILNE-THOMPSON, L. M., 1962, *Theoretical Hydrodynamics*, Macmillan, London.
[3] KINSMAN, B., 1965, *Wind Waves*, Prentice Hall, Englewood Cliffs, New Jersey.
[4] STOKER, J. J., 1957, *Water Waves*, Interscience, New York.
[5] LONGUET-HIGGINS, M. S., 1952, 'On the statistical heights of sea waves'. *J. Marine Res.*, **11**, no. 3, 245–266.

# 8 Generalised frequency analysis

As we have just noted, it is necessary to consider a more general form of frequency analysis. In particular we wish to obtain the Fourier transform of multidimensional parameters[1].

## 8.1 Fourier series in two dimensions

In section 6.1 we discussed a periodic function $g(t)$ and showed how a Fourier series representation can be found for it. We now consider a more general finite periodic function $g(x, t)$ with a 'spatial periodicity' of wavelength $\lambda$ as well as a temporal periodicity of period $T$. The function may be represented by

$$g(x, t) = \tfrac{1}{2}A_{00} + \sum_{m=0}^{\infty} \sum_{n=0}^{\infty} [A_{mn} \cos mk_0x \cos n\omega_0 t$$
$$+ B_{mn} \sin mk_0 x \sin n\omega_0 t + C_{mn} \sin mk_0 x \cos n\omega_0 t$$
$$+ D_{mn} \cos mk_0 x \sin n\omega_0 t]$$

where the 'fundamental frequency' is $\omega_0 = 2\pi/T$ and the 'fundamental wave number' is $k_0 = 2\pi/\lambda$. In the summation $m = 0 = n$ is excluded.

The various coefficients are given by

$$\left.\begin{matrix} A_{mn} \\ B_{mn} \\ C_{mn} \\ D_{mn} \\ A_{00} \end{matrix}\right\} = \frac{4}{T\lambda} \int_{-T/2}^{T/2} \int_{-\lambda/2}^{\lambda/2} g(x, t) \left\{\begin{matrix} \cos mk_0 x \cos n\omega_0 t \\ \sin mk_0 x \sin n\omega_0 t \\ \sin mk_0 x \cos n\omega_0 t \\ \cos mk_0 x \sin n\omega_0 t \\ \tfrac{1}{2} \end{matrix}\right\} dx\, dt$$

for $m = 0, 1, 2, \ldots$ and $n = 0, 1, 2, \ldots$ but excluding $m = 0 = n$ (which is specially catered for). This is merely a statement of the Fourier theorem in a more general form than that given previously. It is assumed that the double integral

$$\int_{-T/2}^{T/2} \int_{-\lambda/2}^{\lambda/2} g(x, t) dx\, dt$$

is finite.

As before, the theorem may be stated in terms of the complex exponential; thus

$$g(x, t) = g(0, 0) + \sum_{m=0}^{\infty} \sum_{n=0}^{\infty} [g(m, n)e^{i(mk_0x - n\omega_0t)}$$
$$+ g(m, -n)e^{i(mk_0x + n\omega_0t)} + g(-m, n)e^{-i(mk_0x + n\omega_0t)}$$
$$+ g(-m, -n)e^{-i(mk_0x - n\omega_0t)}].$$

Again, $m = 0 = n$ is excluded in the summation. Notice that we have again preferred to write the multipliers in the form $g(m, n)$, $g(m, -n)$, etc. rather than in the more natural form $g_{mn}$, $g_{m,-n}$, etc. The new coefficients are related to the old through the relations

$$g(m, -n) = \tfrac{1}{4}\{A_{mn} - B_{mn} - i(C_{mn} + D_{mn})\}$$
$$g(-m, n) = \tfrac{1}{4}\{A_{mn} - B_{mn} + i(C_{mn} + D_{mn})\}$$
$$g(-m, -n) = \tfrac{1}{4}\{A_{mn} + B_{mn} + i(C_{mn} - D_{mn})\}$$
$$g(m, n) = \tfrac{1}{4}\{A_{mn} + B_{mn} - i(C_{mn} - D_{mn})\}$$
$$g(0, 0) = \tfrac{1}{2}A_{00}$$

for $m = 0, 1, 2, \ldots$ and $n = 0, 1, 2, \ldots$ but excluding $m = 0 = n$.

The complex series can be rearranged into the more compact form

$$g(x, t) = \sum_{m=-\infty}^{\infty} \sum_{n=-\infty}^{\infty} g(m, n)e^{i(mk_0x - n\omega_0t)}$$

where the coefficients $g(m, n)$ are expressed in the integral form

$$g(m, n) = \frac{1}{T\lambda} \int_{-T/2}^{T/2} \int_{-\lambda/2}^{\lambda/2} g(x, t)e^{-i(mk_0x - n\omega_0t)} dx\, dt$$

for $m = 0, \pm 1, \pm 2, \ldots.$ and $n = 0, \pm 1, \pm 2, \ldots$

### 8.1.1 Fourier series in three dimensions

The procedure from two to three dimensions is straightforward. The trebly-periodic finite function $g(x, y, t)$ may be expressed in the form of a trigonometric series

$$g(x, y, t) = \tfrac{1}{2} A_{000} + \sum_{l=0}^{\infty} \sum_{m=0}^{\infty} \sum_{n=0}^{\infty} [A_{lmn} \cos lk_{0x}x \cos mk_{0y}y \cos n\omega_0 t$$

$$+ B_{lmn} \cos lk_{0x}x \cos mk_{0y}y \sin n\omega_0 t + \ldots$$

$$+ H_{lmn} \sin lk_{0x}x \sin mk_{0y}y \sin n\omega_0 t]$$

excluding $l = 0 = m = n$ in the summation. Here

$$\omega_0 = \frac{2\pi}{T}; \quad k_{0x} = \frac{2\pi}{\lambda_x}; \quad k_{0y} = \frac{2\pi}{\lambda_y}$$

where $\lambda_x$ is the wavelength in the $OX$ direction and $\lambda_y$ is the wavelength in the direction $OY$. The coefficients are of the form

$$A_{000} = \frac{2}{T\lambda_x\lambda_y} \int_{-T/2}^{T/2} \int_{-\lambda_y/2}^{\lambda_y/2} \int_{-\lambda_x/2}^{\lambda_x/2} g(x, y, t)\, dx\, dy\, dt$$

$$A_{lmn} = \frac{8}{T\lambda_x\lambda_y} \int_{-T/2}^{T/2} \int_{-\lambda_y/2}^{\lambda_y/2} \int_{-\lambda_x/2}^{\lambda_x/2} g(x, y, t) \times$$

$$\cos lk_{0x}x \cos mk_{0y}y \cos n\omega_0 t\, dx\, dy\, dt$$

etc., where $l = 0 = m = n$ is excluded.

If the Fourier series is rearranged it can be shown that a complex form can be used. That is

$$g(x, y, t) = \sum_{l=-\infty}^{\infty} \sum_{m=-\infty}^{\infty} \sum_{n=-\infty}^{\infty} g(l, m, n) e^{i(lk_{0x}x + mk_{0y}y - n\omega_0 t)}$$

where

$$g(l, m, n) = \frac{1}{T\lambda_x\lambda_y} \int_{-T/2}^{T/2} \int_{-\lambda_y/2}^{\lambda_y/2} \int_{-\lambda_x/2}^{\lambda_x/2} g(x, y, t) \times$$

$$e^{-i(lk_{0x}x + mk_{0y}y - n\omega_0 t)}\, dx\, dy\, dt.$$

The new coefficients are in fact related to the old by relationships of the form

$$g(l, m, n) = \frac{1}{8} \{A_{lmn} + E_{lmn} + F_{lmn} - G_{lmn}$$

$$+ i(B_{lmn} - C_{lmn} - D_{lmn} - H_{lmn})\} \ldots \text{ etc.}$$

## 8.2 Fourier integral

The *non*-periodic function $g(x, t)$ can only be represented as a Fourier series if the spatial periodicity $\lambda$ and period $T$ are assumed to be infinite. Thus, if $\lambda$ and $T$ are both very large, we take

$$\frac{2\pi}{\lambda} = \delta k_0; \quad \frac{2\pi}{T} = \delta \omega_0$$

so that

$$g(x, t) = \sum_{m=-\infty}^{\infty} \sum_{n=-\infty}^{\infty} g(m, n) e^{i(m \delta k_0 x - n \delta \omega_0 t)}$$

where

$$g(m, n) = \frac{\delta \omega_0 \delta k_0}{4\pi^2} \int_{-T/2}^{T/2} \int_{-\lambda/2}^{\lambda/2} g(x, t) e^{-i(m \delta k_0 x - n \delta \omega_0 t)} \, dx \, dt.$$

As $\lambda$ and $T$ approach infinity,

$$\delta k_0 \to dk, \qquad m\delta k_0 \to k,$$

$$\delta \omega_0 \to d\omega, \qquad n\delta \omega_0 \to \omega,$$

and the summations become integrals such that

$$g(x, t) = \frac{1}{4\pi^2} \int_{-\infty}^{\infty} \int_{-\infty}^{\infty} e^{i(kx - \omega t)} d\omega dk \int_{-\infty}^{\infty} \int_{-\infty}^{\infty} g(x, t) e^{-i(kx - \omega t)} dx \, dt.$$

This result may be written in the form

$$g(x, t) = \int_{-\infty}^{\infty} \int_{-\infty}^{\infty} G(k, \omega) e^{i(kx - \omega t)} dk \, d\omega$$

where

$$G(k, \omega) = \frac{1}{4\pi^2} \int_{-\infty}^{\infty} \int_{-\infty}^{\infty} g(x, t) e^{-i(kx - \omega t)} dx \, dt \, .$$

This states that the non-periodic function $g(x, t)$ can be expressed in the form of a Fourier integral by means of the Fourier transform $G(k, \omega)$. This latter is, in general, a continuous complex function in the wave number and frequency domain.

In the same way, for the non-periodic function $g(x, y, t)$ we let

$$\frac{2\pi}{\lambda_x} = \delta k_{0x}$$

$$\frac{2\pi}{\lambda_y} = \delta k_{0y}$$

$$\frac{2\pi}{T} = \delta \omega_0$$

whence

$$g(x, y, t) = \frac{1}{8\pi^3} \sum_{l=-\infty}^{\infty} \sum_{m=-\infty}^{\infty} \sum_{n=-\infty}^{\infty} e^{i(l\delta k_{ox}x + m\delta k_{oy}y - n\delta\omega_o t)} \times \delta k_{0x}\delta k_{0y}\delta\omega_0 \times$$

$$\int_{-T/2}^{T/2} \int_{-\lambda_y/2}^{\lambda_y/2} \int_{-\lambda_x/2}^{\lambda_x/2} g(x, y, t)e^{-i(l\delta k_{ox}x + m\delta k_{oy}y - n\delta\omega_o t)}dx\, dy\, dt.$$

As $T$, $\lambda_{0x}$, $\lambda_{0y}$ all tend to infinity,

$$\delta k_{0x} \to dk_x, \quad l\delta k_{0x} \to k_x,$$

$$\delta k_{0y} \to dk_y, \quad m\delta k_{0y} \to k_y,$$

$$\delta \omega_0 \to d\omega, \quad n\delta\omega_0 \to \omega,$$

while the summations become integrations so that

$$g(x, y, t) = \int_{-\infty}^{\infty} \int_{-\infty}^{\infty} \int_{-\infty}^{\infty} G(k_x, k_y, \omega)e^{i(k_x x + k_y y - \omega t)}dk_x dk_y d\omega$$

where

$$G(k_x, k_y, \omega) = \frac{1}{8\pi^3} \int_{-\infty}^{\infty} \int_{-\infty}^{\infty} \int_{-\infty}^{\infty} g(x, y, t)e^{-i(k_x x + k_y y - \omega t)}dx\, dy\, dt.$$

The continuous complex function $G(k_x, k_y, \omega)$ is the Fourier transform of the given function $g(x, y, t)$.

It is common to condense this last result somewhat by writing

$$\mathbf{r} = x\mathbf{i} + y\mathbf{j}; \quad d\mathbf{r} = dx\, dy;$$

$$\mathbf{k} = k_x\mathbf{i} + k_y\mathbf{j}; \quad d\mathbf{k} = dk_x\, dk_y.$$

The Fourier transform pair is now written as

$$g(\mathbf{r}, t) = \int_{-\infty}^{\infty} \int_{-\infty}^{\infty} \int_{-\infty}^{\infty} G(\mathbf{k}, \omega)e^{i(\mathbf{k}.\mathbf{r} - \omega t)}\, d\mathbf{k}\, d\omega$$

where

$$G(\mathbf{k}, \omega) = \frac{1}{8\pi^3} \int_{-\infty}^{\infty} \int_{-\infty}^{\infty} \int_{-\infty}^{\infty} g(\mathbf{r}, t)e^{-i(\mathbf{k}.\mathbf{r} - \omega t)}\, d\mathbf{r}\, dt.$$

But this is merely a shorter way of writing the foregoing results.

As in the simpler one-dimensional case we discussed in section 6.2, it is necessary that $g(x, t)$ and $g(\mathbf{r}, t)$ shall meet certain requirements if they are to possess Fourier transforms. The derivation of these requirements is not easy and we merely note that it is sufficient that:

(a) $g(x, t), g(\mathbf{r}, t)$ must be continuous,

(b) $\int_{-\infty}^{\infty}\int_{-\infty}^{\infty} |g(x, t)|\, dx dt, \int_{-\infty}^{\infty}\int_{-\infty}^{\infty}\int_{-\infty}^{\infty} |g(\mathbf{r}, t)|\, d\mathbf{r}\, dt$ must exist,

(c) $\int_{-\infty}^{\infty}\int_{-\infty}^{\infty} \{g(x, t)\}^2\, dx\, dt, \int_{-\infty}^{\infty}\int_{-\infty}^{\infty}\int_{-\infty}^{\infty} \{g(\mathbf{r}, t)\}^2\, d\mathbf{r}\, dt$

must be finite.

## 8.3 Application of the Fourier integral to random processes

Just as we did in section 6.3, consider a single realisation $X^{(1)}(\mathbf{r}, t)$ of a random process $X(\mathbf{r}, t)$. In general this observed realisation will not possess a Fourier transform since it does not meet the requirement just mentioned. We can, however, truncate $X^{(1)}(\mathbf{r}, t)$ in such a fashion that

$$X_T(\mathbf{r}, t) = \begin{cases} X^{(1)}(\mathbf{r}, t) & \text{if } -\lambda_x/2 < x < \lambda_x/2,\ -\lambda_y/2 < y < \lambda_y/2, \\ & \qquad -T/2 < t < T/2 \\ 0 & \text{otherwise} \end{cases}$$

and $X_T(\mathbf{r}, t)$ does then possess a Fourier transform. The Fourier transform pair is now of the form

$$G_T(\mathbf{k}, \omega) = \frac{1}{8\pi^3} \int_{-\infty}^{\infty}\int_{-\infty}^{\infty}\int_{-\infty}^{\infty} X_T(\mathbf{r}, t) e^{-i(\mathbf{k}.\mathbf{r} - \omega t)}\, d\mathbf{r}\, dt$$

$$X_T(\mathbf{r}, t) = \int_{-\infty}^{\infty}\int_{-\infty}^{\infty}\int_{-\infty}^{\infty} G_T(\mathbf{k}, \omega) e^{i(\mathbf{k}.\mathbf{r} - \omega t)}\, d\mathbf{k}\, d\omega.$$

The Fourier transform of the truncated realisation may be used in expressions for elementary parametric averages. Thus the mean value of $X_T(\mathbf{r}, t)$ is

$$\langle X_T(\mathbf{r}, t)\rangle = \frac{1}{\lambda_x \lambda_y T} \int_{-T/2}^{T/2}\int_{-\lambda_y/2}^{\lambda_y/2}\int_{-\lambda_x/2}^{\lambda_x/2} X_T(x, y, t)\, dx\, dy\, dt$$

$$= \frac{1}{\lambda_x \lambda_y T} \int_{-\infty}^{\infty}\int_{-\infty}^{\infty}\int_{-\infty}^{\infty} X_T(\mathbf{r}, t)\, d\mathbf{r}\, dt$$

$$= \frac{8\pi^3}{\lambda_x \lambda_y T} G_T(\mathbf{0}, 0)$$

so that

$$\langle X^{(1)}(\mathbf{r}, t)\rangle = \lim_{\substack{T\to\infty \\ \lambda_x\to\infty \\ \lambda_y\to\infty}} \left\{\frac{8\pi^3}{\lambda_x\lambda_y T} G_T(0, 0)\right\}.$$

The mean square of $X_T(\mathbf{r}, t)$ is particularly important, as we shall show. It is

$$\langle X_T^2(\mathbf{r}, t)\rangle = \frac{1}{\lambda_x\lambda_y T}\int_{-T/2}^{T/2}\int_{-\lambda_y/2}^{\lambda_y/2}\int_{-\lambda_x/2}^{\lambda_x/2} X_T^2(x, y, t)\,dx\,dy\,dt$$

$$= \frac{1}{\lambda_x\lambda_y T}\int_{-\infty}^{\infty}\int_{-\infty}^{\infty}\int_{-\infty}^{\infty} X_T^2(\mathbf{r}, t)\,d\mathbf{r}\,dt$$

$$= \frac{1}{\lambda_x\lambda_y T}\int_{-\infty}^{\infty}\int_{-\infty}^{\infty}\int_{-\infty}^{\infty} X_T(\mathbf{r}, t)\times$$

$$\left\{\int_{-\infty}^{\infty}\int_{-\infty}^{\infty}\int_{-\infty}^{\infty} G_T(\mathbf{k}, \omega)\,.\,e^{i(\mathbf{k}.\mathbf{r}-\omega t)}\,d\mathbf{k}\,d\omega\right\}d\mathbf{r}\,dt$$

$$= \frac{1}{\lambda_x\lambda_y T}\int_{-\infty}^{\infty}\int_{-\infty}^{\infty}\int_{-\infty}^{\infty} G_T(\mathbf{k}, \omega)\times$$

$$\left\{\int_{-\infty}^{\infty}\int_{-\infty}^{\infty}\int_{-\infty}^{\infty} X_T(\mathbf{r}, t)e^{i(\mathbf{k}.\mathbf{r}-\omega t)}\,d\mathbf{r}\,dt\right\}d\mathbf{k}\,d\omega$$

$$= \frac{8\pi^3}{\lambda_x\lambda_y T}\int_{-\infty}^{\infty}\int_{-\infty}^{\infty}\int_{-\infty}^{\infty} G_T(\mathbf{k}, \omega)G_T^*(\mathbf{k}, \omega)\,d\mathbf{k}\,d\omega$$

where $G_T^*(\mathbf{k}, \omega)$ is the complex conjugate of $G_T(\mathbf{k}, \omega)$. It follows that

$$\langle X^{(1)2}(\mathbf{r}, t)\rangle = \int_{-\infty}^{\infty}\int_{-\infty}^{\infty}\int_{-\infty}^{\infty} \lim_{\substack{\lambda_x\to\infty \\ \lambda_y\to\infty \\ T\to\infty}} \left\{\frac{8\pi^3}{\lambda_x\lambda_y T}|G_T(\mathbf{k}, \omega)|^2\right\}d\mathbf{k}\,d\omega$$

and by generalising the results of section 6, we may write it as

$$\langle X^{(1)2}(\mathbf{r}, t)\rangle = \int_{-\infty}^{\infty}\int_{-\infty}^{\infty}\int_{-\infty}^{\infty} S_{XX}(\mathbf{k}, \omega)\,d\mathbf{k}\,d\omega\,.$$

We now see that the mean square spectral density is a real and even function of the parameters $k_x$, $k_y$ and $\omega$ given by

$$S_{XX}(\mathbf{k},\omega)=\lim_{\substack{\lambda_x\to\infty\\ \lambda_y\to\infty\\ T\to\infty}}\left\{\frac{8\pi^3}{\lambda_x\lambda_y T}\,|G_T(\mathbf{k},\omega)|^2\right\}.$$

The auto-correlation function of the realisation $X^{(1)}(\mathbf{r},t)$ may be found from the parametric average $\langle X_T(\mathbf{r},t)X_T(\mathbf{r}+\boldsymbol{\rho},t+\tau)\rangle$, using the same limiting process. We have

$$R_T(\boldsymbol{\rho},\tau)=\frac{1}{\lambda_x\lambda_y T}\int_{-T/2}^{T/2}\int_{-\lambda_y/2}^{\lambda_y/2}\int_{-\lambda_x/2}^{\lambda_x/2} X_T(x,y,t)\times X_T(x+\rho_x,y+\rho_y,\;t\;+\tau)\,dx\,dy\,dt$$

$$=\frac{1}{\lambda_x\lambda_y T}\int_{-\infty}^{\infty}\int_{-\infty}^{\infty}\int_{-\infty}^{\infty} X_T(\mathbf{r},t)X_T(\mathbf{r}+\boldsymbol{\rho},t+\tau)\,d\mathbf{r}\,dt$$

$$=\frac{1}{\lambda_x\lambda_y T}\int_{-\infty}^{\infty}\int_{-\infty}^{\infty}\int_{-\infty}^{\infty} X_T(x,y,t)\left\{\int_{-\infty}^{\infty}\int_{-\infty}^{\infty}\int_{-\infty}^{\infty} G_T(k_x,k_y,\omega)\times e^{i[k_x(x+\rho_x)+k_y(y+\rho_y)-\omega(t+\tau)]}\,dk_x dk_y d\omega\right\}dx\,dy\,dt$$

$$=\frac{1}{\lambda_x\lambda_y T}\int_{-\infty}^{\infty}\int_{-\infty}^{\infty}\int_{-\infty}^{\infty} G_T(k_x,k_y,\omega)\left\{\int_{-\infty}^{\infty}\int_{-\infty}^{\infty}\int_{-\infty}^{\infty} X_T(x,y,t)\times e^{i(k_x x+k_y y-\omega t)}\,dx\,dy\,dt\right\}e^{i(k_x\rho_x+k_y\rho_y-\omega\tau)}\,dk_x dk_y d\omega.$$

The triple integral within the curly brackets is equal to $8\pi^3 G_T^*(k_x,k_y,\omega)$ so that

$$R_T(\boldsymbol{\rho},\tau)=\frac{8\pi^3}{\lambda_x\lambda_y T}\int_{-\infty}^{\infty}\int_{-\infty}^{\infty}\int_{-\infty}^{\infty}|G_T(k_x,k_y,\omega)|^2\,e^{i(k_x\rho_x+k_y\rho_y-\omega\tau)}\,dk_x dk_y d\omega$$

whence, in the limit, as $\lambda_x$, $\lambda_y$ and $T\to\infty$,

$$R_{XX}(\boldsymbol{\rho},\tau)=\int_{-\infty}^{\infty}\int_{-\infty}^{\infty}\int_{-\infty}^{\infty} S_{XX}(\mathbf{k},\omega)e^{i(\mathbf{k}.\boldsymbol{\rho}-\omega\tau)}\,d\mathbf{k}\,d\omega.$$

These various results apply, of course, to a single realisation of a random process. The parametric averages that we have found, namely $\langle X^{(1)}(\mathbf{r},t)\rangle$, $\langle X^{(1)2}(\mathbf{r},t)\rangle$ and $R_{XX}(\boldsymbol{\rho},\tau)$, represent meaningful expectations of the parent random process $X(\mathbf{r}, t)$ only if that process is ergodic in these quantities. Moreover, $S_{XX}(\mathbf{k}, \omega)$ is the mean square spectral density of $X(\mathbf{r}, t)$ only under that condition.

### 8.3.1 Some properties of the auto-correlation function

In the last section we found expressions for the three-dimensional auto-correlation function and the mean square spectral density of an ergodic random process. These quantities are of great importance and it is therefore worthwhile to reflect briefly on their properties.

We may first notice that, had the random process produced a *two*-dimensional realisation $X^{(1)}(x, t)$ we should have found the results

$$R_{XX}(\rho, \tau) = \int_{-\infty}^{\infty} \int_{-\infty}^{\infty} S_{XX}(k, \omega)\, e^{i(k\rho - \omega\tau)}\, dk\, d\omega$$

$$S_{XX}(k, \omega) = \lim_{\substack{T \to \infty \\ \lambda \to \infty}} \left\{ \frac{4\pi^2}{T\lambda} |G_T(k, \omega)|^2 \right\}$$

where $G_T(k, \omega)$ is the Fourier transform of the truncated realisation $X_T(x, t)$.

It will be noticed that in both the two and three-dimensional cases, the mean square spectral density is simply the Fourier transform of the auto-correlation function. In other words, if the mean square spectral density function $S$ is given, the auto-correlation function $R$ can be found because

$$S_{XX}(k, \omega) = \frac{1}{4\pi^2} \int_{-\infty}^{\infty} \int_{-\infty}^{\infty} R_{XX}(\rho, \tau) e^{-i(k\rho - \omega\tau)}\, d\rho\, d\tau$$

in two dimensions, and

$$S_{XX}(\mathbf{k}, \omega) = \frac{1}{8\pi^3} \int_{-\infty}^{\infty} \int_{-\infty}^{\infty} \int_{-\infty}^{\infty} R_{XX}(\boldsymbol{\rho}, \tau) e^{-i(\mathbf{k}.\boldsymbol{\rho} - \omega\tau)}\, d\boldsymbol{\rho}\, d\tau$$

in three. That the auto-correlation can possess a Fourier transform (i.e. that it does meet the requirements mentioned in section 8.2) can be seen from physical considerations; for if one or more of the arguments of the auto-correlation function $R$ is made large, the expected value that $R$ represents will be vanishingly small. Just as we found in section 6.3.1 (where we introduced the one dimensional problem), we should therefore have been justified in writing the mean square spectral density function $S$ and the auto-correlation function $R$ as a Fourier transform pair without having recourse to the truncated realisation and its Fourier transform.

It is convenient to push the physical reasoning a little further. Since the random process $X(x, t)$ or $X(\mathbf{r}, t)$ is both stationary and homogeneous, it follows that $R_{XX}(\rho, \tau)$ or $R_{XX}(\boldsymbol{\rho}, \tau)$ is a real and even function. Consequently we have confirmed that the quantity $S_{XX}(k, \omega)$ or $S_{XX}(\mathbf{k}, \omega)$ is real and even. Taking the former case, then, we may write

$$R_{XX}(\rho, \tau) = \int_{-\infty}^{\infty} \int_{-\infty}^{\infty} S_{XX}(k, \omega) \cos(k\rho - \omega\tau)\, dk\, d\omega$$

$$= 4 \int_{0}^{\infty} \int_{0}^{\infty} S_{XX}(k, \omega) \cos(k\rho - \omega\tau)\, dk\, d\omega$$

or, alternatively,

$$R_{XX}(\rho, \tau) = \int_{0}^{\infty} \int_{0}^{\infty} \Phi_{XX}(k, \omega) \cos(k\rho - \omega\tau)\, dk\, d\omega$$

where $\Phi_{XX}(k, \omega)$ is the 'two dimensional one sided mean square spectral density function' defined by

$$\Phi_{XX}(k, \omega) = \begin{cases} 4S_{XX}(k, \omega) & \text{for } k \geqslant 0, \omega \geqslant 0 \\ 0 & \text{otherwise.} \end{cases}$$

The foregoing argument may be extended so as to cover the three-dimensional problem. That is, we may write

$$R_{XX}(\boldsymbol{\rho}, \tau) = 8 \int_{0}^{\infty} \int_{0}^{\infty} \int_{0}^{\infty} S_{XX}(\mathbf{k}, \omega) \cos(\mathbf{k} \cdot \boldsymbol{\rho} - \omega\tau)\, d\mathbf{k}\, d\omega$$

$$= \int_{0}^{\infty} \int_{0}^{\infty} \int_{0}^{\infty} \Phi_{XX}(\mathbf{k}, \omega) \cos(\mathbf{k} \cdot \boldsymbol{\rho} - \omega\tau)\, d\mathbf{k}\, d\omega$$

where

$$\Phi_{XX}(\mathbf{k}, \omega) = \begin{cases} 8S_{XX}(\mathbf{k}, \omega) & \text{for } k_x \geqslant 0, k_y \geqslant 0, \omega \geqslant 0 \\ 0 & \text{otherwise.} \end{cases}$$

The quantity $\Phi_{XX}(\mathbf{k}, \omega)$ is now the 'three-dimensional one sided mean square spectral density function'. It is this function which may possibly be measured physically, whereas the functions $S_{XX}(k, \omega)$, $S_{XX}(\mathbf{k}, \omega)$ are defined in the variable ranges of $(-\infty, \infty)$ only for mathematical convenience.

## Reference

[1] CHAMPENEY, D. C., 1973, *Fourier Transforms and their Physical Applications*, Academic Press, London.

# 9 Wave spectral density functions

Let us agree to regard the elevation of the surface of the sea as a random process $\zeta(\mathbf{r}, t)$. And let us also assume that this process is ergodic both as regards space and time statistics. From our previous discussion in chapters 7 and 8 we see that the auto-correlation function and the one-sided spectral density function of the random process $\zeta(\mathbf{r}, t)$ may be written

$$R_{\zeta\zeta}(\boldsymbol{\rho}, \tau) = \int_0^\infty \int_0^\infty \int_0^\infty \Phi_{\zeta\zeta}(\mathbf{k}, \omega)\cos(\mathbf{k} \cdot \boldsymbol{\rho} - \omega\tau)\, d\mathbf{k}\, d\omega$$

and

$$\Phi_{\zeta\zeta}(\mathbf{k}, \omega) = \frac{2}{\pi^3} \int_0^\infty \int_0^\infty \int_0^\infty R_{\zeta\zeta}(\boldsymbol{\rho}, \tau)\cos(\mathbf{k} \cdot \boldsymbol{\rho} - \omega\tau)\, d\boldsymbol{\rho}\, d\tau.$$

The quantity $\Phi_{\zeta\zeta}(\mathbf{k}, \omega)\delta\mathbf{k}\delta\omega$ is defined as the contribution to the mean square elevation having a wave number between $\mathbf{k}$ and $\mathbf{k} + \delta\mathbf{k}$ and a frequency between $\omega$ and $\omega + \delta\omega$.

The spectral density $\Phi_{\zeta\zeta}(\mathbf{k}, \omega)$ is, generally speaking, the most useful function of its type, since averages of higher order are unwieldy to use. Unfortunately it is exceedingly difficult to obtain and it is necessary to examine less demanding alternatives. This, we shall now do.

Before proceeding, it will be as well to refer to a matter of usage. The one-sided mean square spectral densities – the quantities that we have denoted by the symbol $\Phi$ – are frequently referred to as 'energy spectra'. Thus $\Phi_{\zeta\zeta}(\mathbf{k}, \omega)$ is called a 'three dimensional wave energy spectrum'. The reason for this is that in section 7.1.1 it was shown that the total energy per unit wavelength of a simple sinusoidal wave is proportional to the average of the squared wave elevation; i.e. energy = $\rho g a^2/2$. It has also been demonstrated that

$$E[\zeta^2(\mathbf{r}, t)] = \tfrac{1}{2} \sum_{i=1}^{n} a_i^2 = R_{\zeta\zeta}(0, 0) = \int_0^\infty \int_0^\infty \int_0^\infty \Phi_{\zeta\zeta}(\mathbf{k}, \omega) d\mathbf{k}\, d\omega$$

and since the total energy per unit wavelength of the superposed waves is given by

$$\rho g E[\zeta^2(\mathbf{r}, t)] = \frac{\rho g}{2} \sum_{i=1}^{n} a_i^2 = \rho g R_{\zeta\zeta}(0, 0)$$

$$= \rho g \int_0^\infty \int_0^\infty \int_0^\infty \Phi_{\zeta\zeta}(\mathbf{k}, \omega) d\mathbf{k}\, d\omega$$

it follows that the wave energy is proportional to either the mean square value of superposed wave elevation or the integrated three dimensional energy spectrum. Following usual practice, then, we shall occasionally refer to 'energy spectra', although electrical engineers may alternatively refer to 'power spectra' whilst in branches of mathematics the term 'variance' is used.

## 9.1 Instantaneous energy spectra

A stereoscopic photograph[1] of the sea's surface permits the wave characteristics to be ascertained at the moment when the picture is taken. From the definition of the wave auto-correlation function as a parametric average, and on the assumption that the process is ergodic in that quantity, it is seen that $R_{\zeta\zeta}(\boldsymbol{\rho}, 0)$ may be obtained – at least in theory. It is expressible in the form

$$R_{\zeta\zeta}(\boldsymbol{\rho}, 0) = \int_0^\infty \int_0^\infty \Phi'_{\zeta\zeta}(\mathbf{k}) \cos(\mathbf{k} \cdot \boldsymbol{\rho})\, d\mathbf{k}$$

where

$$\Phi'_{\zeta\zeta}(\mathbf{k}) = \frac{2}{\pi^2} \int_0^\infty \int_0^\infty R_{\zeta\zeta}(\boldsymbol{\rho}, 0) \cos(\mathbf{k} \cdot \boldsymbol{\rho}) d\boldsymbol{\rho}.$$

If one could make the necessary calculations[2], then, the 'wave number energy spectrum' $\Phi'_{\zeta\zeta}(\mathbf{k})$ could be determined. To see how $\Phi'_{\zeta\zeta}(\mathbf{k})$ is related to the more general quantity $\Phi_{\zeta\zeta}(\mathbf{k}, \omega)$ we note that if $\boldsymbol{\rho} = 0$

$$R_{\zeta\zeta}(0, 0) = \int_0^\infty \int_0^\infty \Phi'_{\zeta\zeta}(\mathbf{k})\, d\mathbf{k}.$$

But

$$R_{\zeta\zeta}(0,0) = E[\zeta^2(\mathbf{r},t)] = \int_0^\infty \int_0^\infty \int_0^\infty \Phi_{\zeta\zeta}(\mathbf{k},\omega)d\mathbf{k}\,d\omega$$

and by comparing these two equations, we discover that

$$\Phi'_{\zeta\zeta}(\mathbf{k}) = \int_0^\infty \Phi_{\zeta\zeta}(\mathbf{k},\omega)d\omega.$$

That is, $\Phi'_{\zeta\zeta}(\mathbf{k})$ gives the contribution to the wave energy from components of wave number $\mathbf{k}$ irrespective of the frequency associated with the wave number. In component form, the result is

$$E[\zeta^2(\mathbf{r},t)] = \int_0^\infty \int_0^\infty \Phi'_{\zeta\zeta}(k_x, k_y)dk_x\,dk_y$$

where the wave components are

$$k_x = k\cos\mu$$

$$k_y = k\sin\mu$$

(see figure 29).

This last result would convey a better picture if the energy spectrum were expressed in terms of $k$ and $\mu$, rather than $k_x$ and $k_y$. By a change of variable it may be shown that

$$E[\zeta^2(\mathbf{r},t)] = \int_{-\pi/2}^{\pi/2} \int_0^\infty \Phi''_{\zeta\zeta}(k,\mu)\,\|J\|dk\,d\mu$$

where

$$|J| = \begin{vmatrix} \dfrac{\partial k_x}{\partial k} & \dfrac{\partial k_y}{\partial k} \\[2ex] \dfrac{\partial k_x}{\partial \mu} & \dfrac{\partial k_y}{\partial \mu} \end{vmatrix} = \begin{vmatrix} \cos\mu & \sin\mu \\[2ex] -k\sin\mu & k\cos\mu \end{vmatrix} = k$$

is the 'Jacobian' of the transformation and where the modulus $\|J\|$ of $|J|$ is used to ensure that the transformed mean square spectral density function is positive (c.f. section 3.3). Thus

$$E[\zeta^2(\mathbf{r},t)] = \int_{-\pi/2}^{\pi/2} \int_0^\infty \Phi''_{\zeta\zeta}(k,\mu)k\,dk\,d\mu = \int_{-\pi/2}^{\pi/2} \int_0^\infty \Phi'^{*}_{\zeta\zeta}(k,\mu)dk\,d\mu$$

where $\Phi'^{*}_{\zeta\zeta}(k, \mu)$ is the 'directional wave number spectrum' and is derived from the wave number spectrum $\Phi'_{\zeta\zeta}(\mathbf{k})$. This spectrum is of importance in oceanographic studies.

For waves of small amplitude in deep water, the wave number satisfies the dispersion relationship

$$k = \frac{\omega^2}{g}$$

so we may use a second transformation:

$$E[\zeta^2(\mathbf{r}, t)] = \int_{-\pi/2}^{\pi/2} \int_0^\infty \Phi'''_{\zeta\zeta}(\omega, \mu) k \, \|J\| d\omega \, d\mu$$

where, now,

$$|J| = \begin{vmatrix} \dfrac{\partial k}{\partial \omega} & \dfrac{\partial \mu}{\partial \omega} \\ \dfrac{\partial k}{\partial \mu} & \dfrac{\partial \mu}{\partial \mu} \end{vmatrix} = \begin{vmatrix} \dfrac{2\omega}{g} & 0 \\ 0 & 1 \end{vmatrix} = \frac{2\omega}{g}.$$

That is to say

$$E[\zeta^2(\mathbf{r}, t)] = \int_{-\pi/2}^{\pi/2} \int_0^\infty \Phi'''_{\zeta\zeta}(\omega, \mu) \frac{2\omega^3}{g^2} \, d\omega \, d\mu = \int_{-\pi/2}^{\pi/2} \int_0^\infty \Phi'^{*}_{\zeta\zeta}(\omega, \mu) d\omega \, d\mu.$$

The quantity $\Phi'^{*}_{\zeta\zeta}(\omega, \mu)$ is the 'directional frequency spectrum' and it is a positive real function that can be represented graphically by a surface as shown in Figure 33.

## 9.2 Local energy spectra

If observation of the sea's elevation is carried out at a single point, rather than at a single instant, then $\boldsymbol{\rho} = \mathbf{0}$, instead of $\tau = 0$. On the basis of the observation, it is possible to determine $R_{\zeta\zeta}(\mathbf{0}, \tau)$ and so establish the Fourier transform pair

$$R_{\zeta\zeta}(\mathbf{0}, \tau) = \int_0^\infty \Phi^{*}_{\zeta\zeta}(\omega) \cos \omega\tau \, d\omega$$

$$\Phi^{*}_{\zeta\zeta}(\omega) = \frac{2}{\pi} \int_0^\infty R_{\zeta\zeta}(\mathbf{0}, \tau) \cos \omega\tau \, d\tau.$$

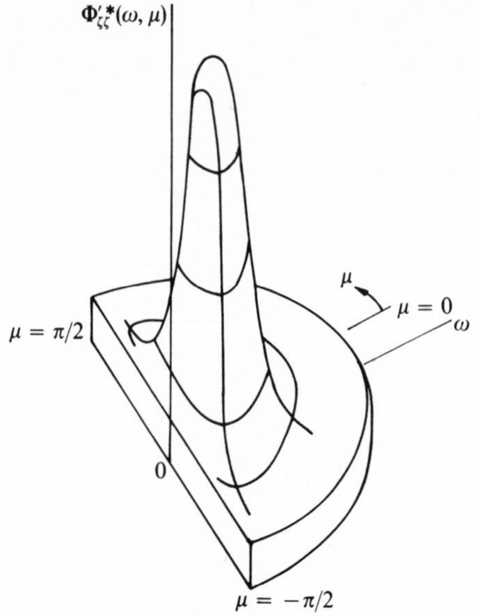


**Figure 33.** Directional frequency spectrum of wave elevation.

From the observed auto-correlation function $R_{\zeta\zeta}(\mathbf{0}, \tau)$ we may discover the mean square value since, with $\tau = 0$,

$$R_{\zeta\zeta}(\mathbf{0}, 0) = E[\zeta^2(\mathbf{r}, t)] = \int_0^\infty \Phi_{\zeta\zeta}^*(\omega)\, d\omega.$$

On comparison with the similar result obtained from the function $R_{\zeta\zeta}(\boldsymbol{\rho}, \tau)$ this shows that

$$\Phi_{\zeta\zeta}^*(\omega) = \int_0^\infty \int_0^\infty \Phi_{\zeta\zeta}(\mathbf{k}, \omega)\, d\mathbf{k}.$$

The one dimensional 'frequency spectrum' $\Phi_{\zeta\zeta}^*(\omega)$ gives the contribution of frequency $\omega$ to the mean square elevation, independently of the wave number associated with the frequency. It is the easiest spectrum to obtain and it is the basis of the commonest way of specifying wave conditions in the ocean. Figure 34 shows curves obtained by Neumann[3] for a fully developed sea for wind speeds of 20, 30 and 40 knots. Notice that, with an increase in the wind speed, the frequency of the component associated with the maximum wave energy decreases. However, the spectrum provides no information on the direction of wave travel.

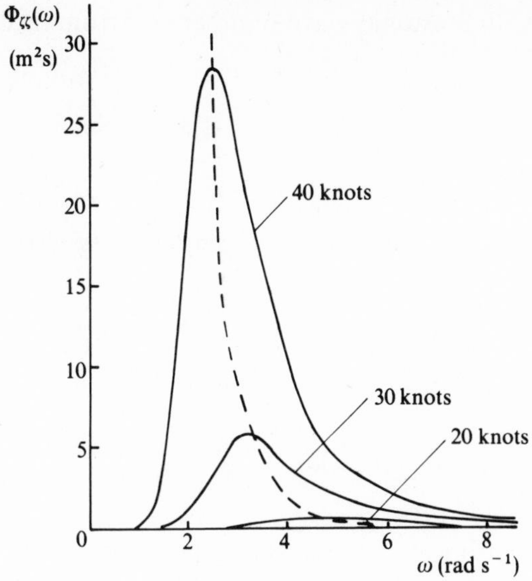


**Figure 34.** Neumann frequency spectra for various wind speeds. The broken line illustrates that higher wind speeds are associated with spectra having maximum values at lower frequencies.

It is of interest to relate the spectrum $\Phi^*_{\zeta\zeta}(\omega)$ with the directional frequency spectrum $\Phi'^*_{\zeta\zeta}(\omega, \mu)$. We have

$$E[\zeta^2(\mathbf{r}, t)] = \int_0^\infty \Phi^*_{\zeta\zeta}(\omega)\, d\omega.$$

When this is compared with the last equation in the previous section, it is seen that

$$\Phi^*_{\zeta\zeta}(\omega) = \int_{-\pi/2}^{\pi/2} \Phi'''_{\zeta\zeta}(\omega, \mu) \frac{2\omega^3}{g^2}\, d\mu = \int_{-\pi/2}^{\pi/2} \Phi'^*_{\zeta\zeta}(\omega, \mu)\, d\mu$$

for $\omega \geqslant 0$.

In section 9.1 the mean square wave value is also given by

$$E[\zeta^2(\mathbf{r}, t)] = \int_0^\infty \left\{ k \int_{-\pi/2}^{\pi/2} \Phi''_{\zeta\zeta}(k, \mu)\, d\mu \right\} dk = \int_0^\infty \int_{-\pi/2}^{\pi/2} \Phi'^*_{\zeta\zeta}(k, \mu)\, dk\, d\mu$$

$$= \int_0^\infty \Phi^{**}_{\zeta\zeta}(k) dk$$

where the 'one dimensional wave number spectrum' is defined by

$$\Phi^{**}_{\zeta\zeta}(k) = k \int_{-\pi/2}^{\pi/2} \Phi''_{\zeta\zeta}(k, \mu) d\mu.$$

From the results of this section and section 9.1 it follows that in deep water we may express the wave frequency spectrum in the form

$$\Phi^{*}_{\zeta\zeta}(\omega) = \frac{2\omega^3}{g^2} \int_{-\pi/2}^{\pi/2} \Phi'''_{\zeta\zeta}(\omega, \mu) d\mu = \frac{2\omega^3}{g^2} \int_{-\pi/2}^{\pi/2} \Phi''_{\zeta\zeta}\left(\frac{\omega^2}{g}, \mu\right) d\mu$$

$$= \frac{2\omega^3}{g^2} \int_{-\pi/2}^{\pi/2} \Phi''_{\zeta\zeta}(k, \mu) d\mu = \frac{2\omega^3}{g^2 k} \Phi^{**}_{\zeta\zeta}(k) = \frac{2\omega}{g} \Phi^{**}_{\zeta\zeta}(k)$$

This transformation expresses the relationship between the one dimensional wave number spectrum, $\Phi^{**}_{\zeta\zeta}(k)$, and the one dimensional wave frequency spectrum $\Phi^{*}_{\zeta\zeta}(\omega)$.

A more direct way of obtaining this relationship is by expressing the mean square wave elevation as a function of either $\Phi^{*}_{\zeta\zeta}(\omega)$ or $\Phi^{**}_{\zeta\zeta}(k)$. That is

$$E[\zeta^2(\mathbf{r}, t)] = \int_0^\infty \Phi^{*}_{\zeta\zeta}(\omega) d\omega = \int_0^\infty \Phi^{**}_{\zeta\zeta}(k) dk.$$

For deep water waves, $\omega^2 = gk$; it follows by a change of variable that

$$\int_0^\infty \Phi^{*}_{\zeta\zeta}(\omega) d\omega = \int_0^\infty [\Phi^{**}_{\zeta\zeta}(k)]_{k=\omega^2/g} \frac{2\omega}{g} d\omega$$

whence

$$\Phi^{*}_{\zeta\zeta}(\omega) = \frac{2\omega}{g} \Phi^{**}_{\zeta\zeta}(k)$$

for $k = \dfrac{\omega^2}{g}$, or

$$\Phi^{*}_{\zeta\zeta}(\omega) = \frac{\Phi^{**}_{\zeta\zeta}(k)}{\left|\dfrac{d\omega}{dk}\right|}$$

for $k = \dfrac{\omega^2}{g}$.

Since $\Phi^{*}_{\zeta\zeta}(\omega)$, $\Phi^{**}_{\zeta\zeta}(k)$ are both positive real quantities, the modulus sign has to be incorporated into the transformation relationship. This last equation expresses a simple transformation between wave spectra which are functions of related variables. It may be written as

$$\Phi^{*}_{\zeta\zeta}(\omega)d\omega = \Phi^{**}_{\zeta\zeta}(k)dk$$

In the literature, the superscripts that we have placed on the $\Phi$ are usually discarded. But their retention is helpful when, as here, the relationships between the different spectra are developed.

### 9.2.1 Wave slope spectra

The auto-correlation function for the wave elevation may be expressed in the component form

$$R_{\zeta\zeta}(\rho_x, \rho_y, \tau) = E[\zeta(x, y, t)\zeta(x + \rho_x, y + \rho_y, t + \tau)]$$

$$= \int_0^\infty \int_0^\infty \int_0^\infty \Phi_{\zeta\zeta}(k_x, k_y, \omega)\cos(k_x\rho_x + k_y\rho_y - \omega\tau)\, dk_x dk_y d\omega$$

as may be seen from section 9.1. By analogy with section 5.4, we see that successive differentiation with respect to $\rho_x$, for example, gives

$$R_{\zeta'\zeta'}(\rho_x, \rho_y, \tau) = -E[\zeta'(x, y, t)\zeta'(x + \rho_x, y + \rho_y, t + \tau)]$$

$$= -\int_0^\infty \int_0^\infty \int_0^\infty \Phi_{\zeta\zeta}(\mathbf{k}, \omega)k_x^2 \cos(\mathbf{k} \cdot \boldsymbol{\rho} - \omega\tau)d\mathbf{k}\, d\omega$$

and

$$E[\zeta'^2(x, y, t)] = \int_0^\infty \int_0^\infty \int_0^\infty k_x^2 \Phi_{\zeta\zeta}(\mathbf{k}, \omega)d\mathbf{k}\, d\omega$$

$$= \int_0^\infty \int_0^\infty \int_0^\infty \Phi_{\zeta'\zeta'}(\mathbf{k}, \omega)d\mathbf{k}\, d\omega.$$

Here, the new random process $\zeta'(\mathbf{r}, t)$ describing the wave slope in the $x$-direction has a wave slope spectrum $\Phi_{\zeta'\zeta'}(\mathbf{k}, \omega)$ related to the wave spectrum by

$$\Phi_{\zeta'\zeta'}(\mathbf{k}, \omega) = k_x^2 \Phi_{\zeta\zeta}(\mathbf{k}, \omega).$$

Generalising this result, we see that the wave slope spectrum may be determined from the wave spectrum by multiplying the latter by the square of the wave number. (In a similar way, relationships between the one dimensional spectrum for the wave slope process are analogous to those obtained in section 9.2; that is,

$$\Phi_{\zeta'\zeta'}(\omega) = \frac{2\omega}{g} \Phi_{\zeta'\zeta'}(k)$$

for $k = \frac{\omega^2}{g}$. However, the above result for the wave slope spectra indicates that

$$\Phi_{\zeta'\zeta'}(k) = k^2 \Phi_{\zeta\zeta}(k)$$

which may be related to the wave spectrum as,

$$\Phi_{\zeta'\zeta'}(\omega) = \frac{2\omega}{g} k^2 \Phi_{\zeta\zeta}(k) = \frac{\omega^4}{g^2} \Phi_{\zeta\zeta}(\omega)$$

for $k = \frac{\omega^2}{g}$ since

$$\Phi_{\zeta\zeta}(\omega) = \frac{2\omega}{g} \Phi_{\zeta\zeta}(k)$$

for $k = \frac{\omega^2}{g}$.

The wave slope spectrum $\Phi_{\zeta'\zeta'}(\omega)$ is sometimes given as a function of log $\omega$ and not $\omega$. By a change of variable it may be shown that

$$\Phi_{\zeta'\zeta'}(\log \omega) = \frac{\Phi_{\zeta'\zeta'}(\omega)}{\left|\dfrac{d(\log \omega)}{d\omega}\right|} = \omega \Phi_{\zeta'\zeta'}(\omega) = \frac{\omega^5}{g^2} \Phi_{\zeta\zeta}(\omega).$$

Such a wave representation is useful when comparing the pitching motions of ships with identical form but of different size.

## 9.3 Simplified three-dimensional energy spectra

Attempts have been made to obtain useful approximations to the general three dimensional spectrum $\Phi_{\zeta\zeta}(k_x, k_y, \omega)$. The easiest approach from a mathematical point of view is to assume that the variables are separable so that if

$$E[\zeta^2(\mathbf{r}, t)] = \int_0^\infty \int_0^\infty \int_0^\infty \Phi_{\zeta\zeta}(\mathbf{k}, \omega) d\mathbf{k}\, d\omega,$$

then

$$\Phi_{\zeta\zeta}(k_x, k_y, \omega) = K(k_x) \,.\, L(k_y) \,.\, \Phi_{\zeta\zeta}(\omega).$$

This has not so far proved particularly rewarding however.

If either the directional frequency spectrum or directional wave number spectrum is treated in this way, so that

$$\Phi'^{*}_{\zeta\zeta}(\omega, \mu) = \Phi^{*}_{\zeta\zeta}(\omega) M(\mu)$$

or

$$\Phi'^{*}_{\zeta\zeta}(k, \mu) = \Phi^{*}_{\zeta\zeta}(k) M(\mu),$$

then the directional function $M(\mu)$ is given by

$$\int_{-\pi/2}^{\pi/2} M(\mu)\, d\mu = 1$$

as may be seen from section 9.2.

Considerable interest has been focussed on the function $M(\mu)$ where $\mu$ is measured from the dominant wind direction. As a result, a number of semi-empirical forms have emerged, such as

$$M(\mu) = \frac{2}{\pi} \cos^2 \mu$$

for $-\pi/2 < \mu < \pi/2$, see ref. [4];

$$M(\mu) = \frac{8}{3\pi} \cos^4 \mu$$

for $-\pi/2 < \mu < \pi/2$, see ref. [5]; and

$$M(\mu) = G(s) \cos^{2s}\{(\mu - \bar{\mu})/2\}$$

for $-\pi/2 < \mu < \pi/2$, see ref. [6], where

$$G(s) = \frac{1}{\displaystyle\int_{-\pi/2}^{\pi/2} \left(\cos \frac{\lambda}{2}\right)^{s} d\lambda}$$

and $s$, $\bar{\mu}$ vary with $\omega$. It has to be admitted however that none of this work is at all conclusive.

## 9.4 Wave formation

Three kinds of disturbance may result in the generation of waves on the surface of the open sea:

1. air/sea interaction at the boundary surface
2. seismic disturbance on the sea bed or at a coastline
3. tidal forces.

The wind generated waves[1] are of the greatest interest to naval architects and a general outline of the basic principles of their generation will be stated here. Fuller mathematical expositions are given in the references.

When a wind blows over the surface of the sea, energy is transferred from the wind to the sea and two principal mechanisms of disturbance have been identified:

(a) the passage of the turbulent pressure spectrum of the atmospheric boundary layer
(b) the generation of viscous shear stresses at the air/sea interface.

The fluid motion caused by the turbulent pressure field can be irrotational, whilst those induced by the tangential shears must have vorticity. In practice, the fluid motion is largely irrotational so that model (a) is preferred; moreover model (a) is theoretically simpler than model (b). However, although the theories are treated separately both mechanisms are relevant in building a sea.

In the theory due to Phillips[7] the first process is taken to be the primary cause of wave formation. It was found that the wave growth develops most rapidly by means of a 'resonance mechanism' which occurs when a component of the surface pressure distribution moves at the same speed as the free surface wave and with the same wave number. Further, Phillips obtained a time dependent expression defining the wave amplitudes in terms of the amplitudes of the pressure fluctuations at all frequencies. The theory explains the effect of a turbulent pressure field on the sea surface, but neglects the interaction between the wave field and pressure field.

The theory of Miles[8] is based on the second effect and predicts an exponential growth of wave energy that is particularly rapid for waves of high frequency. The air flow is assumed inviscid and incompressible and, in the absence of waves, has the specified mean shear flow velocity. The disturbances in the air flow caused by the surface waves are assumed small but the turbulent pressure fluctuations which are needed to maintain a mean shear flow velocity are neglected in the perturbation analysis. In effect the Miles theory considers the interaction between the mean wind flow and the waves, but it is only valid when there exists a disturbance on the sea surface, whereas the resonance mechanism provides an explanation of a disturbance generated from a calm water reference condition. In the formation of waves both effects are expected to be present and Miles[9] has discussed a mathematical model incorporating both effects acting simultaneously.

Unfortunately, measured rates of growth of waves in the ocean have been found to be larger than those predicted by either the resonance mechanism of Phillip's theory or the instability mechanism of Miles' theory. It may be that such essentially linear theories do not adequately describe the complicated mechanism of wave generation and a more refined mathematical model must be postulated. A non-linear mechanism for the generation of waves has been suggested by Longuet-Higgins[10] in which very short waves (that are readily generated by a shear instability mechanism on the forward slopes of long waves) start to break, so that some of their energy is transferred to the longer waves. The latter then increase their energy at a rate that is proportional to the square of the time.

Non-linear theories are very complicated and it is evident that the detailed mechanism of the interaction between air and sea which leads to wave formation is not fully understood. Further the development of satisfactory theories has been hindered by the scarcity of experimental data on wave growth.

### 9.4.1 Growth of wind-generated waves

Consider an area of the surface of an open ocean which is initially stationary and smooth being subjected to a uni-directional, steady wind. The subsequent formation of waves may be regarded as a process of energy transfer from the atmosphere to the sea. In the sea this energy appears as surface waves, and as wind-induced currents. The overall rate at which energy is transferred depends upon the wind velocity (or 'strength'), the area of the sea in contact with the wind (typified by the length of contact, or the 'fetch' of the wind) and, probably, by the size of the waves already generated. The total energy transferred also depends upon the time during which the wind blows over the surface – that is upon its 'duration'.

It was shown in section 7.1.1 that the energy in a wave train is related to the wave elevation. Therefore the typical wave elevation in an area of generation must increase with time, wind speed and fetch, provided that the waves retain the energy they receive from the wind. The waves are therefore developing unsteadily. Energy dissipation may occur through wave-breaking, however, as the higher waves become unstable. Thus the growth of waves is limited and an equilibrium condition is attained when the rates of energy transfer and energy dissipation are of equal magnitude. The wave is now referred to as 'fully developed' and the history of this process is indicated by the curves of Figures 35a and b.

In Figure 35c the growth of the energy spectrum is shown for a sea subjected to a steady wind of 30 knots. Because the high frequency waves develop more rapidly they will achieve a steady state earlier than waves of lower frequency. Thus the peaks of the energy spectra move towards lower frequencies as the duration increases, until the sea becomes 'fully developed'.

Since the transport of mass is negligibly small for deep-water waves, the motion of water particles affected by Coriolis forces is negligible in comparison with gravity and pressure forces. Waves, unlike currents, are therefore generated in the direction of the shear force applied to the surface and this coincides with the direction of the wind.

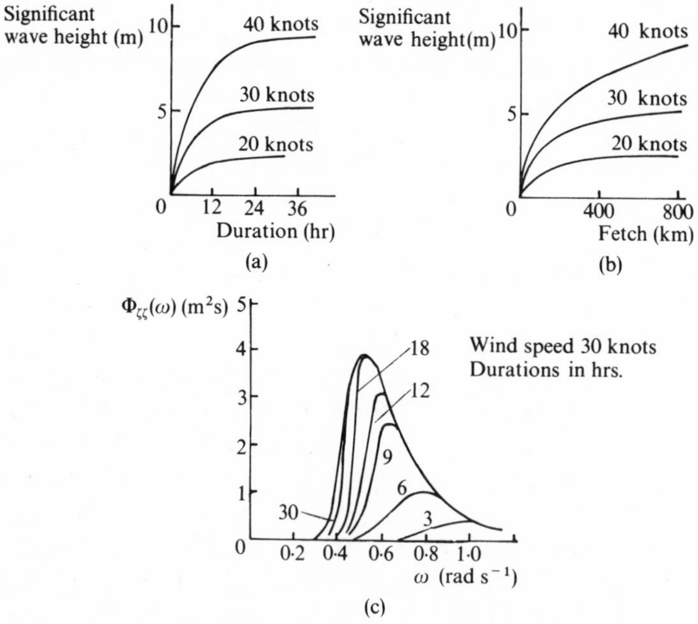


**Figure 35.** Curves illustrating the growth of waves for various wind speeds, showing the effects of (a) duration, and (b) fetch on the significant wave height $h_{1/3}$. Diagram (c) shows the growth of the spectrum for a 30 knot wind with increasing duration.

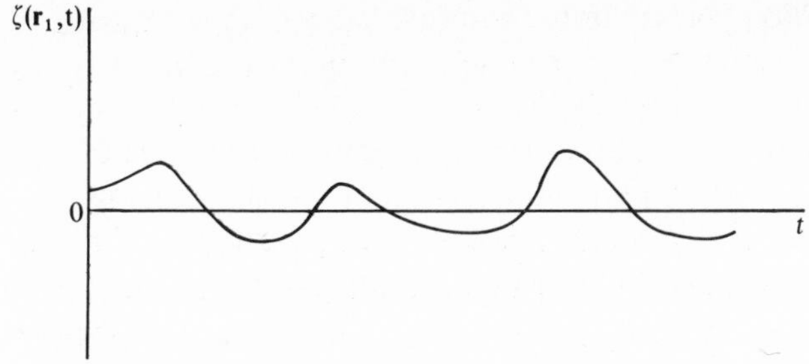


**Figure 36.** Possible wave record at a given $\mathbf{r}_1$ in the sea.

### 9.4.2 Practical representation of crested seas

Consider the wave height record $\zeta(\mathbf{r}_1, t)$ taken at an arbitrary position $\mathbf{r}_1$ in a fully developed sea surface as shown in Figure 36. If the irregularity of the wave is only in the dominant wind direction, so that all the infinitely long, uni-directional wave crests with varying separation remain parallel to each other, the sea is referred to as a 'swell' or a 'long-crested' sea.

At the particular position $\mathbf{r}_1$, $\zeta(\mathbf{r}_1, t)$ is a time dependent random process. At another position $\mathbf{r}_2$, $\zeta(\mathbf{r}_2, t)$ is again a time dependent random process. Since a long-crested sea is assumed homogeneous $\zeta(\mathbf{r}_1, t)$ and $\zeta(\mathbf{r}_2, t)$ are random processes with identical statistical relationships. Thus,

$$\zeta(\mathbf{r}_1, t) = \zeta(\mathbf{r}_2, t) = \zeta(t)$$

so that regardless of the position of measurement, a long crested sea's wave elevation may be considered as a realisation of the time dependent random process $\zeta(t)$. Such a wave disturbance is usually described by the one-dimensional wave energy spectrum $S_{\zeta\zeta}(\omega)$ or $\Phi_{\zeta\zeta}(\omega)$ as discussed in section 9.2.

If, on the other hand, irregularities are apparent along the wave crests at right angles to the direction of the wind, which may be observed in storm areas, there is said to exist a 'short-crested' or 'confused' sea. The position of measurement is now important and the wave elevation $\zeta(\mathbf{r}, t)$ is dependent on position and time. To describe a short-crested sea, we assume that it is composed of an infinite number of waves of small elevation/length ratio coming from a broad sector of direction $\mu$ either side of the average wind direction. From the theory developed, we see that it is desirable at least to use either

$$S_{\zeta\zeta}(\omega, \mu) \quad \text{or,} \quad S_{\zeta\zeta}(k, \mu).$$

Alternatively, it is desirable to use

$$\Phi_{\zeta\zeta}(\omega, \mu) \quad \text{or} \quad \Phi_{\zeta\zeta}(k, \mu).$$

As an aid to visualising the irregular nature of the ocean surface, it is common to employ contour plots as in Figure 37. These plots show the basic difference between the short and long crested seas. The energy spectrum in (a) for an ideal short-crested sea has a wide range of values of $\omega$ and direction $\mu$, whereas the long-crested sea in (b) has a narrow range of variables.

Within an area of atmospheric disturbance, however, where waves are being generated, description of the surface of the sea is likely to be extremely difficult. In this region the waves are produced by a forced oscillation of the surface of the sea and their characteristics are strongly dependent on the nature of the forcing disturbance.

### 9.4.3 Classification of wind generated waves

In general, oscillations of the sea surface are performed with a wide range of frequencies and wave number. A typical general energy spectrum is sketched in Figure 38 which indicates the character of motion in different frequency bands, and the dominant controlling forces.

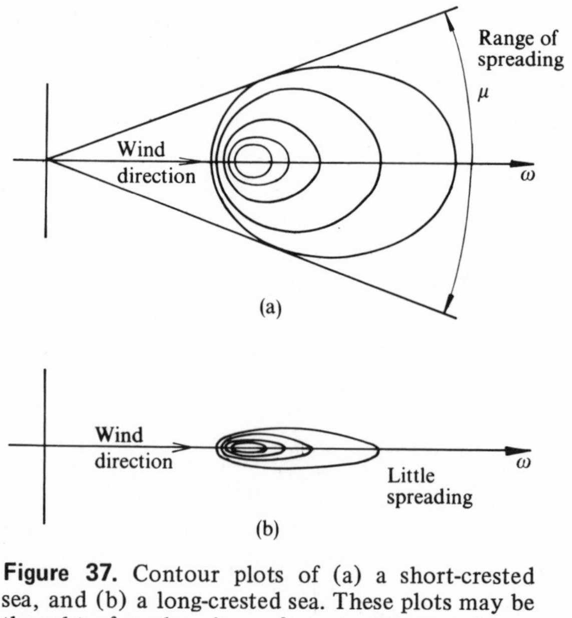


**Figure 37.** Contour plots of (a) a short-crested sea, and (b) a long-crested sea. These plots may be thought of as plan views of spectra like that shown in Figure 33.

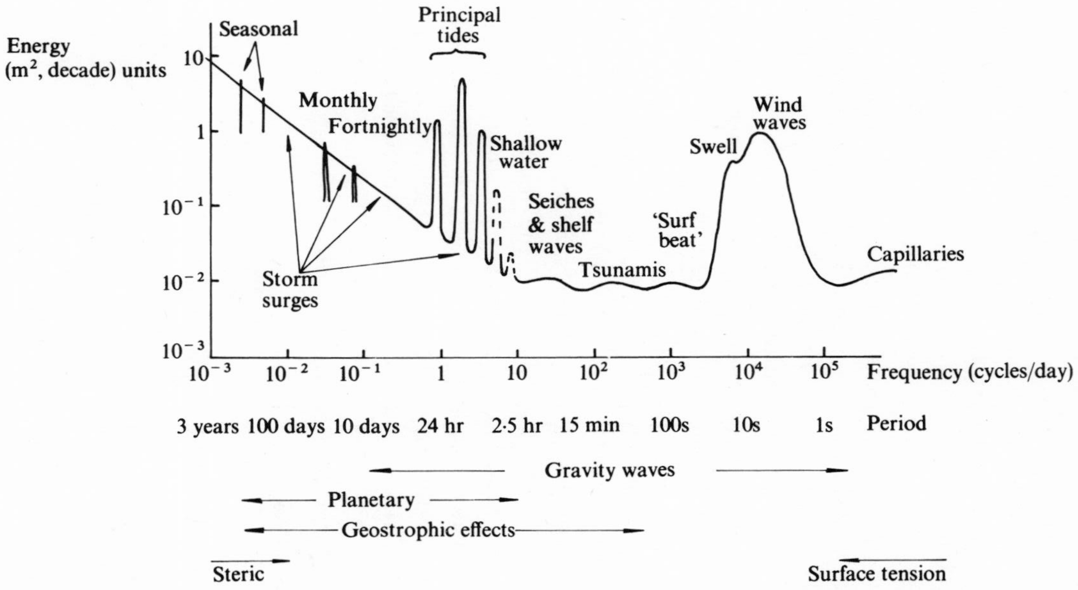


**Figure 38.** Sketch illustrating the way in which the mean square spectral density of wave elevation is distributed.

*Wind waves*

Wind waves are the most important to naval architects and they make a substantial contribution to the overall picture shown in Figure 38. The smallest of the wind-generated waves are the ripples or the capillary waves, which are first to appear when a wind begins to blow over a smooth sea. Surface tension effects play an important part in controlling the formation of such waves (which, incidentally, are essential for the mathematical model of Miles to be valid). These capillary waves are usually less than 1 centimetre high, are of the same order in length, and they travel with low velocity, typically $0{\cdot}3\ \mathrm{m\,s^{-1}}$. They are often present in very large numbers and appear as a fine corrugation on the surface of longer waves.

The formation of larger waves in deep water is controlled primarily by gravitational forces. These waves are, in fact, often referred to as gravity waves and they may be subdivided into the categories of 'sea' and 'swell'. The word 'sea' describes the ocean surface produced when the influence of the wind continues to be exerted on the formed waves. The term 'swell' on the other hand refers to the sea state after the waves have left the atmospheric disturbance. A sea is shorter in length, steeper and far more confused than a swell.

Since the energy of deep-water waves is transmitted from a generating area at the group velocity $c/2$ (see section 7.1.2), waves with an initial frequency of 0·05 Hz spread outwards at 15·6 m (or 30 knots) whilst waves of 0·1 Hz leave at 7·8 m. The resultant swell from a given storm area arrives at a distant locality with a steadily increasing frequency.

*Surf beats*

Surf beating is a motion of very low amplitude with a frequency of the order 0·01 – 0·02 Hz (i.e. with a one or two minute period) produced as a side effect from swell waves when they break on a shore. As the swell moves into shallower water it slows down with the result that it changes form. Prior to breaking, the wave crests become squashed together and the waves increase in height and steepness. Large amounts of energy are expended when the wave breaks on the shore.

*Tsunamis*

'Tsunami' is a Japanese word for sea waves produced by earthquakes. A typical such wave has a frequency of $10^{-3}$ Hz (i.e. a 10–20 minute period) and a wave length of 100 km. At such wave numbers the ocean is effectively 'shallow' and waves are no longer dispersive but travel at an unique speed $\sqrt{gd}$ where $d$ is the ocean depth (see section 7.1.2). A typical speed of such a wave is 800 km $h^{-1}$. An elaborate warning system has been organised in the Pacific Ocean, where tsunamis are most common and dangerous to shore dwellers. Although only a few centimetres high in mid ocean, these waves contain tremendous energy and can build up to amplitudes of 30 m when slowed down by a shelving beach or focussed by off shore topography. Tsunamis echo around the oceans with decreasing energy levels for many days after their initial generation.

*Shelf waves*

Since depths on continental shelves may be some fifty times less than those of the deep oceans, there occurs an abrupt change in the wave velocity at the edge of the shelf. This causes long waves on the shelf to reflect back and forth between the shelf edge and the shoreline with the result that a wave energy 'reservoir' is formed. The relationships between wave number and frequency are dependent on the shelf geometry although the frequencies of the waves are similar to those of the tsunamis. The shelf waves or edge waves tend to travel along the shelf with amplitude large at the shoreline and diminishing to zero out at sea. Resonances may occur, producing a characteristic peak in the wave energy spectrum at a given location. For example, a particularly sharp resonance occurs at island shelves, when the circumference of the island coincides with one or more wavelengths.

*Storm surges*

Storm surges are variations in the sea level caused by strong winds and pressure systems acting over large areas of shallow water. The frequency of the wave motion is of order $10^{-5}$ Hz, i.e. comparable with that of tides, and like the latter these variations can reach heights of several metres. Wave motions with frequencies of the order of $10^{-5}$ Hz (1 cycle/day) or smaller are strongly affected by the geostrophic or Coriolis force due to the rotation of the earth. Such surges occur in the North Sea and they sometimes have disastrous effects in the low coastal regions of Eastern England, The Netherlands and Germany.

*Tides* [11]

The predictability and astronomical origin of tides are unique by comparison with the other motions of the sea. The tidal generating forces occurring over surface regions of the earth are due to the imbalance between the gravitational attraction of the moon and sun and the orbital centrifugal force. Equilibrium of forces occurs only at the centre of the earth. The region of the earth nearest the moon experiences a gravitational attraction greater than the orbital centrifugal forces whilst on the other side of the earth the opposite occurs. Although the forces are small, they cause a water disturbance resulting in two tidal bulges at the near and far regions on the surface of the earth, and these are separated by two tidal depressions. The characteristics of the tides are influenced more by the proximity of the moon than the size of the sun, resulting in the height of an average solar tide which is approximately half that of an average lunar tide.

Due to the passage of the moon, a particular surface region experiences two high and two low tides during each tidal day of 24 hours 50 minutes. This is called a semi-diurnal tide cycle and is common in Europe and the Atlantic Coast of the United States. Since the lunar day exceeds the solar day by 50 minutes, the tidal cycle at a particular surface region will be offset by 50 minutes each day from the previous day.

Since the moon's orbit around the earth is not in exactly the same plane as the earth's orbit around the sun, some complication in the tidal variation occurs. As well as semi-diurnal tides there also exist diurnal and mixed tide cycles. The former, having one low and one high water per tidal day, occur in the Gulf of Mexico and in South-east Asia. The latter occur along the Pacific coast of the United States and display a combination of higher high water and lower high water as well as higher low water and lower low water.

When the earth, moon and sun are aligned so that there exists the greatest influence on the earth, the distance between the high and low semi-diurnal tide is a maximum. These are the so-called 'spring tides' which appear about twice monthly. They are interspersed by 'neap tides' when this influence is a minimum.

*Steric and other motions*

Below the frequency of surges and tides, the spectrum rises continuously, as shown in figure 38. There are some low frequency tides with periods ranging from 14 days to 19 years but they are scarcely detectable. Slow changes in barometric pressure produce changes in the sea level at about a centimetre per millibar but the slowest oscillations, known as 'steric motions', are principally caused by changes in temperature or in the climate. The water column behaves like a giant thermometer, such that 1 centimetre per degree centigrade change in the surface layer is typical. Naturally, the ocean level is also affected by the changes in the polar ice caps. From very limited data, the indications are that a general rise in sea level of about 0·1 m occurred during the last century. However, it is probable that over many centuries such a rise will prove to be merely a phase of a long period oscillation.

### 9.4.4 Compilation of wave data

Visually obtained wave data are collected by observers in merchant ships. Information is also gathered at fixed positions in the oceans. Over one million observations of wave data have been compiled by Hogben and Lumb[12] covering the fifty different sea areas shown in figure 39. This information was collected from about 500 voluntary observation ships during the period 1953 to 1961. For each area, tables are given showing the number of observations in each of twelve direction classes (at intervals of 30° in wave direction) for every combination of wave height (in 0·5 m intervals) and period (in 2 s intervals). A seasonal breakdown is also incorporated. A modified form of these wave data is given in Table II and shows the percentage probability of encountering a wave of known height in the Northern North Atlantic, North Atlantic and World Wide.

**Table II**

| Sea State Code | Description of sea | Wave height observed (m) | World wide | North Atlantic | Northern North Atlantic |
|---|---|---|---|---|---|
| 0 | Calm (glassy) | 0 | | | |
| 1 | Calm (rippled) | 0–0·1 | 11·2486 | 8·3103 | 6·0616 |
| 2 | Smooth (wavelets) | 0·1–0·5 | | | |
| 3 | Slight | 0·5–1·25 | 31·6851 | 28·1996 | 21·5683 |
| 4 | Moderate | 1·25–2·5 | 40.1944 | 42·0273 | 40·9915 |
| 5 | Rough | 2·5–4·0 | 12·8005 | 15·4435 | 21·2383 |
| 6 | Very Rough | 4·0–6·0 | 3·0253 | 4·2938 | 7·0101 |
| 7 | High | 6·0–9·0 | 0·9263 | 1·4968 | 2·6931 |
| 8 | Very High | 9·0–14·0 | 0·1190 | 0·2263 | 0·4346 |
| 9 | Phenomenal | Over 14·0 | 0·0009 | 0·0016 | 0·0035 |

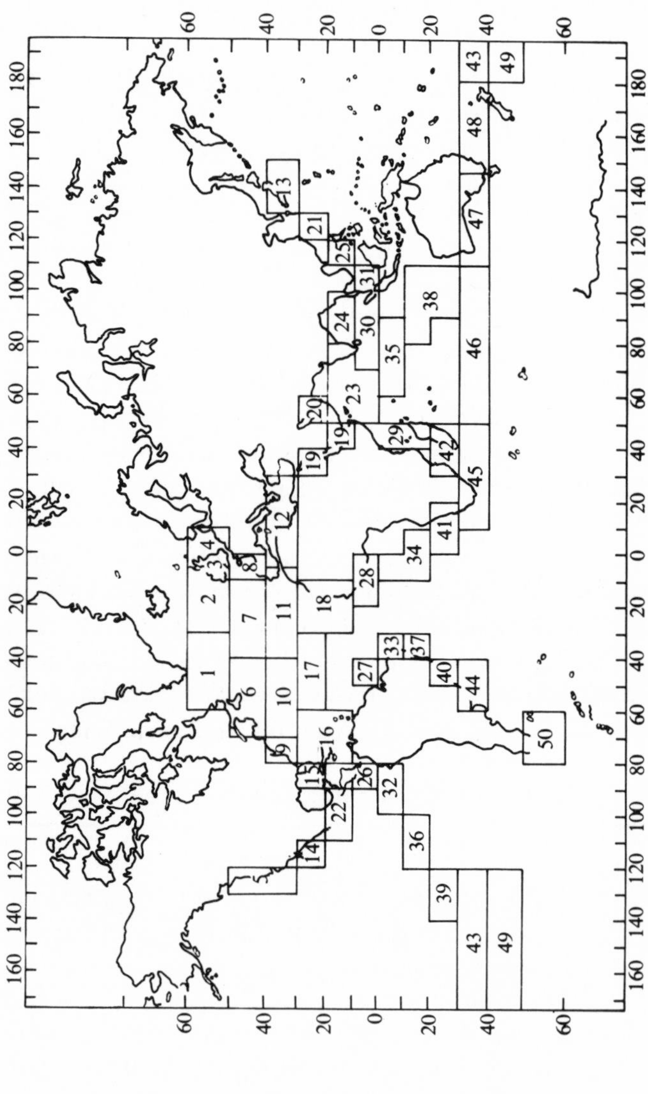

**Figure 39.** Map of the world showing areas for which wave data have been collected by Hogben and Lumb.

Further data for the North Atlantic and North Pacific have been published by the U.S. Naval Oceanographic Office[13] and the U.S. Department of Commerce.

Lofft and Price[14] summarised the wave data of Hogben and Lumb for the following three main areas:

1. World wide
2. North Atlantic latitudes 20–50 degrees North – map areas 1, 2, 6, 7, 8, 9, 10, 11, 16, 17 and 18
3. Northern North Atlantic – map areas 1, 2, 6, 7 and 8.

Table II shows the sea state code, wave height and percentage frequency of occurrence of waves for these areas. A further breakdown of these data is given in Table IIIa–f illustrating the frequency of occurrence of waves observed in specific wave height and wave period ranges. From Table II we see, for example, that in the Northern North Atlantic 40·9915 percent of the waves occur in the 1·25–2·50 m wave height group or the sea condition 'moderate'. Table IIIe shows how this percentage is further broken down. For example, 18·1186 percent of all the observed waves in this region lie in the 1·25–2·5 m wave group and have an observed wave period of 6·5 s.

The data contained in Tables IIIa–f may be formed into a histogram or density function of the wave height. The histogram is obtained by forming a series of rectangles on each wave height band such that the area of each rectangle is proportional to the percentage frequency of occurrence or to the number of waves in the corresponding wave height band. As a rule, this number is plotted as a fraction of the total number of observations so that the total area beneath the rectangles is unity. Figures 40a, b are the histograms of the data from Tables II and IIIf respectively, recorded in the Northern North Atlantic. In each case, it is seen that the area of each rectangle is proportional to the frequency of occurrence of the corresponding range of wave height. Alternatively, we may express the information as a probability distribution function giving the probability of exceeding a particular visual wave height as shown in Figure 41. That is to say, we assume axiomatically that the observations are 'typical'.

The shape of the histogram gives an indication of the irregularity of the record. A narrow and high histogram implies that the data are concentrated within a small area and the record concerned is fairly regular. This may be verified by measuring the amplitude of a sinusoidal wave at a fixed periodic interval over the record. The

**Table IIIa**

*World-wide*

| Sea state | Wave height (m) | *Observed wave period (s)* 2·5 | 6·5 | 8·5 | 10·5 | 12·5 | 14·5 | 16·5 | 18·5 | 20·5 | Over 21 | Totals |
|---|---|---|---|---|---|---|---|---|---|---|---|---|
| 0, 1, 2 | Up to 0·50 | 10·0687 | 0·5583 | 0·1544 | 0·0717 | 0·0272 | 0·0117 | 0·0087 | 0·0076 | 0·0959 | 0·2444 | 11·2486 |
| 3 | 0·5–1·25 | 22·2180 | 6·9233 | 1·4223 | 0·4355 | 0·1650 | 0·0546 | 0·0268 | 0·0123 | 0·0394 | 0·3878 | 31·6850 |
| 4 | 1·25–2·5 | 8·6839 | 18·1608 | 9·0939 | 2·9389 | 0·8933 | 0·2816 | 0·0889 | 0·0222 | 0·0079 | 0·0230 | 40·1944 |
| 5 | 2·5–4 | 0·5444 | 3·0436 | 4·4913 | 2·9163 | 1·2398 | 0·4010 | 0·1220 | 0·0322 | 0·0050 | 0·0049 | 12·8005 |
| 6 | 4–6 | 0·0691 | 0·3883 | 0·8437 | 0·8611 | 0·5214 | 0·2244 | 0·0866 | 0·0229 | 0·0045 | 0·0033 | 3·0253 |
| 7 | 6–9 | 0·0223 | 0·0928 | 0·2164 | 0·2632 | 0·1842 | 0·0908 | 0·0419 | 0·0081 | 0·0032 | 0·0034 | 0·9263 |
| 8 | 9–14 | 0·0014 | 0·0067 | 0·0186 | 0·0316 | 0·0258 | 0·0177 | 0·0094 | 0·0044 | 0·0023 | 0·0011 | 0·1190 |
| 9 | 14+ | | 0·0001 | 0·0001 | 0·0003 | 0·0003 | | | | 0·0001 | | 0·0009 |
| Totals | | 41·6078 | 29·1739 | 16·2408 | 7·5187 | 3·0571 | 1·0814 | 0·3844 | 0·1098 | 0·1583 | 0·6679 | 100·0001 |

**Table IIIb**

*World-wide*

| Wave height (*m*) | 2·5 | 6·5 | 8·5 | 10·5 | 12·5 | 14·5 | 16·5 | 18·5 | 20·5 | Over 21 | Totals |
|---|---|---|---|---|---|---|---|---|---|---|---|
| | *Observed wave period (s)* | | | | | | | | | | |
| 0–1 | 24·0470 | 4·6416 | 0·9954 | 0·3316 | 0·1253 | 0·0440 | 0·0245 | 0·0147 | 0·1041 | 0·5480 | 30·8762 |
| 1–2 | 15·5208 | 17·0941 | 6·1091 | 1·7475 | 0·5498 | 0·1784 | 0·0626 | 0·0175 | 0·0194 | 0·0910 | 41·3902 |
| 2–3 | 1·3763 | 6·0543 | 6·0000 | 2·6736 | 0·8712 | 0·2668 | 0·0778 | 0·0188 | 0·0054 | 0·0057 | 17·3499 |
| 3–4 | 0·2008 | 1·2153 | 2·1165 | 1·6245 | 0·7848 | 0·2611 | 0·0817 | 0·0226 | 0·0026 | 0·0030 | 6·3129 |
| 4–5 | 0·0506 | 0·3278 | 0·6969 | 0·6998 | 0·4151 | 0·1726 | 0·0687 | 0·0196 | 0·0033 | 0·0020 | 2·4564 |
| 5–6 | 0·0187 | 0·0604 | 0·1469 | 0·1614 | 0·1063 | 0·0509 | 0·0180 | 0·0033 | 0·0012 | 0·0014 | 0·5685 |
| 6–7 | 0·0158 | 0·0587 | 0·1275 | 0·1551 | 0·1039 | 0·0490 | 0·0215 | 0·0039 | 0·0010 | 0·0011 | 0·5375 |
| 7–8 | 0·0032 | 0·0240 | 0·0622 | 0·0702 | 0·0501 | 0·0249 | 0·0120 | 0·0026 | 0·0008 | 0·0012 | 0·2512 |
| 8–9 | 0·0028 | 0·0102 | 0·0266 | 0·0380 | 0·0311 | 0·0169 | 0·0084 | 0·0018 | 0·0013 | 0·0011 | 0·1382 |
| 9–10 | 0·0013 | 0·0064 | 0·0182 | 0·0308 | 0·0247 | 0·0174 | 0·0093 | 0·0041 | 0·0022 | 0·0012 | 0·1156 |
| 10–11 | | 0·0003 | 0·0002 | 0·0006 | 0·0006 | 0·0003 | 0·0001 | 0·0001 | | | 0·0022 |
| 11+ | | 0·0001 | 0·0001 | 0·0004 | 0·0007 | | | 0·0001 | | | 0·0014 |
| Totals | 41·2373 | 29·4932 | 16·2996 | 7·5335 | 3·0636 | 1·0823 | 0·3846 | 0·1090 | 0·1414 | 0·6557 | 100·0002 |

**Table IIIc**

*North Atlantic*

| Sea state | Wave height (*m*) | Observed wave period (*s*) 2·5 | 6·5 | 8·5 | 10·5 | 12·5 | 14·5 | 16·5 | 18·5 | 20·5 | Over 21 | Totals |
|---|---|---|---|---|---|---|---|---|---|---|---|---|
| 0, 1, 2 | Up to 0·5 | 7·1465 | 0·5284 | 0·1608 | 0·0792 | 0·0318 | 0·0147 | 0·0140 | 0·0083 | 0·0876 | 0·2390 | 8·3103 |
| 3 | 0·5–1·25 | 18·6916 | 6·8744 | 1·4616 | 0·4616 | 0·1798 | 0·0616 | 0·0355 | 0·0158 | 0·0392 | 0·3785 | 28·1996 |
| 4 | 1·25–2·5 | 8·6495 | 19·1889 | 9·7609 | 3·0633 | 0·9056 | 0·2969 | 0·0967 | 0·0269 | 0·0104 | 0·0290 | 42·0281 |
| 5 | 2·5–4 | 0·6379 | 3·7961 | 5·4827 | 3·4465 | 1·4291 | 0·4597 | 0·1368 | 0·0420 | 0·0067 | 0·0060 | 15·4435 |
| 6 | 4–6 | 0·0863 | 0·5646 | 1·2574 | 1·2199 | 0·7040 | 0·3005 | 0·1143 | 0·0338 | 0·0076 | 0·0054 | 4·2938 |
| 7 | 6–9 | 0·0258 | 0·1468 | 0·3486 | 0·4417 | 0·3052 | 0·1451 | 0·0614 | 0·0126 | 0·0039 | 0·0057 | 1·4968 |
| 8 | 9–14 | 0·0018 | 0·0114 | 0·0362 | 0·0614 | 0·0520 | 0·0320 | 0·0163 | 0·0082 | 0·0046 | 0·0024 | 0·2263 |
| 9 | 14+ | | 0·0002 | 0·0001 | 0·0006 | 0·0006 | | | | 0·0001 | | 0·0016 |
| Totals | | 35·2394 | 31·1109 | 18·5083 | 8·7742 | 3·6081 | 1·3105 | 0·4750 | 0·1476 | 0·1601 | 0·6660 | 100·0000 |

**Table IIId**

*North Atlantic*

| Wave height (*m*) | *Observed wave period (s)* | | | | | | | | | | |
|---|---|---|---|---|---|---|---|---|---|---|---|
| | 2·5 | 6·5 | 8·5 | 10·5 | 12·5 | 14·5 | 16·5 | 18·5 | 20·5 | Over 21 | Totals |
| 0–1 | 18·6846 | 4·5036 | 1·0144 | 0·3511 | 0·1381 | 0·0512 | 0·0341 | 0·0179 | 0·0976 | 0·5336 | 25·4262 |
| 1–2 | 14·4152 | 17·6097 | 6·4484 | 1·7936 | 0·5534 | 0·1852 | 0·0721 | 0·0218 | 0·0213 | 0·0913 | 41·2120 |
| 2–3 | 1·5051 | 6·9322 | 6·7253 | 2·9229 | 0·9292 | 0·2935 | 0·0825 | 0·0230 | 0·0064 | 0·0074 | 19·4275 |
| 3–4 | 0·2466 | 1·5878 | 2·7234 | 1·9934 | 0·9298 | 0·3039 | 0·0945 | 0·0299 | 0·0039 | 0·0032 | 7·9164 |
| 4–5 | 0·0666 | 0·4775 | 1·0347 | 0·9763 | 0·5518 | 0·2304 | 0·0907 | 0·0290 | 0·0058 | 0·0037 | 3·4665 |
| 5–6 | 0·0197 | 0·0868 | 0·2222 | 0·2434 | 0·1522 | 0·0701 | 0·0237 | 0·0048 | 0·0019 | 0·0018 | 0·8266 |
| 6–7 | 0·0191 | 0·0903 | 0·2038 | 0·2534 | 0·1622 | 0·0723 | 0·0271 | 0·0064 | 0·0017 | 0·0017 | 0·8380 |
| 7–8 | 0·0045 | 0·0384 | 0·0996 | 0·1212 | 0·0873 | 0·0425 | 0·0189 | 0·0037 | 0·0009 | 0·0017 | 0·4187 |
| 8–9 | 0·0023 | 0·0172 | 0·0450 | 0·0671 | 0·0570 | 0·0304 | 0·0155 | 0·0026 | 0·0014 | 0·0024 | 0·2409 |
| 9–10 | 0·0018 | 0·0110 | 0·0353 | 0·0602 | 0·0495 | 0·0315 | 0·0163 | 0·0080 | 0·0043 | 0·0024 | 0·2203 |
| 10–11 | | 0·0003 | 0·0005 | 0·0010 | 0·0014 | 0·0005 | | 0·0002 | | | 0·0039 |
| 11+ | | 0·0003 | 0·0003 | 0·0008 | 0·0018 | 0·0001 | | | 0·0001 | | 0·0034 |
| Totals | 34·9655 | 31·3551 | 18·5529 | 8·7844 | 3·6137 | 1·3116 | 0·4754 | 0·1473 | 0·1453 | 0·6492 | 100·0004 |

**Table IIIe**

*N. North Atlantic*

| Sea state | Wave height (*m*) | Observed wave period (*s*) 2·5 | 6·5 | 8·5 | 10·5 | 12·5 | 14·5 | 16·5 | 18·5 | 20·5 | Over 21 | Totals |
|---|---|---|---|---|---|---|---|---|---|---|---|---|
| 0, 1, 2 | Up to 0·5 | 5·1870 | 0·3864 | 0·1364 | 0·0785 | 0·0254 | 0·0125 | 0·0122 | 0·0077 | 0·0655 | 0·1500 | 6·0616 |
| 3 | 0·5–1·25 | 13·9308 | 5·4071 | 1·2508 | 0·4336 | 0·1534 | 0·0515 | 0·0223 | 0·0158 | 0·0282 | 0·2738 | 21·5673 |
| 4 | 1·25–2·5 | 7·4338 | 18·1186 | 10·4412 | 3·4524 | 1·0344 | 0·3308 | 0·1043 | 0·0356 | 0·0153 | 0·0251 | 40·9915 |
| 5 | 2·5–4 | 0·7898 | 5·0003 | 7·5983 | 4·9232 | 2·0100 | 0·6557 | 0·1850 | 0·0592 | 0·0091 | 0·0077 | 21·2383 |
| 6 | 4–6 | 0·1290 | 0·9294 | 2·0748 | 1·9919 | 1·1577 | 0·4774 | 0·1824 | 0·0500 | 0·0099 | 0·0076 | 7·0101 |
| 7 | 6–9 | 0·0398 | 0·2517 | 0·6354 | 0·7925 | 0·5554 | 0·2644 | 0·1158 | 0·0227 | 0·0055 | 0·0099 | 2·6931 |
| 8 | 9–14 | 0·0033 | 0·0209 | 0·0691 | 0·1201 | 0·1039 | 0·0564 | 0·0302 | 0·0178 | 0·0082 | 0·0047 | 0·4346 |
| 9 | 14+ |  | 0·0005 | 0·0002 | 0·0014 | 0·0012 |  |  |  | 0·0002 |  | 0·0035 |
| Totals |  | 27·5135 | 30·1149 | 22·2062 | 11·7936 | 5·0414 | 1·8487 | 0·6522 | 0·2088 | 0·1419 | 0·4788 | 100·0000 |

**Table IIIf**

*N. North Atlantic*

| Wave height (*m*) | Observed wave period (*s*) 2·5 | 6·5 | 8·5 | 10·5 | 12·5 | 14·5 | 16·5 | 18·5 | 20·5 | Over 21 | Totals |
|---|---|---|---|---|---|---|---|---|---|---|---|
| 0–1 | 13·7204 | 3·4934 | 0·8559 | 0·3301 | 0·1127 | 0·0438 | 0·0249 | 0·0172 | 0·0723 | 0·3584 | 19·0291 |
| 1–2 | 11·4889 | 15·5036 | 6·4817 | 1·8618 | 0·5807 | 0·1883 | 0·0671 | 0·0254 | 0·0203 | 0·0763 | 36·2941 |
| 2–3 | 1·5944 | 7·8562 | 8·0854 | 3·7270 | 1·1790 | 0·3713 | 0·1002 | 0·0321 | 0·0091 | 0·0082 | 22·9629 |
| 3–4 | 0·3244 | 2·2487 | 4·0393 | 2·9762 | 1·3536 | 0·4477 | 0·1307 | 0·0428 | 0·0050 | 0·0040 | 11·5724 |
| 4–5 | 0·1027 | 0·7838 | 1·6998 | 1·5882 | 0·9084 | 0·3574 | 0·1443 | 0·0433 | 0·0072 | 0·0049 | 5·6400 |
| 5–6 | 0·0263 | 0·1456 | 0·3749 | 0·4038 | 0·2493 | 0·1200 | 0·0382 | 0·0067 | 0·0027 | 0·0027 | 1·3702 |
| 6–7 | 0·0277 | 0·1477 | 0·3614 | 0·4472 | 0·2804 | 0·1301 | 0·0504 | 0·0113 | 0·0011 | 0·0032 | 1·4605 |
| 7–8 | 0·0084 | 0·0714 | 0·1882 | 0·2199 | 0·1634 | 0·0785 | 0·0353 | 0·0069 | 0·0018 | 0·0034 | 0·7772 |
| 8–9 | 0·0037 | 0·0325 | 0·0856 | 0·1252 | 0·1119 | 0·0558 | 0·0303 | 0·0045 | 0·0027 | 0·0033 | 0·4555 |
| 9–10 | 0·0034 | 0·0204 | 0·0674 | 0·1173 | 0·0983 | 0·0550 | 0·0303 | 0·0173 | 0·0079 | 0·0047 | 0·4220 |
| 10–11 | | 0·0005 | 0·0012 | 0·0023 | 0·0031 | 0·0012 | | 0·0005 | | | 0·0088 |
| 11+ | | 0·0005 | 0·0007 | 0·0019 | 0·0035 | 0·0002 | | | 0·0005 | | 0·0073 |
| Totals | 27·3003 | 30·3043 | 22·2415 | 11·8009 | 5·0143 | 1·8493 | 0·6517 | 0·2080 | 0·1306 | 0·4691 | 100·0000 |

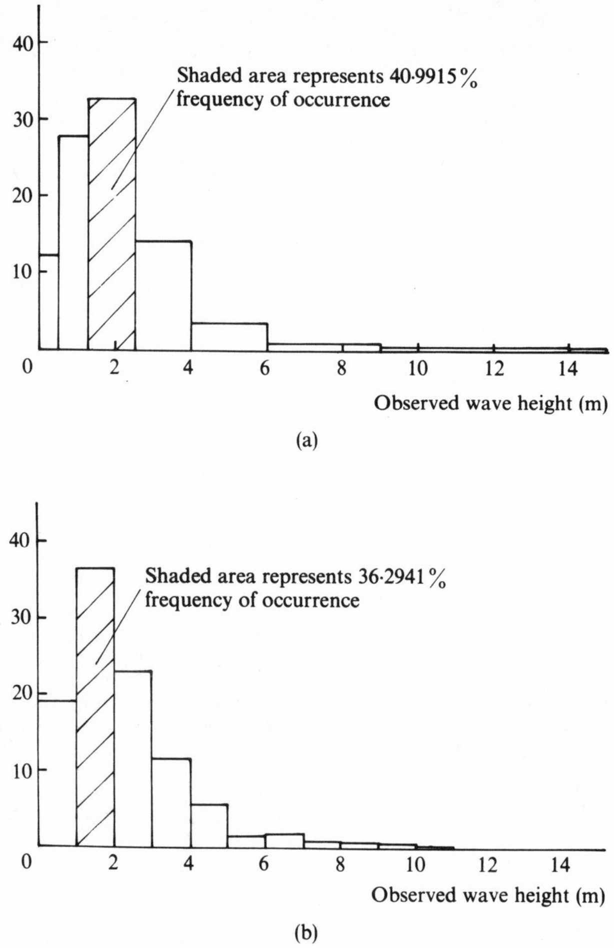


**Figure 40.** Histograms of wave data for (a) Table II, and (b) Table III(f).

measured amplitude is the same at each instant which in the resulting histogram would be represented by a spike (or delta function) at the amplitude. On the other hand, a wide and low histogram implies that the record concerned will be fairly irregular.

From such histograms the expectations of the wave quantity (wave height, period, wave length, etc.) may be determined. The term 'significant value' is used in this context, being defined as the average value of the third highest observations of the wave quantity measured. The 'significant wave height' was introduced by Svedrup and Munk[15]. Evaluation of significant wave height is subjective, since the definition is physically fictitious. However, it does seem to

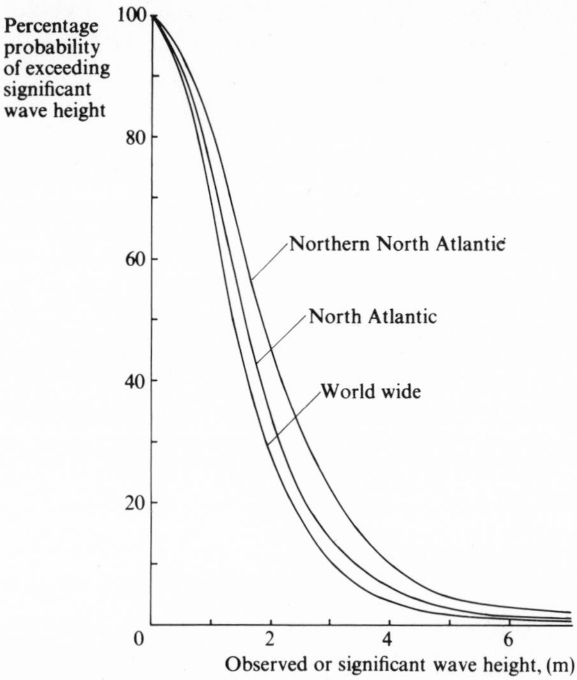


**Figure 41.** Probability distribution function of exceeding a visual wave height $h_{1/3}$ in three areas. Notice that these curves are like Figure 7(a) but, being complementary, are reversed.

correspond to the value an experienced observer will estimate as the average wave height.

Hogben and Lumb compared the visual observations of wave height and period with measurements of the waves. They concluded that visual observations of wave height were reasonably reliable and well correlated with measured waves, i.e. the significant wave height, but the visual estimates of period displayed poor correlation with measured mean periods.

## 9.5 Theories of wave spectra

As mentioned in section 9.4, it is generally believed that waves are created by the wind. Phillips and Miles found that there exists a relationship between the pressure in a turbulent wind and the wave frequency. The shorter waves grow until they 'break' and so start to limit their energy by dissipation. After the wind has blown for a long time the sea becomes 'fully developed' and $\Phi_{\zeta\zeta}(\omega)$ acquires a maximum at a frequency $\omega_0$ that is inversely proportional to the wind speed.

Using a dimensional argument, Phillips produced evidence to show that in the high frequency range, where wave breaking is important, the asymptotic form of the sea spectrum (as $\omega \to \infty$) is given by

$$\Phi_{\zeta\zeta}(\omega) = Ag^2\omega^{-5}$$

where $A$ is a constant of the order of $1{\cdot}35 \times 10^{-2}$. Unfortunately experimental support for this is sparce but it has been shown[10] that the constant has a very slight variation depending on fetch and wind speed. However, a simple spectrum which can be used in certain applications is the cut-off equilibrium spectrum

$$\Phi_{\zeta\zeta}(\omega) = \begin{cases} Ag^2\omega^{-5} & \text{for } \omega > \omega_c \\ 0 & \text{otherwise,} \end{cases}$$

where $\omega_c$ is some suitable cut-off frequency value. Several different forms have been proposed for the wave spectrum $\Phi_{\zeta\zeta}(\omega)$ as a function of the wind speed. They may be summarised as follows:

1. *The Darbyshire wave spectrum*

$$\Phi_{\zeta\zeta}(\omega) = A \exp - \left\{\frac{(\omega - \omega_0)^2}{0{\cdot}054(\omega - \omega_0 + 0{\cdot}265)}\right\}^{1/2}$$

$$= 0 \qquad \text{when } \omega - \omega_0 < -0{\cdot}265,$$

where

$$A = 0{\cdot}186 \times 10^{-5}\, V^4,$$

$V$ being the wind speed, and $\omega_0$ being given by

$$\omega_0 = 6{\cdot}284\,(1{\cdot}94V^{1/2} + 2{\cdot}5 \times 10^{-7}\, V^4)^{-1}.$$

In these formulae, $\Phi_{\zeta\zeta}(\omega)$ is measured in $m^2$, A is in $m^2$ and $\omega$ is in rad $s^{-1}$.

2. *The British Towing Tank Panel wave spectrum*

$$\Phi_{\zeta\zeta}(\omega) = A \exp - \left\{\frac{(\omega - \omega_0)^2}{0{\cdot}065(\omega - \omega_0 + 0{\cdot}265)}\right\}^{1/2}$$

$$= 0 \qquad \text{when } 1{\cdot}65 < \omega - \omega_0 < -0{\cdot}265.$$

In this case, $V$ again represents the wind speed; but now $A$ and $\omega_0$

have the values

$$A = 21{\cdot}5\,(0{\cdot}0625V - 0{\cdot}442)^2,$$

and

$$\omega_0 = 6{\cdot}142\,(0{\cdot}1545V + 7{\cdot}389)^{-1}.$$

In these formulae, $\Phi_{\zeta\zeta}(\omega)$ is measured in $m^2$, A is in $m^2$ and $\omega$ is in rad $s^{-1}$.

3. *Neumann wave spectrum*

The spectral density function developed by Neumann is given by

$$\Phi_{\zeta\zeta}(\omega) = \frac{AB}{\omega^6}\,e^{-B/\omega^2}$$

where $A$ and $B$ are constants depending on the wind speed V. The spectrum is illustrated in Figure 34. As the wind speed increases the frequency corresponding to the maximum value of the spectral density is reduced.

4. *The Pierson-Moskowitz wave spectrum*

The Neumann spectrum has been superseded by the Pierson and Moskowitz[16] spectrum, which was obtained semi-empirically by the analysis of extensive wave data relating to fully developed sea conditions in the North Atlantic. The spectrum is

$$\Phi_{\zeta\zeta}(\omega) = \frac{8{\cdot}1 \times 10^{-3} g^2}{\omega^5}\,e^{-0{\cdot}74(g/V\omega)^4}$$

where $V$ is the wind speed at a height of 19·5 m.

The two main spectrum forms used at present are those adopted by the International Towing Tank Conference (ITTC) and the International Ship Structure Congress (ISSC). They are both based on the Pierson-Moskowitz spectrum and have the general form

$$\Phi_{\zeta\zeta}(\omega) = \frac{A}{\omega^5}\,e^{-B/\omega^4}$$

where $\omega$ is the circular frequency of the waves in rad $s^{-1}$, $A$ and $B$ being constants.

The maximum value of this spectrum occurs when $\omega_m = (0{\cdot}8B)^{1/4}$, with a value

$$\Phi_{\zeta\zeta}(\omega_m) = A(0{\cdot}8B)^{-5/4}e^{-5/4}$$

Alternatively, in accordance with section 9.2, the Pierson-Moskowitz spectrum for fully developed deep water waves may be expressed in terms of the wave number spectrum given by

$$\Phi_{\zeta\zeta}(k) = \frac{\Phi_{\zeta\zeta}(\omega)}{\left|\dfrac{dk}{d\omega}\right|}.$$

Since $\omega = \sqrt{kg}$, this gives

$$\Phi_{\zeta\zeta}(k) = \frac{A^*}{k^3}\, e^{-B^*/k^2}$$

where the constants are $A^* = A/2g^2$ and $B^* = B/g^2$, and the wave numbers are in appropriate units.

5. *The I.T.T.C. wave spectrum*

If the only information available is the significant wave height, $h_{1/3}$, then it is assumed that

$$A = 8{\cdot}1 \times 10^{-3} g^2, \qquad B = \frac{3{\cdot}11}{h_{1/3}^2}$$

in the Pierson and Moskowitz spectrum, where $h_{1/3}$ is in metres and $g$ is the gravitational constant in m s$^{-2}$. The spectrum is illustrated in Figure 42 with a maximum value of $0{\cdot}25e^{-5/4}h_{1/3}^{5/2}$ at a frequency of $1{\cdot}26h_{1/3}^{-1/2}$.

6. *The I.S.S.C. wave spectrum*

If further information is available on the characteristic wave period, then it is assumed that

$$A = \frac{173\, h_{1/3}^2}{T_1^4}, \qquad B = \frac{691}{T_1^4}$$

in the Pierson and Moskowitz spectrum, where the characteristic period

$$T_1 = 2\pi \frac{m_0}{m_1}$$

and

$$m_n = \int_0^\infty \omega^n \Phi_{\zeta\zeta}(\omega)\, d\omega.$$

Data suggest that the period can be taken approximately as the average observed period $T$. It is this form of the wave spectrum which is usually used in calculations of ship strength.

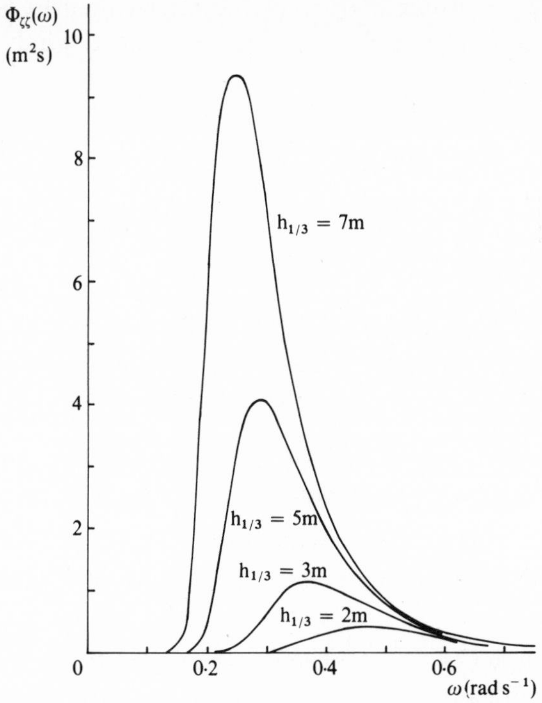


**Figure 42.** The standard wave spectrum of the International Towing Tank Conference for various values of the significant wave height $h_{1/3}$.

This wave spectrum has a maximum value of $0{\cdot}065e^{-5/4}h_{1/3}^2T_1$ at a frequency of $4{\cdot}85T_1^{-1}$. The I.T.T.C and I.S.S.C. wave spectra have maxima occurring at the same value of frequency only when $T_1 = 3{\cdot}86h_{1/3}^{1/2}$.

When no wave data are available, but only the wind speed is known, the I.T.T.C. recommend that the approximate relationship of Table IV should be assumed to exist between wind speed and significant wave height.

**Table IV**

| Wind speed (knots) | Significant wave height (m) |
|---|---|
| 20 | 3·1 |
| 30 | 5·1 |
| 40 | 8·1 |
| 50 | 11·0 |
| 60 | 14·6 |

Sometimes the strength of the wind is classified by the Beaufort Scale (see, e.g. [17]) which is related to the wind speed as in Table V.

**Table V**

| Beaufort number | Description of wind | Wind speed (knots) |
|---|---|---|
| 0 | Calm | 0–1 |
| 1 | Light air | 2–3 |
| 2 | Light breeze | 4–7 |
| 3 | Gentle breeze | 8–11 |
| 4 | Moderate breeze | 12–16 |
| 5 | Fresh breeze | 17–21 |
| 6 | Strong breeze | 22–27 |
| 7 | Moderate gale | 28–33 |
| 8 | Fresh gale | 34–40 |
| 9 | Strong gale | 41–48 |
| 10 | Whole gale | 49–56 |
| 11 | Storm | 57–65 |
| 12 | Hurricane | More than 65 |

Figure 43 shows the relationship between the Beaufort Scale, wind speed and significant wave height implied by Tables IV & V. Due to poor correlation, there is no equivalent relationship between wind speed and wave period.

For a directional wave spectrum the ITTC and ISSC recommend the use of a cosine power spreading function of the form

$$\Phi_{\zeta\zeta}(\omega, \mu) = A(n) \cos^n \mu \; \Phi_{\zeta\zeta}(\omega)$$

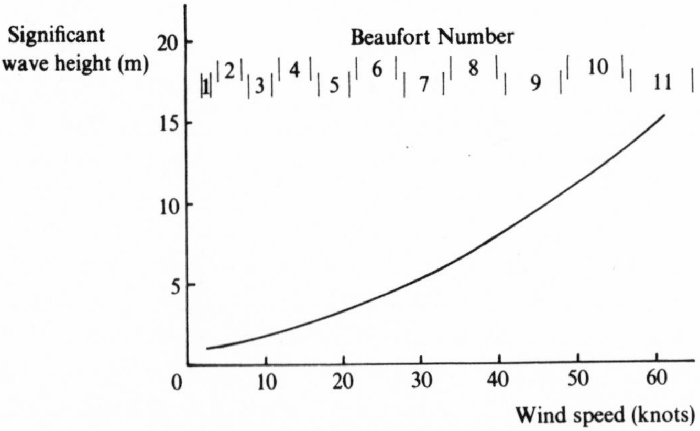


**Figure 43.** Dependence of the significant wave height $h_{1/3}$ on wind speed and Beaufort number (see Tables IV and V).

where $|\mu| \leqslant \pi/2$ and $\mu$ is the wave direction referred to the predominant wind direction; $A(n)$ is a normalizing factor given as

$$n = 2, \qquad A(n) = \frac{2}{\pi}$$

for the ITTC spectrum and

$$n = 4, \qquad A(n) = \frac{8}{3\pi}$$

for the ISSC spectrum. But these values are likely to be modified as more information becomes available.

## 9.6 Some properties of local wave spectra

Almost inevitably we have been forced to consider local spectra and stationary random processes. That is to say we have dwelt at some length on mean square spectral densities of the type $\Phi(\omega)$ and, hence, with auto-correlation functions

$$R(\tau) = \int_0^\infty \Phi(\omega)\cos \omega\tau \, d\omega.$$

Now a random process $\zeta(t)$ is associated with probability density functions of all orders, and in particular that of the second order determines $R_{\zeta\zeta}(\tau)$ because

$$R_{\zeta\zeta}(\tau) = E[\zeta(t)\ \zeta(t+\tau)] = \int_{-\infty}^{\infty}\int_{-\infty}^{\infty} x_1 x_2 f_{\zeta\zeta}(x_1 : x_2; \tau)\, dx_1 dx_2.$$

If we postulate $R_{\zeta\zeta}(\tau)$, or its equivalent $\Phi_{\zeta\zeta}(\omega)$, we do *not* determine a probability density function uniquely. On the contrary, there are infinitely many possible functions $f_{\zeta\zeta}(x_1 : x_2; \tau)$, so that it is left open to us to make a choice. In practice, the Gaussian function is used almost invariably and it is a remarkable property of a Gaussian process that if its mean square spectral density is known, the whole set of probability density functions of all orders can be deduced.

It is not easy to deduce results for a random process whose energy spectrum $\Phi(\omega)$ is selected arbitrarily, even if the process is known to be Gaussian. This is unfortunately the case with the spectra that we have referred to hitherto. There are two special cases, however, that are frequently referred to, since they are convenient for reference purposes, *viz*., 'narrow band' and 'wide band' spectra.

Before examining these special cases it will be convenient to study one feature of mean square spectral densities. It was mentioned in section 4.1.1 that if *all* the moments of a probability density function $f(x)$ about the axis $x = 0$ were known, then the function $f(x)$ could be deduced. Equally the shape of a wave spectrum $\Phi_{\zeta\zeta}(\omega)$ can be described by quantities

$$m_n = \int_0^\infty \omega^n \Phi_{\zeta\zeta}(\omega) d\omega \qquad n = 0, 1, 2, \ldots$$

referred to as the moments of the wave spectrum. In particular

$$m_0 = \int_0^\infty \Phi_{\zeta\zeta}(\omega) d\omega = E[\zeta^2(t)] = \sigma^2$$

which is the mean square value (or variance) of the wave elevation at the point under consideration. Thus $m_0$, the area under the wave spectrum, represents the total energy present in the wave surface.

Further, the moments

$$m_2 = \int_0^\infty \omega^2 \Phi_{\zeta\zeta}(\omega) d\omega = E[\dot{\zeta}^2(t)]$$

and

$$m_4 = \int_0^\infty \omega^4 \Phi_{\zeta\zeta}(\omega) d\omega = E[\ddot{\zeta}^2(t)]$$

are the mean square values of the velocity and acceleration of the wave elevation respectively. The moments $m_0$, $m_2$ and $m_4$ are therefore positive quantities. We shall now discuss the importance of these moments in our analysis.

### 9.6.1 Zero crossings

Although the probability density functions of all orders describe a random process $\zeta(t)$ with realisations $\zeta^{(j)}(t)$ ($j = 1, 2, \ldots$) certain statistical properties are not specified by these density functions. The probability density function gives no direct information either on the distribution of 'zero crossings' or on the distribution of the peaks or maxima of the random process in the time interval $(t_1 \leqslant t \leqslant t_2)$.

Let us first investigate the expected number of crossings of a Gaussian random process $\zeta(t)$ at an arbitrary level $\zeta_0$ in the interval $(\tau = t_2 - t_1)$ as shown in Figure 44a. A 'counting functional' is needed for crossings of level $\zeta_0$ and this may be constructed by considering the unit step function

$$1(t) = \begin{cases} 1 & \text{for } t \geqslant 0 \\ 0 & \text{for } t < 0. \end{cases}$$

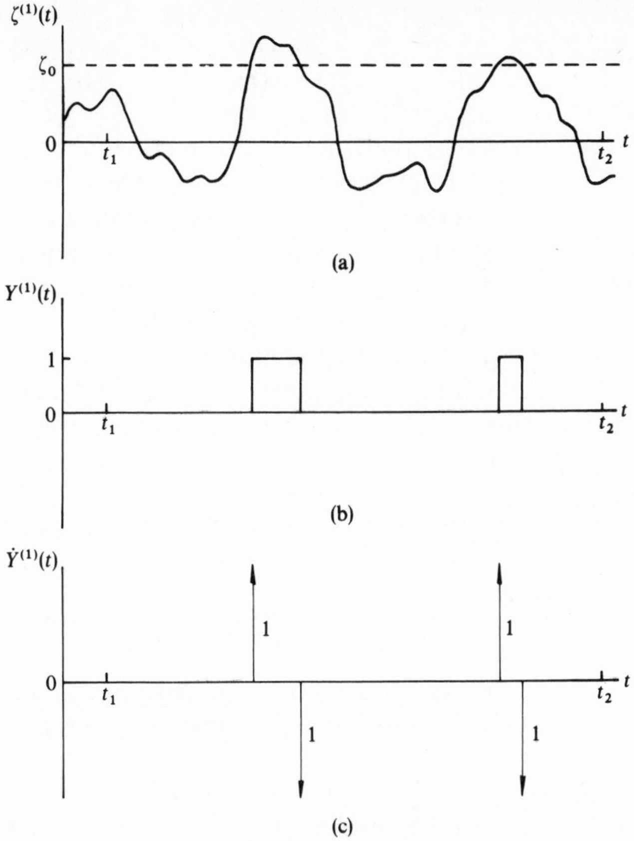


**Figure 44.** Related realisations of random processes: (a) wave elevation $\zeta^{(1)}(t)$, (b) the derived process $Y^{(1)}(t)$ showing intervals for which $\zeta^{(1)}(t) > \zeta_0$, and (c) the derived process $\dot{Y}^{(1)}(t)$ showing the occurrence of up-crossings and down-crossings of $\zeta^{(1)}(t)$ at the level $\zeta_0$.

A realisation of a new random process $Y(t)$ is given by

$$Y^{(1)}(t) = 1\{\zeta^{(1)}(t) - \zeta_0\}$$

having values

$$Y^{(1)}(t) = \begin{cases} 1 & \text{for } \zeta^{(1)}(t) \geqslant \zeta_0 \\ 0 & \text{for } \zeta^{(1)}(t) < \zeta_0 \end{cases}$$

as shown in Figure 44b. Since $\zeta^{(1)}(t)$ and $Y^{(1)}(t)$ are typical realisations of the random processes $\zeta(t)$ and $Y(t)$ respectively, they are deterministic and so may be differentiated such that

$$\dot{Y}^{(1)}(t) = \dot{\zeta}^{(1)}(t)\delta\{\zeta^{(1)}(t) - \zeta_0\}$$

since

$$\frac{d}{dt}\, 1(t) = \delta(t).$$

Due to the presence of the Dirac delta function, $\dot{Y}^{(1)}(t)$ vanishes everywhere except when $\zeta^{(1)}(t) = \zeta_0$, at which point there exists a spike of unit area directed positively or negatively depending on whether the slope of $\zeta^{(1)}(t)$ at $\zeta_0$ is upward or downward (c.f. Figure 44c).

The required counting functional $n(\zeta_0, t)$ for this typical realisation of the random process $\zeta(t)$ is the number of crossings per second at time $t$. The counting functional derived by Middleton[18] includes both upward and downward crossings and is given by

$$n(\zeta_0, t) = |\dot{\zeta}^{(1)}(t)|\delta\,\{\zeta^{(1)}(t) - \zeta_0\}.$$

The total number of crossings in the interval $(t_1, t_2)$ is therefore

$$N(\zeta_0, t_1, t_2) = \int_{t_1}^{t_2} |\dot{\zeta}^{(1)}(t)|\,\delta\{\zeta^{(1)}(t) - \zeta_0\}dt.$$

Thus the average or expected number of crossings in this typical realisation $\zeta^{(1)}(t)$ of the Gaussian random process $\zeta(t)$ is given by

$$E[N(\zeta_0, t_1, t_2)] = \int_{t_1}^{t_2} E[\,|\dot{\zeta}(t)|\delta\{\zeta(t) - \zeta_0\}\,]\,dt.$$

$$= \int_{t_1}^{t_2} \int_{-\infty}^{\infty} \int_{-\infty}^{\infty} |\dot{\zeta}|\delta\{\zeta - \zeta_0\} f_{\zeta\dot{\zeta}}(\zeta, \dot{\zeta})d\zeta\, d\dot{\zeta}\, dt$$

$$= (t_2 - t_1) \int_{-\infty}^{\infty} |\dot{\zeta}| f_{\zeta\dot{\zeta}}(\zeta_0, \dot{\zeta})d\dot{\zeta}.$$

For stationary Gaussian wave random processes $\zeta(t)$ and $\dot{\zeta}(t)$ with zero mean values, the joint probability density function derived in section 4.4 is of the form

$$f_{\zeta\dot{\zeta}}(\zeta, \dot{\zeta}) = \frac{1}{2\pi\sqrt{\{m_0 m_2(1-\rho^2)\}}} \times \exp\left[\frac{-1}{2(1-\rho^2)}\left(\frac{\zeta^2}{m_0} - \frac{2\rho\zeta\dot{\zeta}}{\sqrt{m_0 m_2}} + \frac{\dot{\zeta}^2}{m_2}\right)\right]$$

where

$$\rho = \frac{E[\zeta(t)\dot{\zeta}(t)]}{\sqrt{m_0 m_2}} = \frac{\langle\zeta(t)\dot{\zeta}(t)\rangle}{\sqrt{m_0 m_2}}$$

since the random processes are ergodic. Using the results of section 5.4 we find that the auto-correlation function

$$R_{\zeta\dot{\zeta}}(0) = \langle \zeta(t)\dot{\zeta}(t)\rangle = 0 = \rho$$

so that

$$f_{\zeta\dot{\zeta}}(\zeta, \dot{\zeta}) = \frac{1}{2\pi\sqrt{m_0 m_2}} \exp\left[-\frac{1}{2}\left(\frac{\zeta^2}{m_0} + \frac{\dot{\zeta}^2}{m_2}\right)\right].$$

Since a positive upcrossing is accompanied by a negative downcrossing at another time, the number of upcrossings per unit time is given by

$$E[N_+(\zeta_0)] = \frac{\tfrac{1}{2}E[N(\zeta_0, t_1, t_2)]}{t_2 - t_1} = \tfrac{1}{2} \int_{-\infty}^{\infty} |\dot{\zeta}| f_{\zeta\dot{\zeta}}(\zeta_0, \dot{\zeta})d\dot{\zeta}$$

$$= \int_{-\infty}^{\infty} \frac{|\dot{\zeta}|}{4\pi\sqrt{m_0 m_2}} \exp\left[-\frac{1}{2}\left(\frac{\zeta_0^2}{m_0} + \frac{\dot{\zeta}^2}{m_2}\right)\right] d\dot{\zeta}$$

$$= \frac{1}{2\pi}\sqrt{\frac{m_2}{m_0}}\, e^{-(\zeta_0^2/2m_0)}$$

as may be found from integral tables. If the level $\zeta_0$ is zero, the expected number of zero upcrossings per unit time is given by

$$E[N_+(0)] = \frac{1}{2\pi}\sqrt{\frac{m_2}{m_0}}.$$

### 9.6.2 Peak or maximum probability density function

A peak or maximum occurs in a typical realisation $\zeta^{(1)}(t)$ of the Gaussian random process $\zeta(t)$ provided that, simultaneously, the realisation $\dot{\zeta}^{(1)}(t)$ of the random process $\dot{\zeta}(t)$ is zero and the realisation $\ddot{\zeta}^{(1)}(t)$ of the random process $\ddot{\zeta}(t)$ is negative (assuming that the continuous process $\zeta(t)$ may be differentiated twice). This suggests that information about the peak distribution of the random process $\zeta(t)$ can be obtained from the joint probability density functions of $\zeta(t)$, $\dot{\zeta}(t)$ and $\ddot{\zeta}(t)$.

Consider the pair of step functions

$$1\{\dot{\zeta}^{(1)}(t)-\dot{\zeta}_0\} = \begin{cases} 1 & \text{for } \dot{\zeta}^{(1)}(t) \geqslant \dot{\zeta}_0 \\ 0 & \text{for } \dot{\zeta}^{(1)}(t) < \dot{\zeta}_0 \end{cases}$$

$$1\{\zeta^{(1)}(t)-\zeta_0\} = \begin{cases} 1 & \text{for } \zeta^{(1)}(t) \geqslant \zeta_0 \\ 0 & \text{for } \zeta^{(1)}(t) < \zeta_0. \end{cases}$$

From them we can conveniently define a realisation $Y^{(1)}(t)$ of a new random process $Y(t)$ by the product

$$Y^{(1)}(t) = 1\{\dot{\zeta}^{(1)}(t)-\dot{\zeta}_0\} \,.\, 1\{\zeta^{(1)}(t)-\zeta_0\}$$

$$\begin{cases} = 1 & \text{for } \dot{\zeta}^{(1)}(t) \geqslant \dot{\zeta}_0,\ \zeta^{(1)}(t) \geqslant \zeta_0 \\ = 0 & \text{otherwise} \end{cases}$$

and this, if differentiated, gives

$$\dot{Y}^{(1)}(t) = \dot{\zeta}^{(1)}(t)1\{\dot{\zeta}^{(1)}(t)-\dot{\zeta}_0\} \,.\, \delta\{\zeta^{(1)}(t)-\zeta_0\} + \ddot{\zeta}^{(1)}(t)\delta\{\dot{\zeta}^{(1)}(t)-\dot{\zeta}_0\} \,.\, 1\{\zeta^{(1)}(t)-\zeta_0\}.$$

As we shall see, this realisation of the random process $\dot{Y}(t)$ provides two convenient counting functionals.

The first term of $\dot{Y}^{(1)}(t)$,

$$\dot{\zeta}^{(1)}(t)1\{\dot{\zeta}^{(1)}(t)-\dot{\zeta}_0\}.\,\delta\{\zeta^{(1)}(t)-\zeta_0\},$$

represents a spike of unit area whenever $\zeta^{(1)}(t) = \zeta_0$ and the slope $\dot{\zeta}^{(1)}(t) \geqslant \dot{\zeta}_0$. This term is the counting functional for the number of crossings per second at time $t$ of level $\zeta_0$ with a velocity $\dot{\zeta}^{(1)}(t) \geqslant \dot{\zeta}_0$ and it may be written as

$$n(\zeta_0, \dot{\zeta}_0, t) = \dot{\zeta}^{(1)}(t)\delta\{\zeta^{(1)}(t)-\zeta_0\}.1\{\dot{\zeta}^{(1)}(t)-\dot{\zeta}_0\}.$$

The total number of upcrossings in the time interval $\tau = t_2 - t_1$ is given by

$$N(\zeta_0, \dot{\zeta}_0, t_1, t_2) = \int_{t_1}^{t_2} \dot{\zeta}^{(1)}(t)\delta\{\zeta^{(1)}(t)-\zeta_0\}.1\{\dot{\zeta}^{(1)}(t)-\dot{\zeta}_0\}\, dt.$$

By the procedure of section 9.6.1 for a zero mean valued Gaussian random process, it follows that the expected number of upcrossings with velocity $\dot{\zeta}_0$ per unit time, is given by

$$E[N(\zeta_0, \dot{\zeta}_0)] = \int_{\dot{\zeta}_0}^{\infty} \dot{\zeta} f_{\zeta\dot{\zeta}}(\zeta_0, \dot{\zeta}) d\dot{\zeta}$$

$$= \frac{1}{2\pi} \sqrt{\frac{m_2}{m_0}} \exp\left[-\frac{1}{2}\left(\frac{\zeta_0{}^2}{m_0} + \frac{\dot{\zeta}_0{}^2}{m_2}\right)\right].$$

Notice that if $\dot{\zeta}_0 = 0$, this result reduces to the one found previously.

The other term,

$$\ddot{\zeta}^{(1)}(t)\delta\{\dot{\zeta}^{(1)}(t) - \dot{\zeta}_0\}.1\{\zeta^{(1)}(t) - \zeta_0\},$$

represents a spike of unit area when $\dot{\zeta}^{(1)}(t) = \dot{\zeta}_0$ and the realisation $\zeta^{(1)}(t) \geqslant \zeta_0$. Now if it is wished to investigate the number of maxima per second, the required counting functional must satisfy the condition that $\dot{\zeta}^{(1)}(t) = 0$ and $\ddot{\zeta}^{(1)}(t) \leqslant 0$. Thus the functional is

$$n(\zeta_0, 0, t) = -\ddot{\zeta}^{(1)}(t)\delta\{\dot{\zeta}(t) - 0\}.\,1\{\zeta^{(1)}(t) - \zeta_0\},$$

where it is assumed that the value of $\zeta^{(1)}(t)$ at the maximum is greater than $\zeta_0$.

The total number of peaks in the time interval $\tau = t_2 - t_1$ is

$$N(\zeta_0, 0, t_1, t_2) = -\int_{t_1}^{t_2} \ddot{\zeta}^{(1)}(t)\delta\{\dot{\zeta}^{(1)}(t) - 0\}.1\,\{\zeta^{(1)}(t) - \zeta_0\}\; dt.$$

Proceeding as before, we find that the expected number of maxima greater than $\zeta_0$ per unit time for a zero mean valued Gaussian random process $\zeta(t)$ is given by

$$E[N(\zeta_0, 0)] = -\int_{-\infty}^{\infty} d\zeta \int_{-\infty}^{\infty} d\dot{\zeta} \int_{-\infty}^{\infty} \ddot{\zeta}\delta(\dot{\zeta} - 0).1\,(\zeta - \zeta_0) f_{\zeta\dot{\zeta}\ddot{\zeta}}(\zeta, \dot{\zeta}, \ddot{\zeta})\, d\ddot{\zeta}$$

$$= -\int_{\zeta_0}^{\infty} d\zeta \int_{-\infty}^{0} \ddot{\zeta} f_{\zeta\dot{\zeta}\ddot{\zeta}}(\zeta, 0, \ddot{\zeta})\, d\ddot{\zeta}$$

where $f_{\zeta\dot{\zeta}\ddot{\zeta}}(\zeta, \dot{\zeta}, \ddot{\zeta})$ is the joint probability density function for the three random variables $\zeta$, $\dot{\zeta}$ and $\ddot{\zeta}$. When $\dot{\zeta} = 0$, the required Gaussian joint probability density function of the type obtained in section 4.4 is

$$f_{\zeta\dot{\zeta}\ddot{\zeta}}(\zeta, 0, \ddot{\zeta}) = \frac{1}{(2\pi)^{3/2}\,|\Delta|^{1/2}} \exp\left[-\frac{1}{2|\Delta|}\left(m_2 m_4 \zeta^2 + 2m_2^2 \zeta\ddot{\zeta} + m_0 m_2 \ddot{\zeta}^2\right)\right]$$

where $|\Delta| = m_2(m_0 m_4 - m_2^2)$ and where the results of section 5.4 have been employed, i.e. $E[\zeta(t)\dot{\zeta}(t)] = 0 = E[\dot{\zeta}(t)\ddot{\zeta}(t)]$.

The expected total number of maxima per unit time regardless of their magnitude is obtained by letting $\zeta_0 \to -\infty$ in the above expression so that

$$E[N(-\infty, 0)] = -\int_{-\infty}^{\infty} d\zeta \int_{-\infty}^{0} \ddot{\zeta} f_{\zeta\dot{\zeta}\ddot{\zeta}}(\zeta, 0, \ddot{\zeta})\, d\ddot{\zeta}$$

$$= \frac{1}{2\pi}\sqrt{\frac{m_4}{m_2}}\,.$$

In this integration it is simplest to integrate first with respect to $\zeta$ and then $\ddot{\zeta}$.

The expected number of maxima per unit time lying in the range $(x, x + \delta x)$ is

$$E[N(x, 0)] = -\int_{x}^{x+\delta x} d\zeta \int_{-\infty}^{0} \ddot{\zeta} f_{\zeta\dot{\zeta}\ddot{\zeta}}(\zeta, 0, \ddot{\zeta}) d\ddot{\zeta} = -\delta x \int_{-\infty}^{0} \ddot{\zeta} f_{\zeta\dot{\zeta}\ddot{\zeta}}(x, 0, \ddot{\zeta}) d\ddot{\zeta}.$$

The probability that a peak lies in the range $(x, x + \delta x)$ is therefore

$$P[x < \text{peak} \leqslant x + \delta x] = f_X(x)\delta x\,.$$

$$= \frac{\text{Expected number of peaks per unit time in range } (x, x+\delta x)}{\text{Expected number of peaks per unit time regardless of magnitude}}$$

$$= \frac{E[N(x, 0)]}{E[N(-\infty, 0)]} = -2\pi\sqrt{\frac{m_2}{m_4}}\,\delta x \int_{-\infty}^{0} \ddot{\zeta} f_{\zeta\dot{\zeta}\ddot{\zeta}}(x, 0, \ddot{\zeta}) d\ddot{\zeta}$$

After substituting for the function $f_{\zeta\dot{\zeta}\ddot{\zeta}}(x, 0, \ddot{\zeta})$ and integrating, the probability density function of the random process $X(t)$ describing the peak distribution of the random process $\zeta(t)$ is found to be

$$f_X(x) = \frac{\epsilon}{\sqrt{(2\pi m_0)}}\, e^{-x^2/2m_0\epsilon^2} + \frac{(1-\epsilon^2)^{1/2}}{m_0}\, x e^{-x^2/2m_0}\left[0{\cdot}5 + \operatorname{erf}\left\{\frac{x}{\epsilon}\left(\frac{1-\epsilon^2}{m_0}\right)^{1/2}\right\}\right]$$

where

$$\epsilon^2 = 1 - \frac{m_2^2}{m_0 m_4}; \quad \operatorname{erf}(x) = \frac{1}{\sqrt{2\pi}}\int_0^x e^{-z^2/2} dz; \quad \operatorname{erf}\begin{Bmatrix}0\\ \infty\end{Bmatrix} = \begin{Bmatrix}0\\ 0{\cdot}5\end{Bmatrix}.$$

The quantity $\epsilon$ is referred to as the band-width and as we shall see it has significance in its own right. It is, in fact, a measure of the width of the spectrum.

The probability density function for the minima is simply the reflection in the zero mean level of the probability density function of the maxima.

### 9.6.3 Duration of excursion probability density function

In problems relating to random oscillations, a knowledge of the probability density function of the excursion is sometimes required. The duration of the excursion is the time interval during which the oscillation exceeds a specific limit and is illustrated in Figure 45.

The arbitrary level $\zeta_0$ is exceeded by the stationary random process $\zeta(t)$ at an upcrossing when the oscillation has positive slope $\dot{\zeta}(t_1) \geqslant 0$ whilst at a downcrossing the oscillation has negative slope $\dot{\zeta}(t_2) \leqslant 0$. This suggests that information about the excursion density function may be obtained from the joint probability density function of $\zeta(t)$, $\dot{\zeta}(t)$ at $t = t_1$ and $\zeta(t)$, $\dot{\zeta}(t)$ at time $t = t_2$. Such an approach to this distribution problem was developed initially by Rice[19] although we shall develop an approximate solution by defining a suitable counting functional. Consider the step functions

$$1\{\zeta^{(1)}(t_1) - \zeta_0\} = \begin{cases} 1 & \text{for } \zeta^{(1)}(t_1) \geqslant \zeta_0 \\ 0 & \text{otherwise} \end{cases}$$

$$1\{\zeta^{(1)}(t_2) - \zeta_0\} = \begin{cases} 1 & \text{for } \zeta^{(1)}(t_2) \geqslant \zeta_0 \\ 0 & \text{otherwise} \end{cases}$$

$$1\{\dot{\zeta}^{(1)}(t_1) - 0\} = \begin{cases} 1 & \text{for } \dot{\zeta}^{(1)}(t_1) \geqslant 0 \\ 0 & \text{otherwise} \end{cases}$$

$$1\{-\dot{\zeta}^{(1)}(t_2) - 0\} = \begin{cases} 1 & \text{for } -\infty < -\dot{\zeta}^{(1)}(t_2) \leqslant 0 \\ 0 & \text{otherwise.} \end{cases}$$

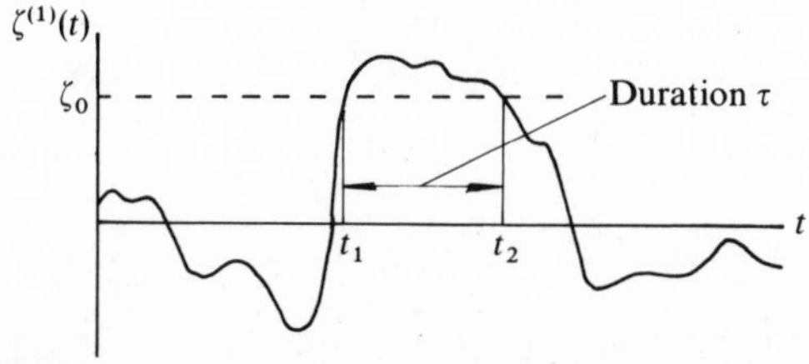


**Figure 45.** The interval between an up-crossing and a down-crossing.

From these functions we can define a realisation of a new random process as

$$Y^{(1)}(t_1, t_2) = 1\{\zeta^{(1)}(t_1) - \zeta_0\}.1\{\dot{\zeta}^{(1)}(t_1) - 0\}.1\{\zeta^{(1)}(t_2) - \zeta_0\}.\ 1\{-\dot{\zeta}^{(1)}(t_2) - 0\}.$$

On forming the differential $d^2 Y^{(1)}(t_1, t_2)/dt_1 dt_2$, four counting functionals are produced one of which is

$$\ddot{Y}^{(1)}(t_1, t_2) = \dot{\zeta}^{(1)}(t_1)\dot{\zeta}^{(1)}(t_2)\delta\{\zeta^{(1)}(t_1) - \zeta_0\}.1\{\dot{\zeta}^{(1)}(t_1) - 0\}.\ \delta\{\zeta^{(1)}(t_2) - \zeta_0\}.1\{-\dot{\zeta}^{(1)}(t_2) - 0\}.$$

This represents a spike of unit area whenever $\zeta^{(1)}(t_1) = \zeta_0 = \zeta^{(1)}(t_2)$ and provided that $\dot{\zeta}^{(1)}(t_1) \geqslant 0$ and $\dot{\zeta}^{(1)}(t_2) < 0$. If we let $t_1 = t$ and $t_2 = t + \tau$ then by proceeding as in the previous sections we discover that the expected number of excursion intervals per unit time is

$$E[N(\tau)] = -\int_{-\infty}^{0}\int_{0}^{\infty} \dot{\zeta}\dot{\zeta}_\tau f_\zeta(\zeta_0, \zeta_0, \dot{\zeta}, \dot{\zeta}_\tau)d\dot{\zeta}\, d\dot{\zeta}_\tau$$

where $\dot{\zeta}_\tau = \dot{\zeta}(t + \tau)$ and $\dot{\zeta} = \dot{\zeta}(t)$.

However, for zero mean valued Gaussian random processes the four variable density function is given in section 4.4 by

$$f_\zeta(\zeta_0, \zeta_0, \dot{\zeta}, \dot{\zeta}_\tau)$$

$$= \frac{1}{4\pi^2 |\boldsymbol{\Delta}|^{1/2}}\, e^{-a\zeta_0^2/|\boldsymbol{\Delta}|} \exp\left[-\frac{1}{|\boldsymbol{\Delta}|}\left[b\zeta_0(\dot{\zeta} - \dot{\zeta}_\tau) + d\dot{\zeta}\dot{\zeta}_\tau + \frac{e}{2}\left(\dot{\zeta}^2 + \dot{\zeta}_\tau^2\right)\right]\right]$$

where $|\boldsymbol{\Delta}|$ is the determinant of the matrix

$$\boldsymbol{\Delta} = \begin{bmatrix} m_0 & m_0(\tau) & 0 & m_1(\tau) \\ m_0(\tau) & m_0 & -m_1(\tau) & 0 \\ 0 & -m_1(\tau) & m_2 & m_2(\tau) \\ m_1(\tau) & 0 & m_2(\tau) & m_2 \end{bmatrix},$$

where the coefficients

$$a = \{m_2 - m_2(\tau)\}\,[\{m_0 - m_0(\tau)\}\{m_2 + m_2(\tau)\} - m_1^2(\tau)]$$

$$b = m_1(\tau)[\{m_0 - m_0(\tau)\}\{m_2 + m_2(\tau)\} - m_1^2(\tau)]$$

$$d = -m_2(\tau)\{m_0^2 - m_0^2(\tau)\} + m_0(\tau)m_1^2(\tau)$$

$$e = m_2\{m_0^2 - m_0^2(\tau)\} - m_0 m_1^2(\tau)$$

$$|\mathbf{\Delta}| = [\{m_0 + m_0(\tau)\}\{m_2 - m_2(\tau)\} - m_1^2(\tau)]\ [\{m_0 - m_0(\tau)\}\{m_2 + m_2(\tau)\} - m_1^2(\tau)]$$

$$= \frac{(e^2 - d^2)}{m_0^2 - m_0^2(\tau)},$$

$$m_0(\tau) = E[\zeta(t)\zeta(t+\tau)] = R_{\zeta\zeta}(\tau) = \int_0^\infty \Phi_{\zeta\zeta}(\omega)\cos\omega\tau\, d\tau$$

with $m_0 = m_0(0) > m_0(\tau)$ for all values of $\tau > 0$,

$$m_1(\tau) = E[\zeta(t)\dot{\zeta}(t+\tau)] = R_{\zeta\dot{\zeta}}(\tau) = -\int_0^\infty \omega\Phi_{\zeta\zeta}(\omega)\sin\omega\tau\, d\tau$$

$$m_2(\tau) = E[\dot{\zeta}(t)\dot{\zeta}(t+\tau)] = R_{\dot{\zeta}\dot{\zeta}}(\tau) = \int_0^\infty \omega^2\Phi_{\zeta\zeta}(\tau)\cos\omega\tau\, d\tau$$

with $m_2 = m_2(0) > m_2(\tau)$ for all values of $\tau > 0$, and where $\Phi_{\zeta\zeta}(\omega)$ is the spectral density function of the oscillation.

The Gaussian density function contains the term $|\mathbf{\Delta}|^{1/2}$ and it is therefore essential that $|\mathbf{\Delta}| > 0$. This property may be easily established, since the terms in square brackets may be expressed in the form

$$\int_0^\infty (1 \pm \cos\omega\tau)\Phi_{\zeta\zeta}(\omega)d\omega \int_0^\infty \omega^2(1 \pm \cos\omega\tau)\Phi_{\zeta\zeta}(\omega)\,d\omega - \left[\int_0^\infty \omega\Phi_{\zeta\zeta}(\omega)\sin\omega\tau\, d\omega\right]^2.$$

If we define

$$r^2(\omega) = (1 \pm \cos\omega\tau)\Phi_{\zeta\zeta}(\omega), \qquad q^2(\omega) = \omega^2(1 \pm \cos\omega\tau)\Phi_{\zeta\zeta}(\omega)$$

such that

$$r(\omega)q(\omega) = \omega\sin\omega\tau\Phi_{\zeta\zeta}(\omega)$$

then by Schwarz's inequality (see next section), it follows that

$$\int_0^\infty r^2(\omega)d\omega \int_0^\infty q^2(\omega)d\omega - \left[\int_0^\infty r(\omega)q(\omega)d\omega\right]^2 > 0$$

whence $|\mathbf{\Delta}| > 0$. Thus $|\mathbf{\Delta}|^{1/2}$ is real and, from the definition of the coefficients, it follows also that $e > d$.

Unfortunately, the general problem with $\zeta_0 \neq 0$ has to be solved numerically[20] although an analytical solution may be obtained when $\zeta_0 = 0$, which is the case considered here.

The probability density function for the interval between zeros of the oscillation, under the condition that the random oscillation intersects the zero level with positive slope at the initial moment ($\tau = 0$), is

$$f(\tau) = \frac{E[N(\tau)]}{E[N_+(\tau)]}$$

where, from section 9.6.1, the number of upcrossings of level $\zeta_0 = 0$ is given by

$$E[N_+(0)] = \frac{1}{2\pi} \sqrt{\frac{m_2}{m_0}}$$

and where

$$E[N(\tau)] = \frac{1}{4\pi^2 |\Delta|^{1/2}} \int_{\dot{\zeta}_\tau = 0}^{-\infty} \int_{\dot{\zeta}=0}^{\infty} \dot{\zeta}\dot{\zeta}_\tau \exp - \left[ -\frac{1}{|\Delta|} \left\{ d\dot{\zeta}\dot{\zeta}_\tau + \frac{e}{2} (\dot{\zeta}^2 + \dot{\zeta}_\tau^2) \right\} \right] d\dot{\zeta}\, d\dot{\zeta}_\tau .$$

Such an integral has been evaluated by Rice and it may be shown that the probability density function for the interval of excursion when the oscillation passed through zero at $\tau = 0$ with positive slope is

$$f(\tau) = \frac{1}{2\pi} \sqrt{\frac{m_0}{m_2}} (e^2 - d^2)^{1/2} \left\{ m_0^2 - m_0^2(\tau) \right\}^{-3/2} \left\{ 1 + H \cot^{-1}(-H) \right\}$$

where

$$H = d(e^2 - d^2)^{-1/2}.$$

This result was initially determined by Rice. Variations of the problem may be solved, such as the probability of the duration between two downcrossings or two upcrossings. These may be obtained in a similar manner to that described in this section.

## 9.7 Special probability density functions

Let us consider the function

$$\beta(y) = \int_0^\infty (y + \omega^2)^2 \Phi_{\zeta\zeta}(\omega) d\omega = y^2 m_0 + 2y m_2 + m_4$$

where $\Phi_{\zeta\zeta}(\omega)$ is a one-sided mean square spectral density and the moments $m_0$, $m_2$ and $m_4$ are positive valued. Since the integrand cannot be negative, the function $\beta(y)$ is positive for all $y$. This implies that no solution of the equation

$$\beta(y) = 0$$

can have two distinct real roots. The solution of this equation is

$$y = \frac{-m_2 \pm \sqrt{(m_2^2 - m_0 m_4)}}{m_0} .$$

Since two distinct, real solutions are impossible, the discriminant cannot be positive, i.e.

$$m_2^2 \leqslant m_0 m_4 .$$

This is called the 'Schwarz Inequality'. Since $m_0$, $m_2$ and $m_4$ are all positive quantities it follows that

$$0 \leqslant \frac{m_2^2}{m_0 m_4} \leqslant 1$$

and the quantity

$$\epsilon^2 = 1 - \frac{m_2^2}{m_0 m_4}$$

lies in the range $0 \leqslant \epsilon \leqslant 1$.

When $\epsilon \to 0$, the probability density function of the peak random variable $X$ of section 9.6.2, reduces to

$$f_X(x) = \frac{x}{m_0} e^{-x^2/2m_0}$$

for $0 < x \leqslant \infty$, and this is the Rayleigh probability density function that is discussed in section 4. The probability of the variable $X$ at time $t$ exceeding the value $x$ is

$$P[X > x] = \int_x^\infty f_X(y)\, dy$$

$$= \int_x^\infty \frac{y}{m_0} e^{-y^2/2m_0}\, dy = e^{-x^2/2m_0} .$$

Thus

$$\frac{x^2}{m_0} = -2 \log_e P[X > x] = -4{\cdot}605 \log_{10} P[X > x]$$

and the relationship is shown in Figure 46 where, for example, the peak value of 1 in 100 oscillations may be expected to exceed ±3·03 times the root mean square value of the variable, or the peak value of 1 in 10 oscillations exceeds ±2·14 times the root mean value of the variable $X$ at time $t$.

On the other hand, for $\epsilon \to 1$, the probability density function of the peaks yields

$$f_X(x) = \frac{1}{\sqrt{(2\pi m_0)}} e^{-x^2/2m_0}$$

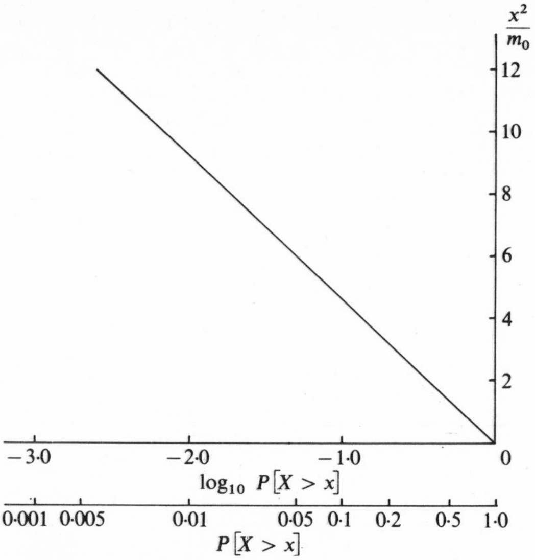


**Figure 46.** Probabilities for a Rayleigh distribution of a narrow band process.

for $-\infty < x < \infty$ and this is the Gaussian probability density function. With the exception of these extreme cases, the probability density function of the peaks for a Gaussian random process is neither a Rayleigh function nor Gaussian.

The probability density function of the peaks for different values of $\epsilon$ were obtained by Cartwright and Longuet-Higgins[21] and are shown in Figure 47. Further, by considering the ratio of the expected number of zero upcrossings per unit time and the total expected number of peaks per unit time regardless of their magnitude, we find that

$$\frac{E[N_+(0)]}{E[N(-\infty, 0)]} = \frac{m_2}{\sqrt{m_0 m_4}} = \sqrt{(1-\epsilon^2)}\,\cdot$$

### 9.7.1 Narrow band spectra

Let us consider the mean square spectral density $\Phi_{\zeta\zeta}(\omega) = \Phi_0 \delta(\omega - \omega_0)$, a Dirac delta function having a spike of area $\Phi_0$ at $\omega = \omega_0$. The moments of the spectrum are given by

$$m_0 = \Phi_0, \qquad m_2 = \omega_0{}^2\Phi_0, \qquad m_4 = \omega_0{}^4\Phi_0$$

and the quantity $\epsilon = 0$.

In general, when $\epsilon \to 0$ the mean square spectral density curve assumes a form somewhat akin to that of Figure 48a. A realisation

$\zeta^{(1)}(t)$ of the random process might have the general form shown in Figure 48b and it will be seen to have a perceptible dominant frequency $\omega_0$. It has also a fluctuating random amplitude and phase angle. Such a process is said to possess a narrow band spectrum.

We have seen in section 9.7 that, corresponding to $\epsilon = 0$, the probability density function of the peaks is

$$f_X(x) = \frac{x}{m_0} e^{-x^2/2m_0}$$

for $x \geqslant 0$. This is the Rayleigh probability density function (see Figure 13).

In a narrow band random process, the probability is very high that each crossing implies a complete 'cycle' and therefore the 'expected radian frequency' per unit time is given by

$$\omega_0 = 2\pi E[N_+(0)]$$

That is to say

$$\omega_0^2 = \frac{m_2}{m_0} = \frac{\int_0^\infty \omega^2 \Phi_{\zeta\zeta}(\omega) d\omega}{\int_0^\infty \Phi_{\zeta\zeta}(\omega) d\omega}$$

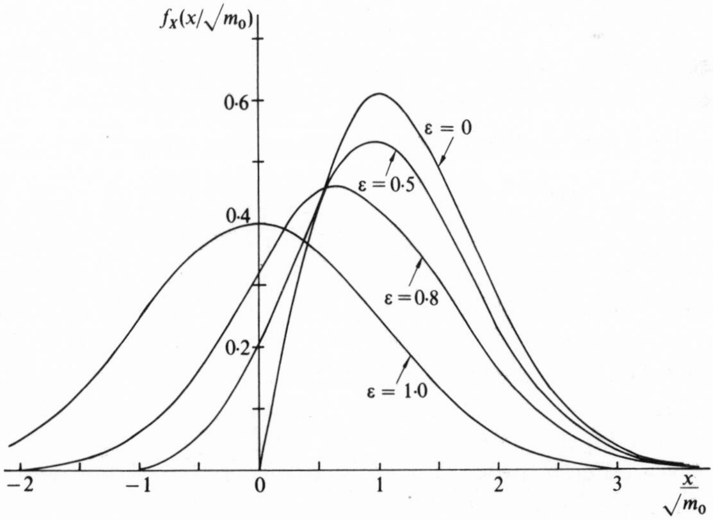


**Figure 47.** Probability density function of the peaks for various values of the band width $\epsilon$.

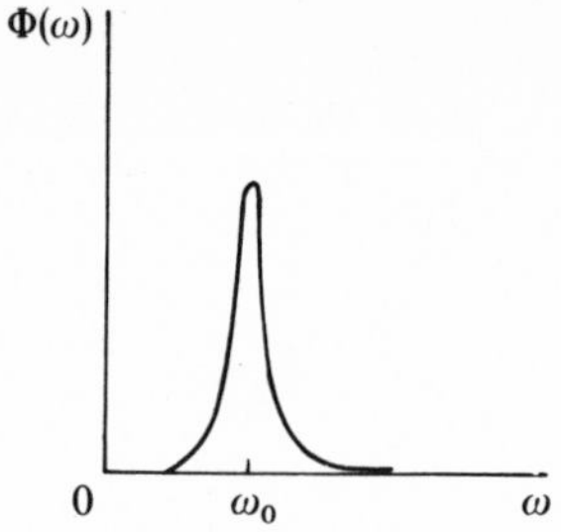


(a)

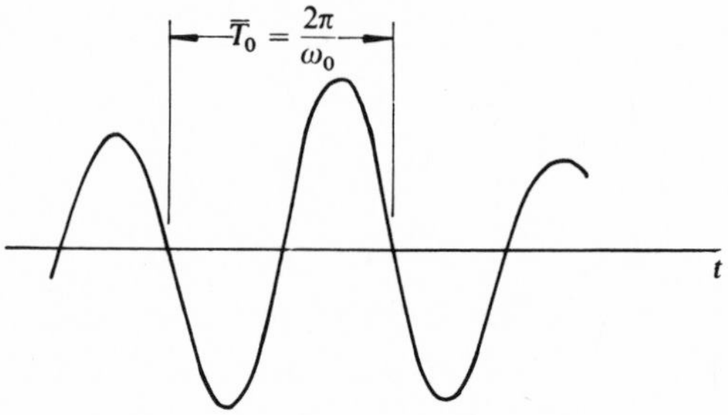


(b)

**Figure 48.** (a) A typical narrow band mean square spectral density function, and (b) a corresponding realisation of the process.

The average period $\bar{T}_0$ measured at zero crossings is given by

$$\bar{T}_0 = \frac{2\pi}{\omega_0} = 2\pi \sqrt{\frac{m_2}{m_0}} .$$

So far we have mainly considered amplitude measurements of the Gaussian wave random process $\zeta(t)$. The theory may be extended to wave height measurements by considering the wave height random process defined by

$$H(t) = 2\zeta(t)$$

with $h = 2a$. By a transformation of variable as described in section 3.3, the wave height probability density function is given by

$$f_H(h) = f_\zeta(a) \bigg/ \left| \frac{dh}{da} \right|$$

so that for a Rayleigh probability density function $f_\zeta(a)$, we have

$$f_H(h) = \frac{h}{4m_0} e^{-h^2/8m_0}$$

for $h \geqslant 0$. This is again a Rayleigh probability density function.

Observations[22] indicate that the Rayleigh distribution of wave height agrees quite closely with reality for seas of low and moderate severity. For seas of great severity the distribution of wave height ceases to follow such simple laws. In order to describe the seaway in a statistical manner, expectations are needed which may be derived by using the Rayleigh probability density function.

### 9.7.2 Statistical values of the 1/nth highest observations

The average value of the $1/n$th highest observations of the wave peaks of a random process $\zeta(t)$ that is described by the Rayleigh probability density function, can be visualised as the distance of the centroid of the shaded area in Figure 49 from the probability density function axis $x = 0$. Thus the probability of exceeding an amplitude $\zeta_{1/n}$ is given by

$$P[X > \zeta_{1/n}] = \frac{1}{n} = \int_{\zeta_{1/n}}^{\infty} f_X(x)\,dx$$

whence

$$\int_{\zeta_{1/n}}^{\infty} \frac{x}{m_0} e^{-x^2/2m_0}\,dx = \frac{1}{n}$$

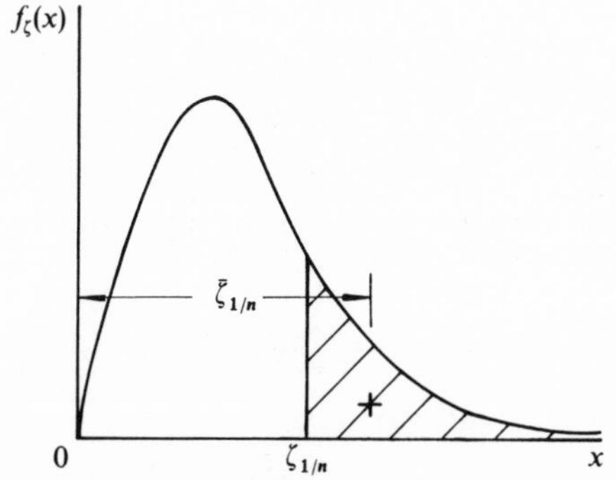


**Figure 49.** Definition of the $1/n$th highest observations.

so that

$$\zeta_{1/n}^2 = 2m_0 \log_e n.$$

The average value of the $1/n$th highest observations is

$$\bar{\zeta}_{1/n} = n \int_{\zeta_{1/n}}^{\infty} x f_X(x) dx = n \int_{\zeta_{1/n}}^{\infty} \frac{x^2}{m_0} e^{-x^2/2m_0}\, dx.$$

That is to say

$$\bar{\zeta}_{1/n} = n\sqrt{2m_0}\left[\frac{1}{n}(\log_e n)^{1/2} + \sqrt{\pi}\{0{\cdot}5 - \operatorname{erf}(2\log_e n)^{1/2}\}\right]$$

where

$$\operatorname{erf} x = \frac{1}{\sqrt{2\pi}} \int_0^x e^{-y^2/2}\, dy$$

(see Table I). Numerical values of $\bar{\zeta}_{1/n}$ are given in Table VI.

**Table VI**

| $1/n$ | $(2m_0)^{-1/2}\bar{\zeta}_{1/n}$ | $1/n$ | $(2m_0)^{-1/2}\bar{\zeta}_{1/n}$ | $1/n$ | $(2m_0)^{-1/2}\bar{\zeta}_{1/n}$ |
|---|---|---|---|---|---|
| 0·01 | 2·359 | 0·4 | 1·347 | 0·8 | 1·031 |
| 0·1 | 1·80 | 0·5 | 1·256 | 0·9 | 0·961 |
| 0·2 | 2·591 | 0·6 | 1·176 | 1·0 | 0·886 |
| 0·3 | 1·454 | 0·7 | 1·102 | | |

Thus the average or mean value of the wave amplitude is given by $n = 1$ and

$$\bar{a} = \int_0^{\infty} x f(x) dx = 1{\cdot}25\sqrt{m_0}$$

whilst the average wave height is

$$\bar{h} = 2\bar{a} = 2{\cdot}5\sqrt{m_0}\,.$$

The 'significant wave amplitude' is the average of the one third highest observations of the wave amplitude and is given by $n = 3$. That is

$$\bar{a}_{1/3} = 2{\cdot}0\sqrt{m_0}$$

so the significant wave height is

$$\bar{h}_{1/3} = 4{\cdot}0\sqrt{m_0}\,.$$

An estimate of the maximum value of the wave amplitude is given by the 'one-tenth highest wave amplitude'. This is the average of the one tenth highest observations such that, for $n = 10$,

$$\bar{a}_{1/10} = 2{\cdot}55\sqrt{m_0}\,.$$

The corresponding wave height is

$$h_{1/10} = 5{\cdot}1\sqrt{m_0}.$$

Suppose that, in a sample of 150 wave height measurements, the numbers of various heights are

| | | | | | | | | |
|---|---|---|---|---|---|---|---|---|
| *Wave height h*(m) | 0·5 | 1·0 | 1·5 | 2·0 | 2·5 | 3·0 | 3·5 | 4·0 |
| *Number of waves* | 15 | 30 | 55 | 21 | 14 | 9 | 5 | 1 |

The average wave height

$$\bar{h} = \frac{0{\cdot}5\times 15+1{\cdot}0\times 30+1{\cdot}5\times 55+2{\cdot}0\times 21+2{\cdot}5\times 14+3{\cdot}0\times 9+3{\cdot}5\times 5+4{\cdot}0\times 1}{150}$$

$$= 1{\cdot}64 \text{ m}.$$

The one third highest waves are the top 50 wave heights measured and the significant wave height is

$$\bar{h}_{1/3} = \frac{3(2{\cdot}0\times 21 + 2{\cdot}5\times 14 + 3{\cdot}0\times 9 + 3{\cdot}5\times 5 + 4{\cdot}0\times 1)}{150}$$

$$= 2{\cdot}61 \text{ m}.$$

The average of the 1/10th highest measurements in the sample is

$$\bar{h}_{1/10} = \frac{10(3{\cdot}0\times 9 + 3{\cdot}5\times 5 + 4{\cdot}0\times 1)}{150}$$

$$= 3{\cdot}23 \text{ m}.$$

### 9.7.3 Expected maximum wave amplitude

Consider an interval containing $N$ waves from which the expected wave amplitude of the highest wave $\zeta_m$ is required. From the previous section, we see that an approximate answer may be obtained by setting $n = N$ and calculating $\bar{\zeta}_{1/N}$. This answer will only be an approximation however, since $\bar{\zeta}_{1/n}$ represents the mean of the $1/N$th highest observations in a large sample $mN$ which may have been obtained by collecting together $m$ samples each of $N$ waves. The $m$ highest observations may not be distributed evenly to each of the samples so that the mean of the highest observations taken one from each group will be less than the mean of the $m$ highest taken from all the $mN$ observations together. Thus it is to be expected that $\zeta_m$ is less than $\zeta_{1/N}$.

From the previous section, it will be seen that the probability of a single wave amplitude exceeding a given amplitude $\zeta_{1/n}$ is, say,

$$P[X > \zeta_{1/n}] = e^{-\zeta_{1/n}^2/2m_0} = g(\zeta_{1/n}).$$

The probability of every wave amplitude $X$ in the $N$ independent waves being less than $\zeta_{1/n}$ is

$$P[X \leqslant \zeta_{1/n}] = F_X(\zeta_{1/n}) = \{1 - g(\zeta_{1/n})\}^N.$$

Alternatively, the probability of at least one wave amplitude exceeding $\zeta_{1/n}$ in the $N$ independent waves is

$$P[X > \zeta_{1/n}] = 1 - F_X(\zeta_{1/n}) = 1 - \{1 - g(\zeta_{1/n})\}^N.$$

The probability that the maximum wave amplitude $\zeta_m$ lies in the range $(\zeta_{1/n}, \zeta_{1/n} + d\zeta_{1/n})$ is the probability that at least one wave amplitude $X$ exceeds $\zeta_{1/n}$ minus the probability that at least one wave amplitude $X$ exceeds $(\zeta_{1/n} + d\zeta_{1/n})$. That is,

$$\begin{aligned} P[\zeta_{1/n} < \zeta_m \leqslant \zeta_{1/n} + d\zeta_{1/n}] &= f_{\zeta_m}(\zeta_{1/n})d\zeta_{1/n} \\ &= P[X > \zeta_{1/n}] - P[X > \zeta_{1/n} + d\zeta_{1/n}] \\ &= F_X(\zeta_{1/n} + d\zeta_{1/n}) - F_X(\zeta_{1/n}) \end{aligned}$$

so that the probability density function of the maximum wave amplitude is

$$f_{\zeta_m}(\zeta_{1/n}) = \frac{dF_X(\zeta_{1/n})}{d\zeta_{1/n}} = N\left\{1 - g(\zeta_{1/n})\right\}^{N-1} \frac{dg(\zeta_{1/n})}{d\zeta_{1/n}}.$$

The expected value of the maximum wave amplitude is

$$\begin{aligned} E[\zeta_m] &= \int_0^\infty \zeta_{1/n} f_{\zeta_m}(\zeta_{1/n}) d\zeta_{1/n} \\ &= \sqrt{2m_0} \int_0^\infty u^{1/2} N e^{-u} (1 - e^{-u})^{N-1}\, du \end{aligned}$$

where $u = \zeta_{1/n}^2/2m_0$. That is

$$E[\zeta_m] = \sqrt{2m_0} \int_0^\infty u^{1/2} \left\{ Ne^{-u} - N(N-1)e^{-2u} + \frac{N(N-1)(N-2)}{2!} e^{-3u} \dots \right\} du.$$

Since

$$\int_0^\infty u^{1/2} e^{-nu}\, du = \frac{1}{2n}\left(\frac{\pi}{n}\right)^{1/2},$$

it follows that

$$E[\zeta_m] = \sqrt{2m_0}\,\frac{\sqrt{\pi}}{2}\left[N - \frac{N(N-1)}{2!\sqrt{2}} + \frac{N(N-1)(N-2)}{3!\sqrt{3}} \cdots\right]$$

which expresses the expected value of the maximum wave amplitude as a function of the size of the wave sample. Longuet-Higgins[23] has evaluated this function for various values of $N$ as shown in Table VII. Unfortunately the computation is tedious for large values of $N$ and so an asymptotic expression has been derived by that author; it is

$$E[\zeta_m] = \sqrt{2m_0}\;\{(\log_e N)^{1/2} + 0{\cdot}2886(\log_e N)^{-1/2}\}.$$

Numerical values are included in Table VII.

**Table VII**

| $N$ | $E[\zeta_m]/\sqrt{2m_0}$ | Asymptotic expression | $\bar{\zeta}_{1/n}/\sqrt{2m_0}$ |
|---|---|---|---|
| 1 | 0·886 | – | 0·707 |
| 2 | 1·146 | – | 1·030 |
| 5 | 1·462 | – | 1·366 |
| 10 | 1·676 | 1·708 | 1·583 |
| 20 | 1·870 | 1·898 | 1·778 |
| 100 | – | 2·280 | 2·172 |
| 1000 | – | 2·738 | 2·642 |
| 10,000 | – | 3·130 | 3·044 |
| 100,000 | – | 3·478 | 3·400 |

### 9.7.4 Mean wave period and wave-length

The average time between successive zero upcrossings is given by

$$\bar{T}_0 = \frac{1}{E[N_+(0)]} = 2\pi\sqrt{\frac{m_0}{m_2}}$$

whilst the average time between successive maxima, regardless of their magnitude, in the wave random process $\zeta(t)$, is given by

$$\bar{T}_c = \frac{1}{E[N(-\infty, 0)]} = 2\pi\sqrt{\frac{m_2}{m_4}}\,.$$

The band-width quantity $\epsilon$ may be defined in terms of the mean wave period since

$$\epsilon^2 = 1 - \frac{m_2^2}{m_0 m_4} = 1 - \frac{\bar{T}_c^2}{\bar{T}_0^2}\,.$$

Thus for a narrow band spectrum (for which $\epsilon = 0$) the mean wave periods $\bar{T}_0$ and $\bar{T}_c$ are equal. On the other hand, for a wide band spectrum (with $\epsilon = 1$) it follows that $\bar{T}_c \ll \bar{T}_0$.

In section 9.2 the wave number and frequency spectra are shown to be related by

$$\Phi_{\zeta\zeta}(\omega) = \frac{2\omega}{g}\,\Phi_{\zeta\zeta}(k)$$

for $k = \dfrac{\omega^2}{g}$. If we define the moments of the wave number spectrum by $m_n'$ then we find

$$m_n' = \int_0^\infty k^n\,\Phi_{\zeta\zeta}(k)dk = \int_0^\infty \frac{\omega^{2n}}{g^n}\,\Phi_{\zeta\zeta}(\omega)d\omega = \frac{m_{2n}}{g^n}$$

since $\omega^2 = kg$. Thus instead of working in the time domain, we can work in the wave number domain, so that the mean wave-length based on zero upcrossings is

$$\bar{\lambda} = 2\pi\sqrt{\frac{m_0'}{m_2'}} = 2\pi g\sqrt{\frac{m_0}{m_4}}$$

and the mean wave length between successive maxima is

$$\bar{\lambda}_c = 2\pi\sqrt{\frac{m_2'}{m_4'}} = 2\pi g\sqrt{\frac{m_4}{m_8}}\,.$$

For simple harmonic waves of period $T$ we have $\omega^2 = gk$ or $(\lambda/T^2) = (g^2/2\pi)$. For zero upcrossings we now have the analogous result

$$\frac{\bar{\lambda}_0}{\bar{T}_0^2} = \frac{g}{2\pi}\cdot\sqrt{\frac{m_2^2}{m_0 m_4}}$$

and for peaks

$$\frac{\bar{\lambda}_c}{\bar{T}_c^2} = \frac{g}{2\pi}\sqrt{\frac{m_4^{3/2}}{m_2^2 m_8}}\,.$$

### 9.7.5 Wide band spectra

Let us consider the energy spectrum

$$\Phi(\omega) = \begin{cases} \Phi_0 & \text{for } 0 < \omega \leqslant \omega_0 \\ 0 & \text{otherwise.} \end{cases}$$

The moments are given by

$$m_0 = \Phi_0 \omega_0; \quad m_2 = \frac{\Phi_0 \omega_0^3}{3}; \quad m_4 = \frac{\Phi_0 \omega_0^5}{5},$$

giving a value of $\epsilon = 2/3$.

The Pierson-Moskowitz spectrum, defined in section 9.5 as

$$\Phi_{\zeta\zeta}(\omega) = \frac{A}{\omega^5} e^{-B/\omega^4},$$

has a zero moment

$$m_0 = \frac{A}{4B} = \frac{1}{16} h_{1/3}^2$$

giving the significant wave height $h_{1/3} = 4\sqrt{m_0}$ which agrees with the value calculated in section 9.6.4. Furthermore, the second and fourth moments are respectively

$$m_2 = \frac{A}{4}\sqrt{\frac{\pi}{B}} \quad \text{and} \quad m_4 = \frac{A}{4}\Gamma(0)$$

where the Gamma function is defined as

$$\Gamma(\beta) = \int_0^\infty x^{\beta - 1} e^{-x}\, dx.$$

This latter function has the properties

$$\Gamma(\beta + 1) = \beta! \quad \Gamma(0+) = \infty, \quad \Gamma(+\infty) = \infty$$

where 0+ indicates that $\beta$ is permitted to approach zero from above.

For this spectrum

$$\epsilon^2 = 1 - \frac{\pi}{\Gamma(0)}$$

and provided the limit 0+ is approached, then $\epsilon \to 1$. When $\epsilon \to 1$ we approach another extreme. The curve $\Phi(\omega)$ is now spread out along the axis as in Figure 50a. A typical realisation $\zeta^{(1)}(t)$ is now like that shown in Figure 50b. The process is now said to be 'wide-band' and

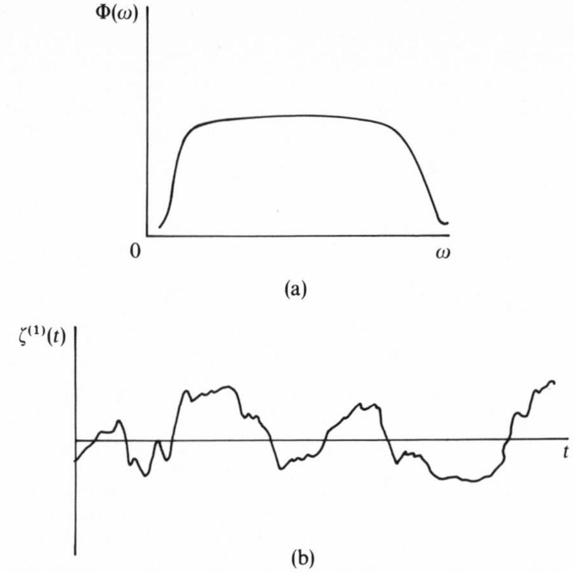


**Figure 50.** (a) A typical wide band mean square spectral density function, and (b) a corresponding realisation of the process.

the random process at the extreme $\epsilon = 1$ is described by the Gaussian density function

$$f_X(x) = \frac{1}{\sqrt{(2\pi m_0)}} e^{-x^2/2m_0}.$$

It is common to introduce two idealised forms of wide-band spectrum. These are illustrated in Figure 51. The spectrum (a) represents 'white noise' (by a somewhat vulnerable analogy with white light). Care has to be used in analysis when this concept is used, since white noise cannot be realised physically because it

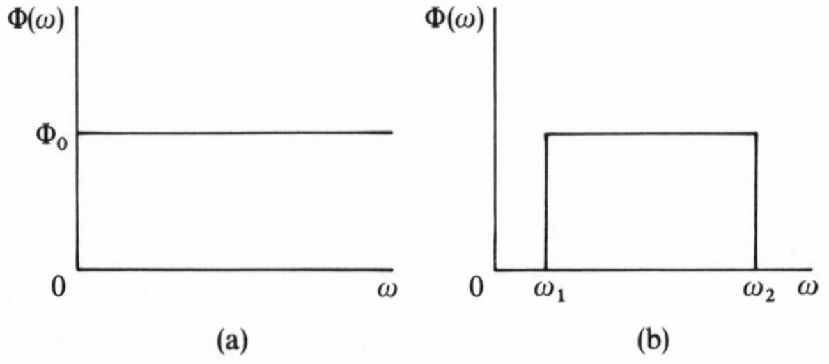


**Figure 51.** (a) The 'white noise' spectrum, and (b) a typical band-limited white noise spectrum.

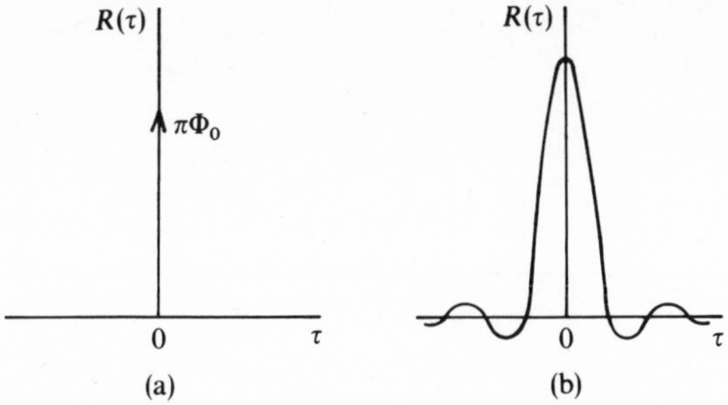


**Figure 52.** (a) The auto-correlation function of the white noise spectrum. (b) A typical auto-correlation function of a band-limited white noise spectrum.

implies an infinite mean square of the random process concerned. The spectrum (b) represents 'band-limited white noise' and this, by contrast, can be quite a good approximation to a real mean square spectrum.

It is of interest briefly to notice the auto-correlation functions corresponding to the spectra of Figure 51. They are shown sketched in Figures 52a and b respectively, the appropriate results being

$$R(\tau) = \pi\Phi_0\delta(\tau)$$

and

$$R(\tau) = \Phi_0 \frac{\sin \omega_2\tau - \sin \omega_1\tau}{\tau}.$$

Since the magnitude of $R(\tau)$ is infinite at $\tau = 0$ and $R(\tau) = 0$ for all other values of $\tau$, white noise has an infinite intensity, but there is no correlation between the values of the random variable at any two instants, no matter how close they are.

As we have mentioned, the probability density functions of all orders may be postulated for a random process with a wide band spectrum. In particular, the process may be Gaussian.

## References


[1] COTE, L. J., DAVIS, J. O., MARKS, W., McGOUGH, R. J., MEHR, E., PIERSON (Jr), W. J., ROPEK, J. F., STEPHENSON, G. and VETTER, R. C., 1960, 'The directional spectrum of a wind generated sea as determined from data by the Stereo Wave Observation Project (SWOP)'. Meteorol. Papers, New York University, Coll. of Eng., New York.

[2] KINSMAN, B., 1965, *Wind Waves*, Prentice Hall, Englewood Cliffs, New Jersey.

[3] KORVIN-KROUKOVSKY, B. V., 1961, *Theory of Seakeeping*, SNAME, New York, p. 37.

[4] St. DENIS, M. and PIERSON, W. J., 1953, 'On the motions of ships in confused seas'. *Trans. SNAME* **61**, 280–357.

[5] St. DENIS, M., 1957, 'On the reduction of motion data from model tests in confused seas'. Proceedings of Symposium on the Behaviour of Ships in Seaway. Netherlands Ship Model Basin, Wageningen, 133–144.

[6] CANHAM, H. J. S., CARTWRIGHT, D. E., GOODRICH, G. J. and HOGBEN, N., 1962, 'Seakeeping trials in O. W. S. Weather Reporter', Trans R. I. N. A., **104**, 447–492.

[7] PHILLIPS, O. M., 1966, *The Dynamics of the Upper Ocean*, C.U.P., Cambridge.

[8] MILES, J. W., 1957, 'On the generation of surface waves by shear flows'. Journal of Fluid Mechanics, **3**, 185–204.

[9] MILES, J. W., 1960, 'On the generation of surface waves by turbulent shear flows'. J. Fluid Mechanics, **7**, 469–478.

[10] LONGUET-HIGGINS, M. S., 1969, 'A non-linear mechanism for the generation of sea waves'. Proc. Roy. Soc., **A310**. 151–159.

[11] LAMB, H., 1945, *Hydrodynamics*, C.U.P., Cambridge.

[12] HOGBEN, N., and LUMB, F. E., 1967, *Ocean Statistics*, H.M.S.O., London.

[13] U.S. Naval Oceanographic Office, 1963, *Oceanographic Atlas of the North Atlantic Ocean; Section IV; Sea and Swell*. Pub. No. 700, Washington.

[14] LOFFT, R. F. and PRICE, W. G., 1973, 'Ocean wave statistics: frequency of occurrence of states'. A.E.W. Tech. Memo. No. 19/73.

[15] SVEDRUP, H. U. and MUNK, W. H., 1947, 'Wind, sea and swell; theory of relations for forecasting', U.S. Navy Hydrodynamic Office Pub. No. 601.

[16] PIERSON, W. J. and MOSKOWITZ, L., 1963, 'A proposed spectral form for fully developed wind seas based on the similarity theory of S. A. Kitaigorodsku', Tech. Rept. U.S. Naval Oceanographic Office Contract No. 62306–1042.

[17] RAWSON, K. J. and TUPPER, E. C., 1968, *Basic Ship Theory*, Longmans, London.

[18] MIDDLETON, D., 1960, *An Introduction to Statistical Communication Theory*, McGraw-Hill, New York, pp. 426–435.

[19] RICE, S. O., 1954, *Mathematical Analysis of Random Noise. Selected Papers on Noise and Stochastic Processes*. Ed., N. Wax, Dover, New York, pp. 133–294.

[20] TIKHONOV, V. I., 1965, *The Distribution of the Duration of Excursions of Normal Fluctuations. Non-linear Transformations of Stochastic Processes*. Ed., P. I. Kusnetsov, R. L. Stratonovick and V. I. Tikhonov, Pergamon, Oxford, pp. 354–367.

[21] CARTWRIGHT, D. E. and LONGUET-HIGGINS, M. S., 1956, 'The statistical distribution of the maxima of a random function', *Proc. Roy. Soc.*, **A237**, 212–232.

[22] JASPER, N. H., 1956, 'Statistical distribution patterns of ocean and wave induced ship stresses and motions with engineering application'. *Trans. SNAME*, **64**, 375–432.

[23] LONGUET-HIGGINS, M. S., 1952, 'On the statistical distribution of the heights of sea waves'. Journal of Marine Research, **XI**, 245–266.

# Part 3 Theory of Seakeeping

# 10 Random vibration of linear systems

A ship that is assumed rigid, has motions which are usually considered to occur in the six degrees of freedom known as surge, sway, heave, roll, pitch and yaw as illustrated in Figure 53. With body axes *Cx, Cy* and *Cz* fixed in the ship and having their origin at the centre of gravity *C*, the translational motions of surge, sway and heave are directed along the body axes *Cx, Cy* and *Cz* respectively. The roll, pitch and yaw motions are variations of orientation about the axes *Cx, Cy* and *Cz* respectively measured from a convenient inertial coordinate system *XYZ*.

None of the motions occur singly; they are coupled in a way that depends on the wave direction. In beam seas, roll and sway motions

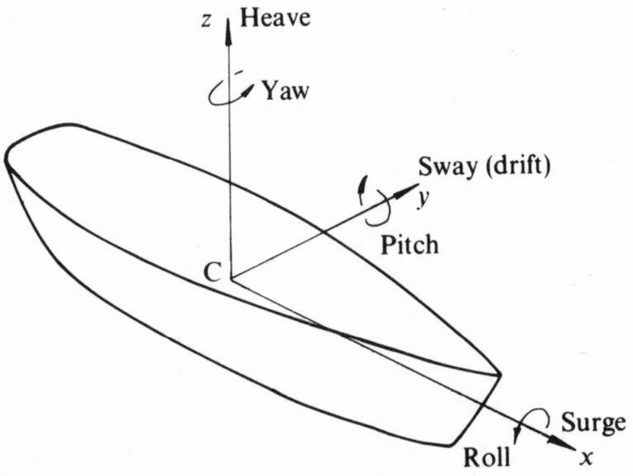


**Figure 53.** Body axes for which translational and rotational motions (of heave, yaw, etc.) may be defined.

are dominant whilst in head seas, pitch and heave motions predominate. Heave, pitch and sometimes surge are treated under the heading of 'seakeeping' and the study of sway and yaw motions has formed the subject of manoeuvring. Roll motion is treated in both. Briefly, a ship has good seakeeping qualities if it is able to perform its intended task safely in the roughest seas with the minimum delay.

The motions excited in the ship are mainly due to the wave disturbance and to the ship's orientation relative to the direction of wave propagation. In effect, the wave disturbance is the 'input' to the ship and the resulting ship motions are the 'output'. Figure 54a shows this relationship diagramatically, the 'black box' serving to illustrate the transformation between input and output. The 'black box' indicates the system which causes the transformation. In the majority of cases to be considered, the system[1] is assumed to be,

a) *linear*; i.e. double the magnitude of the input doubles the magnitude of the output; treble the input trebles the magnitude of the output, etc.

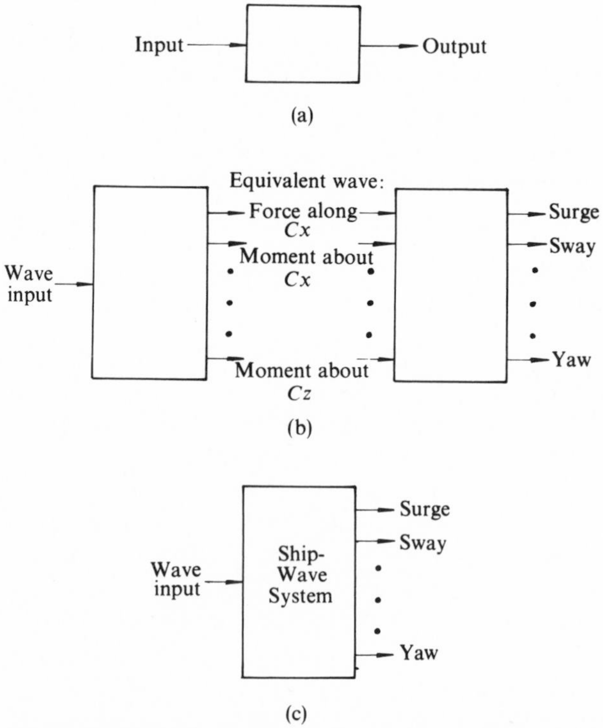


**Figure 54.** Block diagrams for possible input–output representations.

b) *time invariant*; i.e. an impulsive input applied at time $t$ produces the same output at time $t + \tau$ as an impulsive input applied at time $t + \sigma$ gives at $t + \sigma + \tau$.

In general, the output may be considered to be a ship's motion, whilst the wave input may take a modified form. Figure 54b indicates that the wave input distributed along the ship's length has been replaced by an equivalent wave loading acting at the centre of gravity, which excites the corresponding ship motion. Alternatively, Figure 54c takes the system as a combined and interacting ship-wave system discarding all information about wave loading. By redefining the 'black box', then, various interpretations may be placed on 'input' and 'output'.

Before tackling problems of seakeeping, it is necessary to examine some elementary aspects of random vibration of simple linear systems. This we shall now do without specific reference to ships or the sea.

## 10.1 Impulsive inputs

In dynamical systems we frequently encounter an input of very large magnitude acting for a very short time and having a time integral that is finite. Such inputs are described as 'impulsive' and a unit impulse is defined to be the limiting form of a rectangular input having magnitude $A$ acting for a time interval $\Delta t$ such that $A \,.\, \Delta t = 1$ always, as $A$ becomes infinite and $\Delta t$ becomes $dt$. The unit impulse denoted by $\delta(t)$ (see section 6.2), has magnitude tending to infinity during time $dt$, but its value is zero outside this time interval.

The output $h(t)$ due to this unit impulse input applied to the linear system at $t = 0$ is such that $h(t) = 0$ for $t < 0$. In other words, it is physically impossible for the unit impulse to have any effect on the output for $t < 0$. If the unit impulse is delayed by a time $t'$ so that it is applied at time $t = t'$ then the output is also delayed by the same period of time; i.e. an input $\delta(t - t')$ produces an output $h(t - t')$.

Consider an input $Q(t)$ that varies in the manner shown in Figure 55a, commencing at the instant $t = -a$. The function may be divided into $n$ steps of width $\Delta t$ such that at time $t = t_1$ the step has width $\Delta t_1$ and height $Q(t_1)$. As the value of $n$ increases, the element $Q(t_1)\Delta t_1$ at $t = t_1$ has the form of a modified impulse, having an elemental output given by

$$dq(t_1) = Q(t_1)\Delta t_1 h(t - t_1)$$

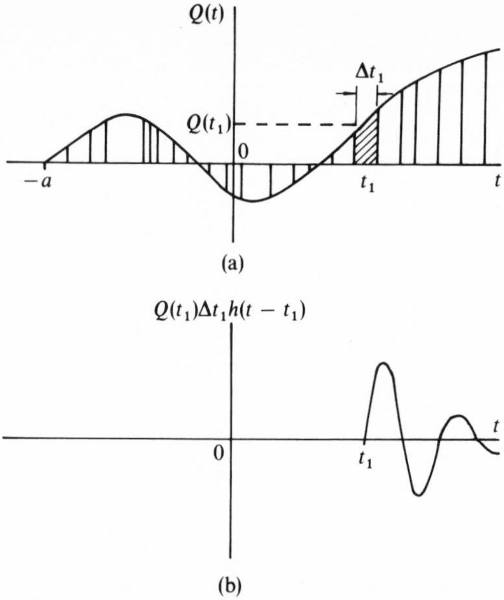


**Figure 55.** Variation of generalised force $Q(t)$, showing (a) its representation as a series of impulses, and (b) the output of the single impulse shown cross-hatched in (a).

as indicated in Figure 55b. Similarly, the element at $t = t_2$ has an elemental output

$$dq(t_2) = Q(t_2)\Delta t_2 h(t - t_2)$$

and by summing all these infinitesimal outputs due to all the elements in the interval $(-a, t)$ we have a resultant output at time $t$ given by

$$q(t) = \int_{-a}^{t} h(t - t')Q(t')dt'$$

which, by the change of variable $\tau = t - t'$, becomes

$$q(t) = \int_{0}^{t+a} h(\tau)Q(t - \tau)d\tau$$

However, since

$$h(\tau) = 0 \qquad \text{for } \tau < 0,$$

and

$$Q(t - \tau) = 0 \qquad \text{for } \tau > t,$$

it follows that the integrand is zero outside the limits of integration, and nothing is added to or subtracted from the integrals if the limits are replaced by $(-\infty, \infty)$. Thus the output,

$$q(t) = \int_{-\infty}^{\infty} h(\tau)Q(t-\tau)d\tau$$

of the linear time invariant system, is determined by a convolution or Duhamel-type integral involving the unit-impulse response and the input.

### 10.1.1 Excitation of a system with one degree of freedom

Consider the well known equation that governs the distortion $q(t)$ of a simple system with but one degree of freedom, *viz.*,

$$\ddot{q} + 2\nu\omega_1\dot{q} + \omega_1^2 q = Q(t).$$

Here $\nu$ is the damping factor, $\omega_1$ is the natural frequency, and $Q(t)$ is the excitation. It will be helpful to think of $Q(t)$ as an 'input' and $q(t)$ as an 'output'. If

$$Q(t) = Q_0 e^{i\omega t}$$

then

$$q(t) = q_0 e^{i\omega t} = H(\omega) Q_0 e^{i\omega t}$$

where the complex 'receptance' $H(\omega)$ *is*

$$H(\omega) = \frac{1}{\omega_1^2 - \omega^2 + 2i\nu\omega_1\omega}$$

so that, if the linear system is stable, a sinusoidal input produces a sinusoidal output of the same frequency, but with a different phase.

If the input is a unit impulse applied at the instant $t = 0$, so that $Q(t) = \delta(t)$, then the output is

$$q(t) = h(t) = \frac{\sin\sqrt{(1-\nu^2)}\omega_1 t}{m\omega_1\sqrt{(1-\nu^2)}}$$

where $m$ is the generalised mass associated with the coordinate $q$. The quantity $h(t)$ is sometimes referred to as the 'unit impulse response'; alternatively, it is sometimes written as $\alpha^{\delta}(t)$ and referred to as an 'impulsive receptance'.

As seen in the previous section, the function $h(t)$ provides a convenient method of determining the response to an input of any form, through the method of Duhamel's integral. Thus

$$q(t) = \int_{-\infty}^{\infty} h(\tau)Q(t-\tau)d\tau$$

where the expression for the output is a convolution integral.

Notice that if $Q(t) = Q_0 e^{i\omega t}$, the convolution integral becomes

$$q(t) = \int_{-\infty}^{\infty} h(\tau) Q_0 e^{i\omega(t-\tau)} d\tau$$

$$= \left\{ \int_{-\infty}^{\infty} h(\tau) e^{-i\omega\tau} d\tau \right\} Q_0 e^{i\omega t}.$$

That is to say

$$H(\omega) = \int_{-\infty}^{\infty} h(t) e^{-i\omega t} dt$$

so that $H(\omega)$ is the Fourier transform of $h(t)$. The inverse transform gives

$$h(t) = \frac{1}{2\pi} \int_{-\infty}^{\infty} H(\omega) e^{i\omega t} d\omega,$$

the functions $H(\omega)$ and $h(t)$ being a Fourier transform pair. On comparing these results with those given in section 6.2 it is seen that the factor $1/2\pi$ appears in the expression for $h(t)$ rather than that for $H(\omega)$. This is of no consequence and, indeed, our previous convention could be preserved by contemplating not a unit impulse, but one of magnitude $2\pi$. Notice also that our assumption that the system is stable ensures that the integral

$$\int_{-\infty}^{\infty} |h(t)| dt$$

is finite, thus ensuring that the Fourier transform pair $H(\omega)$ and $h(t)$ do exist.

### 10.1.2 Step input and step response

If instead of applying a unit impulse, we impose a step input such that

$$Q(t) = U(t) = \begin{cases} 1 & \text{for } t \geqslant 0 \\ 0 & \text{for } t < 0, \end{cases}$$

then the indicial output response, $u(t)$, is

$$u(t) = \int_{-\infty}^{\infty} h(\tau) U(t-\tau) d\tau$$

$$= \int_{-\infty}^{t} h(\tau) d\tau.$$

However, the output response to the unit impulse $Q(t) = \delta(t)$ is given in the previous section as

$$q(t) = h(t)$$

so by differentiation it follows that

$$h(t) = \frac{du}{dt} = \lim_{\delta t \to 0} \frac{u(t + \delta t) - u(t - \delta t)}{2\delta t}.$$

We have therefore found a relationship between the indicial and impulsive responses. Difficulties arise at $t = 0$ since, for physically realisable systems (or by the 'principle of causality'), $h(t) = 0$ and $u(t) = 0$ for $t < 0$. If $u(t)$ is discontinuous at $t = 0$ such that

$$u(0) = \lim_{\beta \to 0} u(0 + \beta) \neq 0$$

then the function $h(t)$ behaves in a similar manner at $t = 0$ to the Dirac delta function $\delta(t)$. Its value jumps to infinity and then decreases to a finite value or zero whereas

$$\lim_{\beta \to 0} \int_{0-\beta}^{0+\beta} h(t)dt = u(0).$$

Thus at $t = 0$ we have

$$h(0) = \lim_{\delta t \to 0} \frac{u(\delta t) - u(-\delta t)}{2\delta t} = u(0)\delta(t)$$

and this result may be incorporated into the response relationship, giving

$$h(t) = u(0)\delta(t) + \frac{du}{dt}.$$

### 10.1.3 Properties of impulse response functions

The Fourier transform of the impulse response function is given in section 10.1 by

$$H(\omega) = H^R(\omega) + iH^I(\omega) = \int_{-\infty}^{\infty} h(t)e^{-i\omega t}dt$$

where $H^R(\omega)$ and $H^I(\omega)$ represent the real and imaginary terms associated with $H(\omega)$. Equating real and imaginary parts gives

$$H^R(\omega) = \int_{-\infty}^{\infty} h(t)\cos \omega t \, dt$$

which is an even function in $\omega$, and

$$H^I(\omega) = -\int_{-\infty}^{\infty} h(t)\sin \omega t \, dt$$

which is an odd function in $\omega$. Alternatively, we may write

$$h(t) = \frac{1}{2\pi}\int_{-\infty}^{\infty} H(\omega)e^{i\omega t}d\omega$$

$$= \frac{1}{2\pi}\int_{-\infty}^{\infty} [\{H^R(\omega)\cos \omega t - H^I(\omega)\sin \omega t\} + i\{H^R(\omega)\sin \omega t + H^I(\omega)\cos \omega t\}]\, d\omega .$$

The contents of the second pair of curly brackets are an odd function of $\omega$ and the corresponding integral value is zero; therefore,

$$h(t) = \frac{1}{2\pi}\int_{-\infty}^{\infty} \{H^R(\omega)\cos \omega t - H^I(\omega)\sin \omega t\}\, d\omega .$$

It follows by the principle of causality (i.e. $h(t) = 0$ for $t < 0$), that

$$h(t) = \frac{1}{2\pi}\int_{-\infty}^{\infty} \{H^R(\omega)\cos \omega t - H^I(\omega)\sin \omega t\}\, d\omega = 0 \qquad \text{for } t < 0,$$

and by replacing $-t$ for $t$ we have

$$h(-t) = \frac{1}{2\pi}\int_{-\infty}^{\infty} \{H^R(\omega)\cos \omega t + H^I(\omega)\sin \omega t\}\, d\omega = 0 \qquad \text{for } t > 0.$$

By considering the terms $h(t) \pm h(-t)$, then, we see that either

$$h(t) = \frac{1}{\pi}\int_{-\infty}^{\infty} H^R(\omega)\cos \omega t \, d\omega = \frac{2}{\pi}\int_{0}^{\infty} H^R(\omega)\cos \omega t \, d\omega$$

or

$$h(t) = -\frac{1}{\pi}\int_{-\infty}^{\infty} H^I(\omega)\sin \omega t \, d\omega = -\frac{2}{\pi}\int_{0}^{\infty} H^I(\omega)\sin \omega t \, d\omega$$

which have the inverse Fourier transforms

$$H^R(\omega) = \int_{0}^{\infty} h(t)\cos \omega t \, dt \quad \text{and} \quad H^I(\omega) = -\int_{0}^{\infty} h(t)\sin \omega t \, dt.$$

Thus the impulse response function may be derived from a knowledge of either $H^R(\omega)$ or $H^I(\omega)$. Alternatively, knowing $h(t)$ then $H^R(\omega)$ and $H^I(\omega)$ may be calculated.

This result is of considerable practical value in the oscillatory testing of ship models. Suppose, for example, that a model is towed at constant speed along a tank and that a sinusoidal sway motion is imposed upon it by a planar motion mechanism. The sinusoidal hydrodynamic force can be measured, and the components of it that are in quadrature and in phase with the sway displacement may be separately identified. It can be shown[2, 3] that the quadrature component gives $H_v^R(\omega)$, the real part of the Fourier transform of $h_v(t)$, the sway force variation following a sudden unit step to port. Equally, the in-phase component gives $H_v^I(\omega)$, the imaginary part of the Fourier transform of $h_v(t)$. The results of actual measurements (e.g. see the work of van Leeuwen[4]) might give curves like those sketched in Figure 56a, b. We now see that such curves are not independent of one another.

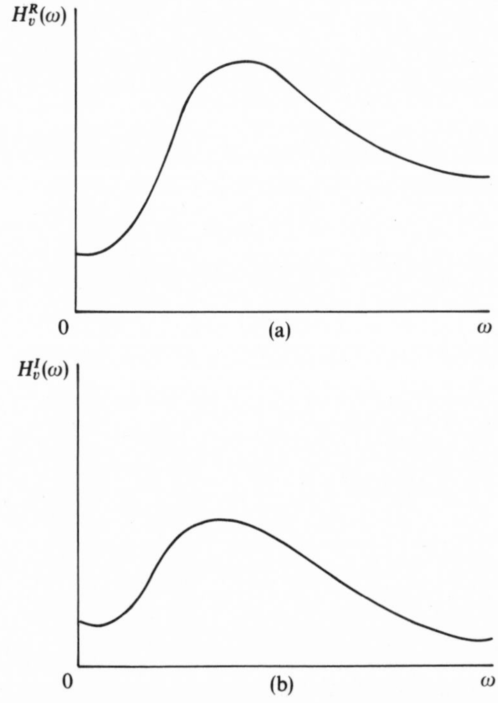


**Figure 56.** Possible forms of the real and imaginary parts of the Fourier transform for an impulse response function.

Suppose a curve of $H^R(\omega)$, as in Figure 56a has been found. After subtracting the value of $H^R(\omega)$ at $\omega = \infty$ the resultant experimental curve may be transformed to a unit response function. Such a derived function $h(t)$ is shown sketched in Figure 57. (The subtracted value of $H^R(\omega)$ contributes a delta function at $t = 0$ in $h(t)$.) Now by an inverse transformation, the other frequency curve $H^I(\omega)$ may be calculated, and it may be compared[3, 4] with the experimental curve of the type in Figure 56b. This may be done either with a view to checking the results or to checking their linearity (for all the foregoing theory rests on the assumption of linearity).

### 10.1.4 The Hilbert transform or Kramers-Kronig relationship

From the previous section it will be seen that the real and imaginary parts of the receptance may be expressed in the form

$$H^R(\omega) = \int_0^\infty h(t)\cos \omega t \, dt$$

$$= -\frac{2}{\pi}\int_0^\infty \cos \omega t \, dt \int_0^\infty H^I(\omega')\sin \omega' t \, d\omega' .$$

Since $H^I(\omega')$ is an odd function of $\omega'$,

$$\int_{-\infty}^\infty H^I(\omega')\sin \omega' t \cos \omega t \, d\omega' = 2\int_0^\infty H^I(\omega')\sin \omega' t \cos \omega t \, d\omega'$$

$$\int_{-\infty}^\infty H^I(\omega')\cos \omega' t \sin \omega t \, d\omega' = 0 .$$

After subtraction, these integrals give

$$\int_{-\infty}^\infty H^I(\omega')\sin(\omega' - \omega)t \, d\omega' = 2\int_0^\infty H^I(\omega')\sin \omega' t \cos \omega t \, d\omega' .$$

Substituting this result into the above expression for $H^R(\omega)$, we find that

$$H^R(\omega) = -\frac{1}{\pi}\int_{-\infty}^\infty H^I(\omega')\int_0^\infty \sin(\omega' - \omega)t \, dt \, d\omega'$$

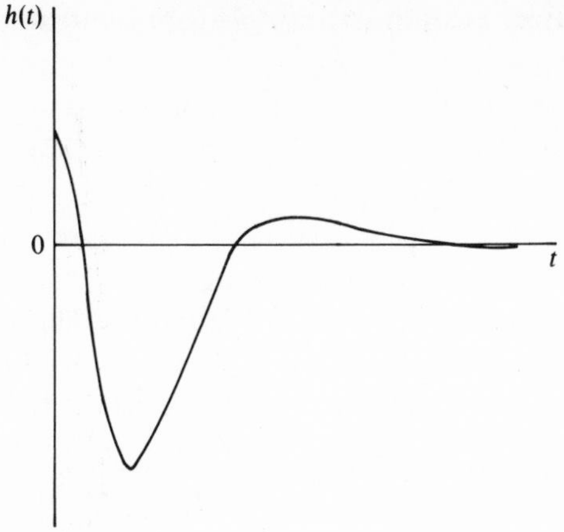


**Figure 57.** Sketch of a possible impulse response function. Note that in practice a function of this sort may be augmented by a delta function at $t = 0$.

or, after integrating with respect to $t$,

$$H^R(\omega) = -\frac{1}{\pi}\int_{-\infty}^{\infty} H^I(\omega') \lim_{t\to\infty}\left\{\frac{1-\cos(\omega'-\omega)t}{(\omega'-\omega)}\right\} d\omega'$$

$$= -\frac{1}{\pi}\int_{-\infty}^{\infty}\frac{H^I(\omega')d\omega'}{\omega'-\omega} + \frac{1}{\pi}\lim_{t\to\infty}\int_{-\infty}^{\infty}\frac{H^I(\omega')\cos(\omega'-\omega)t}{\omega'-\omega}d\omega'.$$

However by the Riemann-Lebesgue[5] lemma, the second integral is zero in the limit as $t \to \infty$. Hence,

$$H^R(\omega) = -\frac{1}{\pi}\int_{-\infty}^{\infty}\frac{H^I(\omega')}{\omega'-\omega}d\omega'.$$

with an inverse transformation (obtained by repeating the above procedure) given by

$$H^I(\omega) = \frac{1}{\pi}\int_{-\infty}^{\infty}\frac{H^R(\omega')}{\omega'-\omega}d\omega'.$$

These results are valid for any physically realisable system where $h(t) = 0$ for $t < 0$. The relationships between the frequency characteristics $H^R(\omega)$ and $H^I(\omega)$ are referred to as the 'Kramers-Kronig relationships' or 'Hilbert transforms'.

## 10.2 General linear system with single input and single output

By referring specifically to a system with only one degree of freedom, we have been able to illustrate the basic ideas of a receptance and its corresponding unit impulse response. There is, however, nothing in those ideas that *requires* the system to be so highly specialised. In fact we may refer to *any* time invariant, stable linear system and think of its 'input' $Q(t)$ as the generalised force applied at some generalised coordinate and its 'output' as the response at the same, or some other, generalised coordinate. In the language of vibration theory,

$H(\omega)$ is a 'direct receptance'
$h(t)$ is an 'impulsive direct receptance'

if the input and output 'correspond' to each other in the Lagrangian sense, while

$H(\omega)$ is a 'cross receptance'
$h(t)$ is an 'impulsive cross receptance'

if they do not so correspond.

The system to which we shall refer henceforth may be represented as in Figure 58. The requirement that we place on it is that it is time invariant, that its motions shall be governed by linear equations and that it shall be stable.

### 10.2.1 Input-output relations for random processes

Suppose that the input $Q(t)$ to the linear system is a random process. The output $q(t)$ is also a random process and we now begin to examine the question of what can be stated about the output in terms of statistical averages of the input.

The expected value of $q(t)$ is

$$E[q(t)] = E\left[\int_{-\infty}^{\infty} h(\tau)Q(t-\tau)d\tau\right].$$

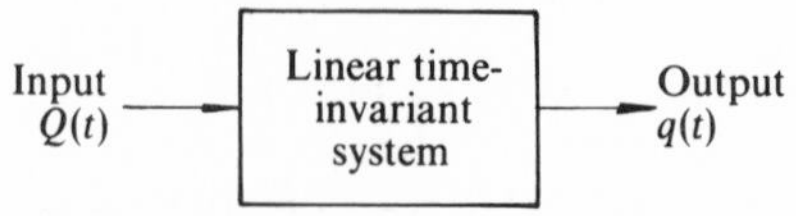


**Figure 58.** Block representation of a time invariant system.

If we reverse the order of integration and averaging we find that

$$E[q(t)] = \int_{-\infty}^{\infty} h(\tau)E[Q(t-\tau)]\,d\tau.$$

If the input process is stationary, $E[Q(t-\tau)]$ is a constant, so that

$$E[q(t)] = \int_{-\infty}^{\infty} h(\tau)d\tau\; E[Q(t)]$$

$$= H(0)E[Q(t)].$$

Instead of finding the expected value, we could find the temporal means of single realisations $Q^{(1)}(t)$ and $q^{(1)}(t)$:

$$\langle q^{(1)}(t)\rangle = \lim_{T\to\infty} \frac{1}{T} \int_{-T/2}^{T/2} q^{(1)}(t)dt$$

$$= \lim_{T\to\infty} \frac{1}{T} \int_{-T/2}^{T/2} \left\{ \int_{-\infty}^{\infty} h(\tau)Q^{(1)}(t-\tau)d\tau \right\} dt$$

$$= \int_{-\infty}^{\infty} h(\tau)\langle Q^{(1)}(t-\tau)\rangle\, d\tau$$

$$= \langle Q^{(1)}(t)\rangle \int_{-\infty}^{\infty} h(\tau)d\tau.$$

That is to say

$$\langle q^{(1)}(t)\rangle = H(0)\langle Q^{(1)}(t)\rangle.$$

If the random process $Q(t)$ is ergodic as regards its mean value,

$$E[Q(t)] = \langle Q^{(1)}(t)\rangle.$$

It follows from the foregoing results, therefore, that

$$E[q(t)] = \langle q^{(1)}(t)\rangle.$$

Under this circumstance it is meaningful to drop the superscript (1) from the time averages and to refer to $\langle Q(t)\rangle$ and $\langle q(t)\rangle$ as the input and output mean values.

We turn next to a more complicated average, the auto-correlation function. We may write

$$q(t)q(t+\tau) = \int_{-\infty}^{\infty} h(\tau_1)Q(t-\tau_1)d\tau_1 \int_{-\infty}^{\infty} h(\tau_2)Q(t+\tau-\tau_2)d\tau_2$$

where the dummy variables have been written as $\tau_1$ and $\tau_2$ so as to preserve their identities. When the integrals concerned are suitably well-behaved, the product of two integrals may be written as a double integral. That is

$$q(t)q(t+\tau) = \int_{-\infty}^{\infty}\int_{-\infty}^{\infty} Q(t-\tau_1)Q(t+\tau-\tau_2)h(\tau_1)h(\tau_2)d\tau_1\, d\tau_2.$$

Consider now the expectation of each side of this equation and interchange the order of integration and averaging on the right hand side. We know that, if the input process is stationary,

$$E[Q(t-\tau_1)Q(t+\tau-\tau_2)] = R_{QQ}(\tau+\tau_1-\tau_2)$$

so that

$$E[q(t)q(t+\tau)] = \int_{-\infty}^{\infty}\int_{-\infty}^{\infty} R_{QQ}(\tau+\tau_1-\tau_2)h(\tau_1)h(\tau_2)d\tau_1\, d\tau_2$$

$$= R_{qq}(\tau).$$

That is to say, the output is also stationary.

This argument can be extended to cover *all* the statistical averages (i.e. the expectations) of the output of the linear stable system. They are unchanged by a shift of the time scale when the input is a stationary process. The output is also stationary, therefore.

Instead of forming the expectations, let us consider the temporal average of single realisations '(1)' of input and output. The temporal auto-correlation function of the output is

$$\langle q^{(1)}(t)q^{(1)}(t+\tau)\rangle$$
$$= \left\langle \int_{-\infty}^{\infty} h(\tau_1)Q^{(1)}(t-\tau_1)d\tau_1 \cdot \int_{-\infty}^{\infty} h(\tau_2)Q^{(1)}(t+\tau-\tau_2)d\tau_2 \right\rangle.$$

The time average implies integration with respect to $t$ so that, by changing the order of integration, we find

$$\langle q^{(1)}(t)q^{(1)}(t+\tau)\rangle$$
$$= \int_{-\infty}^{\infty} h(\tau_1)\left\{ \int_{-\infty}^{\infty} h(\tau_2)\langle Q^{(1)}(t-\tau_1)Q^{(1)}(t+\tau-\tau_2)\rangle d\tau_2\right\} d\tau_1.$$

The fact that this expression depends on $\tau$ and not on $t$ confirms that we may again write

$$R_{qq}^{(1)}(\tau) = \int_{-\infty}^{\infty}\int_{-\infty}^{\infty} R_{QQ}^{(1)}(\tau + \tau_1 - \tau_2)h(\tau_1)h(\tau_2)d\tau_1\, d\tau_2$$

where $R_{qq}^{(1)}(\tau)$ and $R_{QQ}^{(1)}(\tau)$ are now the output and input *temporal* auto-correlation functions of the known realisation '(1)'.

It can be shown in a similar fashion that *all* the temporal averages of output are related to those of a stationary input in the same way as are the expectations.

In general, temporal averages differ from realisation to realisation and none is equal to the corresponding expectation of the random process. But if $Q(t)$ is an ergodic process in its expected value, then the corresponding expectation and the temporal average of *any* realisation are equal. It follows now that the same will be true for the process $q(t)$. In other words, an ergodic input produces an ergodic output.

The mean square spectral densities of $Q(t)$ and $q(t)$ are respectively

$$S_{QQ}(\omega) = \frac{1}{2\pi}\int_{-\infty}^{\infty} R_{QQ}(\tau)e^{-i\omega\tau}d\tau$$

and

$$S_{qq}(\omega) = \frac{1}{2\pi}\int_{-\infty}^{\infty} R_{qq}(\tau)e^{-i\omega\tau}d\tau$$

$$= \frac{1}{2\pi}\int_{-\infty}^{\infty}\left\{\int_{-\infty}^{\infty}\int_{-\infty}^{\infty} R_{QQ}(\tau + \tau_1 - \tau_2)h(\tau_1)h(\tau_2)d\tau_1\, d\tau_2\right\}e^{-i\omega\tau}d\tau$$

$$= \int_{-\infty}^{\infty} h(\tau_1)e^{i\omega\tau_1}d\tau_1 \int_{-\infty}^{\infty} h(\tau_2)e^{-i\omega\tau_2}d\tau_2 \times$$

$$\left\{\frac{1}{2\pi}\int_{-\infty}^{\infty} R_{QQ}(\tau + \tau_1 - \tau_2)e^{-i\omega(\tau+\tau_1-\tau_2)}d(\tau + \tau_1 - \tau_2)\right\}$$

$$= H^*(\omega)\,.\,H(\omega)\,.\,S_{QQ}(\omega)$$

$$= |H(\omega)|^2 S_{QQ}(\omega)$$

since $H^*(\omega)$ is the complex conjugate of $H(\omega)$. If the mean square spectral density of the input is known, therefore, that of the output may be calculated provided the modulus of the receptance is known. Notice that by using the spectra, we disclaim all interest in phase relationships between input and output. For a physically realisable system we may express this result as

$$\Phi_{qq}(\omega) = |H(\omega)|^2 \Phi_{QQ}(\omega)$$

where

$$\left.\begin{aligned}\Phi_{qq}(\omega) &= 2S_{qq}(\omega)\\ \Phi_{QQ}(\omega) &= 2S_{QQ}(\omega)\end{aligned}\right\}$$

for $\omega \geqslant 0$, and

$$\Phi_{qq}(\omega) = \Phi_{QQ}(\omega) = 0$$

for $\omega < 0$. Notice that all the spectral densities $S$ and $\Phi$ are real functions of $\omega$;

The mean square of the output is

$$E[q^2(t)] = \int_{-\infty}^{\infty} S_{qq}(\omega)d\omega = \int_{-\infty}^{\infty} |H(\omega)|^2 S_{QQ}(\omega)d\omega.$$

Alternatively, for realisation '(1)',

$$\langle q^{(1)2}(t)\rangle = \int_{-\infty}^{\infty} |H(\omega)|^2 S_{QQ}^{(1)}(\omega)d\omega.$$

If the process is ergodic in the mean square, the two averages are equal and we may legitimately drop the superscript (1).

Since there are now two related random processes under discussion, we may consider the correlation between them, through the cross-correlation function. For simplicity, suppose that $Q(t)$ (and hence $q(t)$) is an ergodic process so that a temporal average need not be identified with any particular realisation. The cross-correlation function is

$$\begin{aligned}R_{Qq}(\tau) &= \langle Q(t)q(t+\tau)\rangle\\ &= \int_{-\infty}^{\infty} h(\tau_1)\langle Q(t)Q(t+\tau-\tau_1)\rangle d\tau_1\\ &= \int_{-\infty}^{\infty} h(\tau_1)R_{QQ}(\tau-\tau_1)d\tau_1.\end{aligned}$$

In words, then, the cross-correlation function is the convolution of the impulsive response and the auto-correlation function of the input process.

If the Fourier transform of both sides of this equation is taken, it is found that

$$\frac{1}{2\pi}\int_{-\infty}^{\infty} R_{Qq}(\tau)e^{-i\omega\tau}d\tau$$

$$=\frac{1}{2\pi}\int_{-\infty}^{\infty} h(\tau_1)e^{-i\omega\tau_1}d\tau_1 \int_{-\infty}^{\infty} R_{QQ}(\tau-\tau_1)e^{-i\omega(\tau-\tau_1)}d(\tau-\tau_1)$$

which reduces to

$$S_{Qq}(\omega)=H(\omega)\,.\,S_{QQ}(\omega).$$

This complex cross spectral density $S_{Qq}(\omega)$ of the linear system is the product of the *complex* receptance and the mean square spectral density of the input. Notice that if $S_{Qq}(\omega)$ and $S_{QQ}(\omega)$ are known for all $\omega$, then $H(\omega)$ may be determined.

Similarly, it may be shown that the cross correlation function of the output-input is given by

$$S_{qQ}(\omega)=H^*(\omega)\,.\,S_{QQ}(\omega)$$

and hence

$$S_{Qq}(\omega)=S^*_{qQ}(\omega)=S_{qQ}(-\omega),$$

forming a complex conjugate pair of functions. Whereas mean square spectral densities are real functions, cross correlation functions are in general complex.

The complex cross-spectral density function of the input-output can be expressed as

$$S_{Qq}(\omega)=C_{Qq}(\omega)-iQ_{Qq}(\omega)$$

$$=|S_{Qq}(\omega)|e^{-i\epsilon_{Qq}(\omega)}$$

where

$$|S_{Qq}(\omega)|=\surd\{C^2_{Qq}(\omega)+Q^2_{Qq}(\omega)\}$$

and

$$\epsilon_{Qq}(\omega)=\tan^{-1}\left[\frac{Q_{Qq}(\omega)}{C_{Qq}(\omega)}\right].$$

The real part $C_{Qq}(\omega)$ is called the 'co-spectral density function' and the imaginary part $Q_{Qq}(\omega)$ is called the 'quadrature spectral density function' (see section 6.4).

Figure 59 illustrates the magnitude and phase relationship described by the cross-spectral density for a pair of input-output random processes. The receptance of the system can be defined as

$$H(\omega)=\frac{S_{Qq}(\omega)}{S_{QQ}(\omega)}=\frac{|S_{Qq}(\omega)|}{S_{QQ}(\omega)}e^{-i\epsilon_{Qq}(\omega)}$$

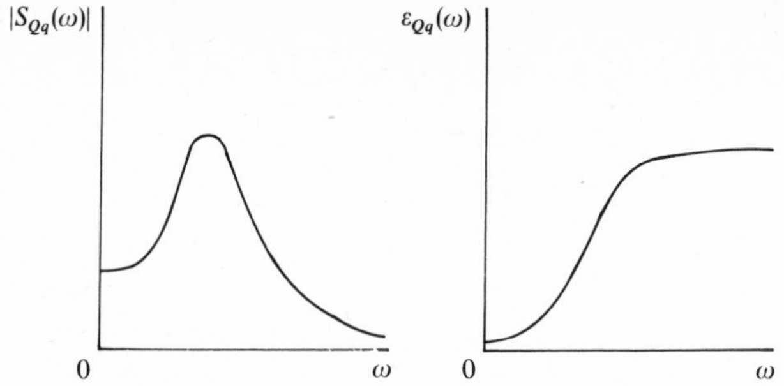


**Figure 59.** Relationships between random processes of input and output showing possible forms of (a) the cross spectral density, and (b) phase difference.

so that the magnitude of the receptance is given by

$$|H(\omega)| = \frac{|S_{Qq}(\omega)|}{S_{QQ}(\omega)}$$

and $\epsilon_{Qq}(\omega)$ describes the probabilistic phase relationship between input and output.

By means of the inverse Fourier transform, we may express the impulse response function in the form

$$h(t) = \frac{1}{2\pi} \int_{-\infty}^{\infty} \frac{S_{Qq}(\omega)}{S_{QQ}(\omega)} e^{i\omega t} d\omega.$$

This is a real function of time.

If the input is a wide band process such that

$$S_{QQ}(\omega) = S_0 \qquad \text{for all } \omega$$

then the cross-correlation function between input and output becomes

$$R_{Qq}(\tau) = \int_{-\infty}^{\infty} S_{Qq}(\omega) e^{i\omega\tau} d\omega$$

$$= S_0 \int_{-\infty}^{\infty} H(\omega) e^{i\omega\tau} d\omega$$

$$= 2\pi S_0 h(\tau),$$

indicating that the input-output cross-correlation function of a linear system under white noise input is proportional to the system's impulse response function.

### 10.2.2 Input-output relations for derivatives

The output $q(t)$ is a random process of generalised displacement. Under certain circumstances we may be interested in its derivatives $\dot{q}(t)$ and $\ddot{q}(t)$, rather than the response itself. Probably the two most important of the relevant statistical averages are the mean square spectral density $S_{\dot{q}\dot{q}}(\omega)$ of the output $\dot{q}(t)$ and the mean square spectral density $S_{\ddot{q}\ddot{q}}(\omega)$ of the output $\ddot{q}(t)$. These we shall now determine under the assumption that the input is ergodic.

Consider first the output random process $\dot{q}(t)$. The autocorrelation function is

$$R_{\dot{q}\dot{q}}(\tau) = \langle \dot{q}(t)\dot{q}(t+\tau)\rangle .$$

But we saw in section 5.4 that

$$\ddot{R}_{qq}(\tau) = -\langle \dot{q}(t-\tau)\dot{q}(t)\rangle = -\langle \dot{q}(t)\dot{q}(t+\tau)\rangle$$

so that

$$R_{\dot{q}\dot{q}}(\tau) = -\ddot{R}_{qq}(\tau).$$

Now since

$$R_{qq}(\tau) = \int_{-\infty}^{\infty}\int_{-\infty}^{\infty} R_{QQ}(\tau+\tau_1-\tau_2)h(\tau_1)h(\tau_2)d\tau_1\, d\tau_2,$$

by differentiating twice with respect to $\tau$ and using the previous results, we find

$$R_{\dot{q}\dot{q}}(\tau) = -\int_{-\infty}^{\infty}\int_{-\infty}^{\infty} \ddot{R}_{QQ}(\tau+\tau_1-\tau_2)h(\tau_1)h(\tau_2)d\tau_1\, d\tau_2.$$

However, we know that

$$R_{QQ}(\tau) = \int_{-\infty}^{\infty} S_{QQ}(\omega)e^{i\omega\tau}d\omega$$

so that

$$\ddot{R}_{QQ}(\tau) = -\int_{-\infty}^{\infty} \omega^2 S_{QQ}(\omega)e^{i\omega\tau}d\omega$$

and the Fourier transform is

$$\omega^2 S_{QQ}(\omega) = -\frac{1}{2\pi}\int_{-\infty}^{\infty} \ddot{R}_{QQ}(\tau)e^{-i\omega\tau}d\tau.$$

The mean square spectral density of the output $\dot{q}(t)$ is given by

$$S_{\dot{q}\dot{q}}(\omega) = \frac{1}{2\pi}\int_{-\infty}^{\infty} R_{\dot{q}\dot{q}}(\tau)e^{-i\omega\tau}d\tau$$

and the foregoing results now give

$$S_{\dot{q}\dot{q}}(\omega) = \int_{-\infty}^{\infty} h(\tau_1)e^{i\omega\tau_1}d\tau_1 \,.\, \int_{-\infty}^{\infty} h(\tau_2)e^{-i\omega\tau_2}d\tau_2 \,.$$

$$\left\{-\frac{1}{2\pi}\int_{-\infty}^{\infty} \ddot{R}_{QQ}(\tau+\tau_1-\tau_2)e^{-i\omega(\tau+\tau_1-\tau_2)}d(\tau+\tau_1-\tau_2)\right\}$$

$$= H^*(\omega)\,.\,H(\omega)\,.\,\omega^2 S_{QQ}(\omega).$$

Therefore

$$S_{\dot{q}\dot{q}}(\omega) = |H(\omega)|^2\omega^2 S_{QQ}(\omega) = \omega^2 S_{qq}(\omega).$$

Similarly the mean square spectral density of the output process $\ddot{q}(t)$ is

$$S_{\ddot{q}\ddot{q}}(\omega) = |H(\omega)|^2\omega^4 S_{QQ}(\omega) = \omega^4 S_{qq}(\omega).$$

### 10.2.3 Gaussian random processes

If the input process $Q(t)$ has a Gaussian probability density function, so too has the output process $q(t)$. The mean and mean square of the input determine the mean and mean square of the output and, hence, of the whole process. A simple illustration that a Gaussian distribution persists under a linear transformation is given in section 4.3.

If on the other hand the input is not Gaussian, the mean and mean square of the output can be obtained as we have shown. But this is not sufficient to determine the output process in much detail.

## 10.3 Multiple input and output

Instead of the system of Figure 58, we now consider the arrangement shown in Figure 60. There are random inputs $Q_1(t)$, $Q_2(t)$, . . . , $Q_m(t)$ and $n$ random outputs $q_1(t)$, $q_2(t)$, . . . , $q_n(t)$. It will be convenient to assemble them in the column matrices

$$\mathbf{Q}(t) = \begin{bmatrix} Q_1(t) \\ Q_2(t) \\ \vdots \\ Q_m(t) \end{bmatrix} \qquad \mathbf{q}(t) = \begin{bmatrix} q_1(t) \\ q_2(t) \\ \vdots \\ q_n(t) \end{bmatrix}$$

It is a fundamental property of linear systems that the separate responses due to a number of excitations may be superposed when those excitations act simultaneously. The following generalisations of the theory merely reflect this property.

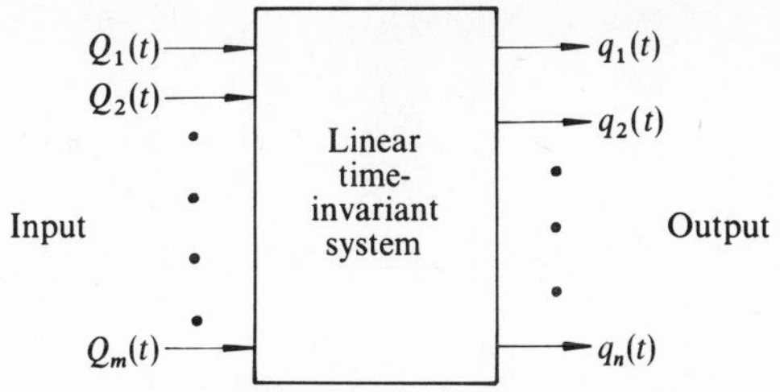


**Figure 60.** Block diagram of a time invariant system with multiple inputs and outputs.

Instead of a single impulse response function (or impulsive receptance) we now have an array of such functions, given by $\mathbf{h}(t)$, the 'impulsive response matrix' (or 'impulsive receptance matrix') which is of order $nxm$. The output matrix is given by the convolution integral

$$\mathrm{q}(t) = \int_{-\infty}^{\infty} \mathrm{h}(\tau)\mathrm{Q}(t-\tau)d\tau.$$

The output auto-correlation matrix is defined as

$$\mathbf{R}_{\mathbf{q}\mathbf{q}}(\tau) = \langle \mathrm{q}(t)\mathrm{q}^T(t+\tau)\rangle$$

where $\mathrm{q}^T(t)$ is the transpose of the matrix $\mathbf{q}(t)$. It is a square matrix of the form

$$\mathbf{R}_{\mathbf{q}\mathbf{q}}(\tau) = \begin{bmatrix} R_{q_1q_1}(\tau) & R_{q_1q_2}(\tau) & \cdots & R_{q_1q_n}(\tau) \\ R_{q_2q_1}(\tau) & R_{q_2q_2}(\tau) & \cdots & R_{q_2q_n}(\tau) \\ \cdots & \cdots & \cdots & \cdots \\ R_{q_nq_1}(\tau) & R_{q_nq_2}(\tau) & \cdots & R_{q_nq_n}(\tau) \end{bmatrix}$$

where

$$R_{q_iq_j}(\tau) = \langle q_i(t)q_j(t+\tau)\rangle$$

for $i, j = 1, 2, \ldots, n$. In terms of the impulsive response matrix, we have

$$\mathbf{R}_{\mathbf{q}\mathbf{q}}(\tau) = \left\langle \int_{-\infty}^{\infty} \mathrm{h}(\tau_1)\mathrm{Q}(t-\tau_1)d\tau_1 \,.\, \int_{-\infty}^{\infty} \mathrm{Q}^T(t+\tau-\tau_2)\mathrm{h}^T(\tau_2)d\tau_2 \right\rangle$$

$$= \int_{-\infty}^{\infty} \mathrm{h}(\tau_1) \int_{-\infty}^{\infty} \langle \mathrm{Q}(t-\tau_1)\mathrm{Q}^T(t+\tau-\tau_2)\rangle \mathrm{h}^T(\tau_2)d\tau_2\, d\tau_1$$

$$= \int_{-\infty}^{\infty} \mathrm{h}(\tau_1) \int_{-\infty}^{\infty} \mathbf{R}_{\mathbf{Q}\mathbf{Q}}(\tau+\tau_1-\tau_2)\mathrm{h}^T(\tau_2)d\tau_2\, d\tau_1.$$

The output mean square spectral density matrix is defined to be

$$\mathbf{S}_{\mathbf{q}\mathbf{q}}(\omega) = \frac{1}{2\pi} \int_{-\infty}^{\infty} \mathbf{R}_{\mathbf{q}\mathbf{q}}(\tau) e^{-i\omega\tau} d\tau.$$

If we now substitute for $\mathbf{R}_{\mathbf{q}\mathbf{q}}(\tau)$, we find that

$$\mathbf{S}_{\mathbf{q}\mathbf{q}}(\omega) = \int_{-\infty}^{\infty} \mathbf{h}(\tau_1) e^{i\omega\tau_1} d\tau_1 \left\{ \frac{1}{2\pi} \int_{-\infty}^{\infty} \mathbf{R}_{\mathbf{Q}\mathbf{Q}}(\tau + \tau_1 - \tau_2) \right.$$

$$\left. e^{-i\omega(\tau+\tau_1-\tau_2)} d(\tau + \tau_1 - \tau_2) \right\} \int_{-\infty}^{\infty} \mathbf{h}^T(\tau_2) e^{-i\omega\tau_2} d\tau_2$$

$$= \mathbf{H}^*(\omega) \mathbf{S}_{\mathbf{Q}\mathbf{Q}}(\omega) \mathbf{H}^T(\omega)$$

where $\mathbf{H}^*(\omega)$ and $\mathbf{H}^T(\omega)$ are respectively the complex conjugate and transpose of the receptance matrix $\mathbf{H}(\omega)$.

Suppose that there are two inputs $Q_1(t)$, $Q_2(t)$ and two outputs $q_1(t)$, $q_2(t)$. We then have

$$\begin{bmatrix} S_{q_1 q_1} & S_{q_1 q_2} \\ S_{q_2 q_1} & S_{q_2 q_2} \end{bmatrix} = \begin{bmatrix} H^*_{q_1 Q_1} & H^*_{q_1 Q_2} \\ H^*_{q_2 Q_1} & H^*_{q_2 Q_2} \end{bmatrix} \begin{bmatrix} S_{Q_1 Q_1} & S_{Q_1 Q_2} \\ S_{Q_2 Q_1} & S_{Q_2 Q_2} \end{bmatrix} \times$$

$$\begin{bmatrix} H_{q_1 Q_1} & H_{q_2 Q_1} \\ H_{q_1 Q_2} & H_{q_2 Q_2} \end{bmatrix}$$

where, for $i, j = 1, 2$,

$S_{q_i q_j}$, etc. are the spectral densities of the outputs $q_i(t)$ and $q_j(t)$
$S_{Q_i Q_j}$, etc. are the spectral densities of the inputs $Q_i(t)$ and $Q_j(t)$
$H_{q_i Q_j}(\omega)$ is the output at $q_i$ for a unit sinusoidal input $Q_j$

To return to the general case, the *mxn* cross-correlation input-output matrix is

$$\mathbf{R}_{\mathbf{Q}\mathbf{q}}(\tau) = \langle \mathbf{Q}(t) \mathbf{q}^T(t + \tau) \rangle$$

$$= \langle \mathbf{Q}(t) \int_{-\infty}^{\infty} \mathbf{Q}^T(t + \tau - \tau_1) \mathbf{h}^T(\tau_1) d\tau_1 \rangle$$

$$= \int_{-\infty}^{\infty} \mathbf{R}_{\mathbf{Q}\mathbf{Q}}(\tau - \tau_1) \mathbf{h}^T(\tau_1) d\tau_1.$$

The cross spectral density matrix is defined as

$$\mathbf{S}_{\mathbf{Q}\mathbf{q}}(\omega) = \frac{1}{2\pi}\int_{-\infty}^{\infty} \mathbf{R}_{\mathbf{Q}\mathbf{q}}(\tau)e^{-i\omega\tau}d\tau$$

so that

$$\mathbf{S}_{\mathbf{Q}\mathbf{q}}(\omega) = \frac{1}{2\pi}\int_{-\infty}^{\infty} \mathbf{R}_{\mathbf{Q}\mathbf{Q}}(\tau - \tau_1)e^{-i\omega(\tau-\tau_1)}d(\tau - \tau_1)\int_{-\infty}^{\infty} \mathbf{h}^T(\tau_1)e^{-i\omega\tau_1}d\tau_1$$

whence

$$\mathbf{S}_{\mathbf{Q}\mathbf{q}}(\omega) = \mathbf{S}_{\mathbf{Q}\mathbf{Q}}(\omega)\,.\,\mathbf{H}^T(\omega).$$

Thus if $\mathbf{S}_{\mathbf{Q}\mathbf{q}}(\omega)$ and $\mathbf{S}_{\mathbf{Q}\mathbf{Q}}(\omega)$ are known for all frequencies, the receptances can be determined since

$$\mathbf{H}^T(\omega) = \mathbf{S}_{\mathbf{Q}\mathbf{Q}}^{-1}(\omega)\mathbf{S}_{\mathbf{Q}\mathbf{q}}(\omega).$$

## 10.4 Dependence on position and time

The vibration of a linear system having a finite number of degrees of freedom is governed by one or more ordinary differential equations. If the system is continuous, like a simply supported beam for instance, the motion is governed by a partial differential equation. In this latter case, since the output may be a deflection and is dependent upon time and position, it will have the form $q(x, t)$ where $x$ is a space variable (like distance from one support). The input might be a transverse force and if its position as well as its magnitude varies, it will be of the form $Q(x, t)$. The essential point is that introduction of the space variable does not necessarily impair the *linearity* of the system.

Consider a linear system with a single input and single output, such that both are functions of position $(x, y)$ and time $t$. For the sake of definiteness, one might think of the small, linear forced vibrations of a flat plate. Then $Q(x, y, t)$ can be a transverse force and $q(x, y, t)$ might be the deflection. Let the system be excited by the sinusoidal input

$$Q(x, y, t) = Q_0 e^{i(k_x x + k_y y - \omega t)}$$

and suppose that this produces a response

$$q(x, y, t) = H(k_x, k_y, \omega)Q_0 e^{i(k_x x + k_y y - \omega t)}$$

where the quantity $H(k_x, k_y, \omega)$ is merely an extended form of receptance.

By adopting the vector notation of section 7.1, we can abbreviate this last result to

$$q(\mathbf{r}, t) = H(\mathbf{k}, \omega) \,.\, Q(\mathbf{r}, t)$$

on the understanding that only sinusoidal relationships are admitted. The receptance $H(\mathbf{k}, \omega)$ gives the response output due to a sinusoidal excitation of unit amplitude, vectorial wave number $\mathbf{k}$ and frequency $\omega$.

Proceeding as before, we now consider a unit impulse $\delta(\mathbf{r}, t)$. Let the input $Q(x, y, t)$ be a stress $\sigma$ acting on an area $\Delta A$ that is centred about the point $x = 0 = y$ (or $\mathbf{r} = 0$) and suppose that it acts at time $t = 0$ for a period $\Delta t$. The impulse delivered to the system is of magnitude $\sigma \Delta A \Delta t = 1$. Now suppose that $\sigma \to \infty$ as $\Delta A \to 0$ and $\Delta t \to 0$ such that the product remains of unit magnitude. That is what we mean by the unit impulse $\delta(\mathbf{r}, t)$. Let the response to the unit impulse be $h(\mathbf{r}, t)$, such that $h(\mathbf{r}, t) = 0$ everywhere before the instant $t = 0$.

An input $Q(\mathbf{r}, t)$ may be thought of as being composed of contributions each of which acts on a small area for a short interval of time. The output at a point $\mathbf{r}$ at time $t$ due to that contribution which acts at the point $\mathbf{r} = \mathbf{r}'$ and at time $t = t'$ is

$$dq(\mathbf{r}, t) = Q(\mathbf{r}', t') \,.\, h(\mathbf{r} - \mathbf{r}', t - t')d\mathbf{r}'dt$$

Summing all such contributions from the instant $t = 0$ when the disturbance commenced, we find that

$$q(\mathbf{r}, t) = \int_{-\infty}^{\infty} \int_{-\infty}^{\infty} \int_{0}^{t} h(\mathbf{r} - \mathbf{r}', t - t')Q(\mathbf{r}', t')d\mathbf{r}'\, dt'.$$

Let $\boldsymbol{\rho} = \mathbf{r} - \mathbf{r}'$, $\tau = t - t'$ so that $d\boldsymbol{\rho} = -\, d\mathbf{r}'$ and $d\tau = -\, dt'$. We now have

$$q(\mathbf{r}, t) = \int_{-\infty}^{\infty} \int_{-\infty}^{\infty} \int_{0}^{t} h(\boldsymbol{\rho}, \tau)Q(\mathbf{r} - \boldsymbol{\rho}, t - \tau)d\boldsymbol{\rho}\, d\tau.$$

Since $h(\boldsymbol{\rho}, \tau) = 0$ when $\tau < 0$, this result may be written without loss of generality in the form

$$q(\mathbf{r}, t) = \int_{-\infty}^{\infty} \int_{-\infty}^{\infty} \int_{-\infty}^{\infty} h(\boldsymbol{\rho}, \tau)Q(\mathbf{r} - \boldsymbol{\rho}, t - \tau)d\boldsymbol{\rho}\, d\tau.$$

It is of interest to consider the particular input

$$Q(\mathbf{r}, t) = Q_0 e^{i(\mathbf{k}.\mathbf{r} - \omega t)}.$$

The foregoing result gives

$$q(\mathbf{r}, t) = Q(\mathbf{r}, t) \int_{-\infty}^{\infty} \int_{-\infty}^{\infty} \int_{-\infty}^{\infty} h(\boldsymbol{\rho}, \tau) e^{-i(\mathbf{k}.\boldsymbol{\rho} - \omega\tau)} d\boldsymbol{\rho}\, d\tau$$

so that a receptance $H(\mathbf{k}, \omega)$ is defined, namely

$$H(\mathbf{k}, \omega) = \int_{-\infty}^{\infty} \int_{-\infty}^{\infty} \int_{-\infty}^{\infty} h(\boldsymbol{\rho}, \tau) e^{-i(\mathbf{k}.\boldsymbol{\rho} - \omega\tau)} d\boldsymbol{\rho}\, d\tau.$$

Referring to section 8.1.1, we see that the inverse Fourier transform is

$$h(\boldsymbol{\rho}, \tau) = \frac{1}{8\pi^3} \int_{-\infty}^{\infty} \int_{-\infty}^{\infty} \int_{-\infty}^{\infty} H(\mathbf{k}, \omega) e^{i(\mathbf{k}.\boldsymbol{\rho} - \omega\tau)} d\mathbf{k}\, d\omega$$

where, it will be noted, the factor $1/(8\pi^3)$ is associated with the frequency dependent function $H(\mathbf{k}, \omega)$ rather than the time dependent function $h(\boldsymbol{\rho}, \tau)$.

### 10.4.1 Input-output relationship for homogeneous, stationary random processes

The introduction of a space variable $\mathbf{r}$ does not alter the general nature of our previous theory. Thus analogous expressions may be found for expected values and parametric averages. Again a homogeneous and stationary input produces a homogeneous and stationary output. It will therefore be sufficient for our purposes merely to list a few of the more relevant results.

Suppose that the input is an ergodic process as regards both position and time. If the angular brackets now signify averaging over position as well as time, we may define an input auto-correlation function as

$$R_{QQ}(\boldsymbol{\rho}, \tau) = \langle Q(\mathbf{r}, t) Q(\mathbf{r} + \boldsymbol{\rho}, t + \tau)\rangle.$$

Notice that we have not indicated a particular realisation here since it is unnecessary to do so, the process being ergodic. Again, choosing to define it as a parametric average rather than an expectation, we can express an output auto-correlation function in the form

$$R_{qq}(\boldsymbol{\rho}, \tau) = \langle q(\mathbf{r}, t) q(\mathbf{r} + \boldsymbol{\rho}, t + \tau)\rangle.$$

These two auto-correlation functions are related by the equation

$$R_{qq}(\boldsymbol{\rho}, \tau) = \int_{-\infty}^{\infty} \int_{-\infty}^{\infty} \int_{-\infty}^{\infty} h(\boldsymbol{\rho}_1, \tau_1) \int_{-\infty}^{\infty} \int_{-\infty}^{\infty} \int_{-\infty}^{\infty} h(\boldsymbol{\rho}_2, \tau_2) \cdot$$
$$R_{QQ}(\boldsymbol{\rho} + \boldsymbol{\rho}_1 - \boldsymbol{\rho}_2, \tau + \tau_1 - \tau_2) d\boldsymbol{\rho}_1\, d\boldsymbol{\rho}_2\, d\tau_1\, d\tau_2.$$

The mean square spectral density of the output is given by

$$S_{qq}(\mathbf{k}, \omega) = \frac{1}{8\pi^3} \int_{-\infty}^{\infty} \int_{-\infty}^{\infty} \int_{-\infty}^{\infty} R_{qq}(\boldsymbol{\rho}, \tau) e^{-i(\mathbf{k}.\boldsymbol{\rho} - \omega\tau)} d\boldsymbol{\rho}\, d\tau$$

which, after substitution for $R_{qq}(\boldsymbol{\rho}, \tau)$ reduces to

$$S_{qq}(\mathbf{k}, \omega) = |H(\mathbf{k}, \omega)|^2 S_{QQ}(\mathbf{k}, \omega).$$

Thus the spectral density of the output at any wave number and frequency is equal to the spectral density of the input at that wave number and frequency multiplied by the square of the modulus of the receptance at that wave number and frequency. The one dimensional spectral density of the output is given by

$$S_{qq}(\omega) = \int_{-\infty}^{\infty} \int_{-\infty}^{\infty} S_{qq}(\mathbf{k}, \omega) d\mathbf{k}.$$

The spectral densities of the time derivatives $\dot{q}(\mathbf{r}, t)$ and $\ddot{q}(\mathbf{r}, t)$ of the output response $q(\mathbf{r}, t)$ are given by

$$S_{\dot{q}\dot{q}}(\mathbf{k}, \omega) = |H(\mathbf{k}, \omega)|^2 \omega^2 S_{QQ}(\mathbf{k}, \omega),$$
$$S_{\ddot{q}\ddot{q}}(\mathbf{k}, \omega) = |H(\mathbf{k}, \omega)|^2 \omega^4 S_{QQ}(\mathbf{k}, \omega).$$

Further, the spectral densities of the spatial derivatives

$$q_x(\mathbf{r}, t) = \frac{\partial q(\mathbf{r}, t)}{\partial x}$$

$$q_y(\mathbf{r}, t) = \frac{\partial q(\mathbf{r}, t)}{\partial y}$$

$$\dot{q}_x(\mathbf{r}, t) = \frac{\partial \dot{q}(\mathbf{r}, t)}{\partial x}$$

$$\cdots\cdots\cdots\cdots\cdots$$

are given by

$$S_{q_x q_x}(\mathbf{k}, \omega) = |H(\mathbf{k}, \omega)|^2 k_x^2 S_{QQ}(\mathbf{k}, \omega),$$
$$S_{q_y q_y}(\mathbf{k}, \omega) = |H(\mathbf{k}, \omega)|^2 k_y^2 S_{QQ}(\mathbf{k}, \omega),$$
$$S_{\dot{q}_x \dot{q}_x}(\mathbf{k}, \omega) = |H(\mathbf{k}, \omega)|^2 k_x^2 \omega^2 S_{QQ}(\mathbf{k}, \omega),$$

$$\cdots\cdots\cdots\cdots\cdots\cdots\cdots\cdots\cdots\cdots\cdots\cdots$$

The single input-output correlation through the cross spectral density function is given by

$$S_{Qq}(\mathbf{k}, \omega) = H(\mathbf{k}, \omega) S_{QQ}(\mathbf{k}, \omega)$$

and the results for a multiple input-output system become

$$\mathbf{S_{qq}}(\mathbf{k}, \omega) = \mathbf{H}^*(\mathbf{k}, \omega)\mathbf{S_{QQ}}(\mathbf{k}, \omega)\mathbf{H}^T(\mathbf{k}, \omega)$$

and

$$\mathbf{S_{Qq}}(\mathbf{k}, \omega) = \mathbf{S_{QQ}}(\mathbf{k}, \omega)\mathbf{H}^T(\mathbf{k}, \omega).$$

These relationships may be expressed in terms of the physically realisable one-sided spectral density function $\Phi(\omega)$ or $\Phi(\mathbf{k}, \omega)$ defined as

$$\Phi(\omega) = \begin{cases} 2S(\omega) & \text{for } 0 \leqslant \omega < \infty \\ 0 & \text{otherwise,} \end{cases}$$

$$\Phi(\mathbf{k}, \omega) = \begin{cases} 8S(\mathbf{k}, \omega) & \text{for } 0 \leqslant \omega < \infty, 0 \leqslant k_x < \infty, 0 \leqslant k_y < \infty \\ 0 & \text{otherwise.} \end{cases}$$

## References


[1] BROWN, B. M., 1961, *The Mathematical Theory of Linear Systems*, Chapman and Hall, London.

[2] van LEEUWEN, G., 1964, 'The lateral damping and added mass of a horizontally oscillating ship model' Technological University, Delft, Report No. 23.

[3] BISHOP, R. E. D., BURCHER, R. K. and PRICE, W. G., 1973, 'The uses of functional analysis in ship dynamics'. *Proc. Roy. Soc.* **A332**, 23–35.

[4] BISHOP, R. E. D., BURCHER, R. K. and PRICE, W. G., 1973, 'Application of functional analysis to oscillatory ship model testing'. *Proc. Roy. Soc.* **A332**, 37–49.

[5] APOSTOL, T. M., 1960, *Mathematical Analysis*, Addison Wesley, Reading, Massachusetts.

# 11 Deterministic theory of ship motions

It has been shown that, in order to obtain data on the random process of ship motion from given data on random wave elevation, it is necessary to have information on the vehicle's response to deterministic excitations. This is typified in the previous chapter by our use of the receptance matrix. The individual receptance elements may either be determined by experimental procedures, in which case the equations of motion are redundant, or by calculation (when the form of the equations of motion becomes crucial).

The experimental method has, as input, a wave excitation or wave loading and an output that is a motion. No insight is provided into how the individual ship characteristics affect the response unless the experiment is repeated numerous times under different conditions. However, this information may be obtained with comparative ease from the hydrodynamic equations of motion, provided that the mathematical model is adequate.

When a ship or other vehicle moves in waves, hydrodynamic forces are exerted on it as a consequence of that motion. Considerable effort has been made to formulate an adequate theory for the motion of a surface ship in steady waves. In these theories, the ship is idealised in one of several ways and the essential differences between four common analytical approaches are as shown in the table[1]. In the table the entries have the following meanings:

$B$ = beam of ship $\qquad \lambda$ = wavelength
$T$ = draught of ship $\qquad \omega$ = frequency of wave.
$L$ = length of ship

**Table VIII**

| | $\frac{B}{L}$ | $\frac{T}{L}$ | $\frac{\lambda}{L}$ | $\omega\left(\frac{L}{g}\right)^{1/2}$ | $F_n = \frac{\bar{U}}{\sqrt{gL}}$ |
|---|---|---|---|---|---|
| *Thin ship theory* [2] | o(β) | o(1) | o(1) | o(1) | zero or o(1) |
| *Flat ship theory* [3] | o(1) | o(β) | o(1) | o(1) | zero or o(1) |
| *Slender ship theory* [4] | o(β) | o(β) | o(1) | o(1) | zero or o(1) |
| *Strip theory* [5] | o(β) | o(β) | o(β) | $o(\beta^{-1/2})$ | zero or o(1) |

The entries o(1) and o(β), where $\beta \ll 1$, describe the magnitudes or order of the ratios referred to. For example $B/L = o(1)$ refers to a ship whose beam and length are of approximately the same magnitude, while $B/L = o(\beta)$ refers to a ship whose beam is much smaller than its length.

Although they are in many respects rigorous, the first two theories do not describe the geometry of the ship adequately, since usually $B \ll L$ and $T \ll L$. The 'slender ship' and 'strip' theories overcome these defects and only differ from each other with respect to the incident waves.

Slender ship theory attempts to account for differences in the flow condition in the fore and aft directions, either arising from wave effects or from the effects of forward motion. But this refinement is achieved at the expense of neglecting any flow interaction between transverse points on the surface of the ship, since the beam is assumed small compared with the wavelength. Further it is assumed that the wavelength of waves striking or radiating from the ship are of the same order as, or greater than, the length of the ship.

The strip theory which assumes two dimensional flow in transverse planes at each section of the ship holds good only if the wavelength is small compared with the ship length. Thus interference between the bow and stern are negligible since they are many wavelengths apart, and the three dimensional hydrodynamic problem is reduced to one in two dimensions. Strip theory is the crudest of the four mentioned approaches but it often gives reasonable agreement with experimental results and is becoming, if anything, the most widely used.

Our purpose in the present chapter will be to examine the necessary deterministic theory in more detail, prior to our discussion of random responses.

## 11.1 Frames of reference

In order to describe the motion of a ship or other vehicle moving in a seaway, it is necessary to define axes of reference. We now consider the right handed systems of axes that we shall require.

The coordinate system $OXYZ$ is fixed with $OXY$ in the mean sea surface and with the axis $OZ$ pointing vertically upwards. At any point $X$, $Y$ the elevation of waves travelling with velocity $c$ at an angle $\mu$ to the axis $OX$ is given by

$$\zeta(X, Y, t) = \zeta_0 \cos(k \cos \mu \,.\, X + k \sin \mu \,.\, Y - \omega t).$$

The fixed frame of reference $OX'Y'Z'$ also has $OX'Y'$ in the mean sea surface, but $OX'$, points in the direction of the ship heading. The ship is assumed to have velocity $\overline{U}$ and a heading that is inclined at an angle $\xi$ to the axis $OX$ as shown in Figure 61. It is easily shown by a simple transformation of coordinates that

$$X = X' \cos \xi - Y' \sin \xi,$$
$$Y = X' \sin \xi + Y' \cos \xi,$$
$$Z = Z'.$$

The system $C^*x^*y^*z^*$ is a set of equilibrium axes which has the undisturbed motion of the ship. Its origin $C^*$ occupies the position that the centre of mass $C$ of the ship would occupy if the waves caused no parasitic motion of the ship, and this frame always remains parallel to $OX'Y'Z'$. At the instant $t = 0$, the origins $O$ and $C^*$

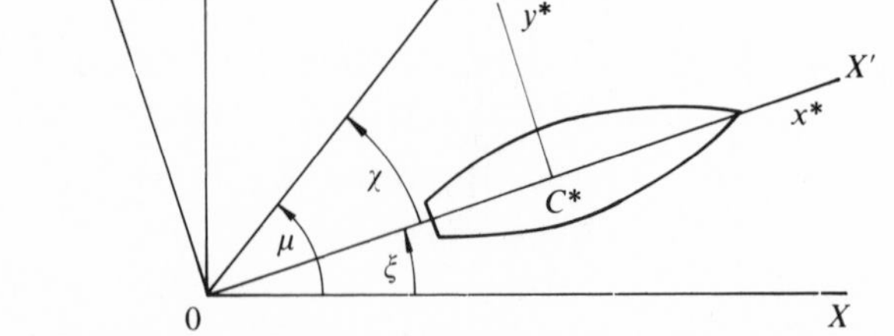

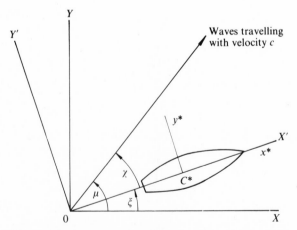

**Figure 61.** Frames of reference used in seakeeping analysis. $OXY$ and $OX'Y'$ are fixed while $C^*x^*y^*$ are 'equilibrium axes'.

coincide so that, at time $t$, the coordinates of any point within the ship are

$$X' = \bar{U}t + x^*, \qquad Y' = y^*, \qquad Z' = z^*$$

when measured with respect to $OX'Y'Z'$, or

$$\begin{aligned} X &= x^* \cos \xi - y^* \sin \xi + \bar{U}t \cos \xi, \\ Y &= x^* \sin \xi + y^* \cos \xi + \bar{U}t \sin \xi, \\ Z &= z^*, \end{aligned}$$

when measured with respect to $OXYZ$.

Notice that in the theory of seakeeping, it is common to use equilibrium axes. This is in contradiction to the theory of directional stability and control, where body axes are usually employed. Body axes are fixed to the ship whereas equilibrium axes are not. It is also worth noting that the axes employed in the theory of directional stability and control usually have the $OZ$ axis and the $Cz$ body axis pointing downwards.

The wave elevation at any point, measured with respect to the equilibrium axes is given by

$$\begin{aligned} \zeta(x^*, y^*, t) &= \zeta_0 \cos[k \cos \mu \{x^* \cos \xi - y^* \sin \xi + \bar{U}t \cos \xi\} + \\ &\qquad k \sin \mu \{x^* \sin \xi + y^* \cos \xi + \bar{U}t \sin \xi\} - \omega t] \\ &= \zeta_0 \cos[kx^* \cos(\mu - \xi) + ky^* \sin(\mu - \xi) \\ &\qquad - \{\omega - \bar{U}k \cos(\mu - \xi)\}t]. \end{aligned}$$

If $\mu - \xi = \chi$, this becomes

$$\zeta(x^*, y^*, t) = \zeta_0 \cos(kx^* \cos \chi + ky^* \sin \chi - \omega_e t)$$

where

$$\omega_e = \omega - \bar{U}k \cos \chi$$

is the 'frequency of encounter'. Let

$$k_x = k \cos \chi, \qquad k_y = k \sin \chi$$

and write $\mathbf{k} = k_x \mathbf{i}^* + k_y \mathbf{j}^*$ where $\mathbf{i}^*$, $\mathbf{j}^*$ are unit vectors in the directions $C^*x^*$, $C^*y^*$ respectively. Then if $\mathbf{r}^* = x^*\mathbf{i}^* + y^*\mathbf{j}^*$ we have

$$\zeta(x^*, y^*, t) = \zeta_0 \cos(\mathbf{k} . \mathbf{r}^* - \omega_e t).$$

For waves in deep water, the dispersion relationship is

$$k = \frac{\omega^2}{g}$$

so that

$$\omega_e = \omega - \frac{\bar{U}\omega^2}{g}\cos(\mu - \xi) = \omega - \frac{\bar{U}\omega^2}{g}\cos\chi.$$

When $\mu - \xi = 0 = \chi$, the vehicle is in a following sea and the frequency of encounter with the stern is

$$\omega_e = \omega - \frac{\bar{U}\omega^2}{g}.$$

But when $\mu - \xi = \pi = \chi$ the ship is in a head sea and the frequency of encounter with the bows is

$$\omega_e = \omega + \frac{\bar{U}\omega^2}{g}.$$

One more system of coordinate axes is sometimes needed. It is the frame of body axes $Axyz$ whose origin $A$ lies at some convenient position in the ship. The axis $Ax$ points forward, $Ay$ points to port and $Az$ points upwards, all three being fixed to the vehicle. This system permits the orientation of the vehicle to be specified.

## 11.2 Equations of motion

Some of the earliest attempts at the derivation of a set of differential equations describing the ship's motion in waves, were by Kriloff[6] and Mitchell[7] at the turn of the century. Unfortunately, this work was not further developed by other naval architects and for the next fifty years the topic of ship's motion lay relatively dormant until the early 1950s. After the theoretical successes of St Denis and Pierson[8], Peters and Stoker[9] and the introduction of seakeeping experimental facilities at several tank establishments, the subject of ship's motion has since become an active field of research.

The majority of this research activity was concerned with the determination of the symmetric responses of heave and pitch (surge being neglected) in head seas. Korvin-Kroukovsky and Jacobs[5] developed a strip theory to calculate these responses. Although much physical insight was built into their theoretical model and the motions predicted agreed quite well with experiments, it may be condemned for the lack of a rational mathematical philosophy. The strip theory has since been modified by Gerritsma and Beukelmann[10] and it is now found that the predicted motions for a ship having various bow shapes and travelling in heads seas even at high speeds, are in good agreement with the corresponding experiment.

Various other strip theories[11] have been developed, some based on slender body theory.

Less work has been done on the prediction of the anti-symmetric motions of roll, sway and yaw in a seaway, although much attention has been given to these motions in determining the stability and control characteristics of a ship in calm water. Tasai[12] has developed a linear strip analysis for the prediction of these motions but difficulties arise in following seas where a nonlinear analysis may be necessary. In beam seas, roll motion is dominant and is sometimes decoupled from the other motions.

For a rigid ship travelling with constant forward speed $\bar{U}$ at an arbitrary angle to regular sinusoidal waves, the resultant motions in the six degrees of freedom are governed by the set of second order linear differential equations

$$(\mathbf{m} + \mathbf{a})\ddot{\mathbf{q}}(t) + \mathbf{b}\dot{\mathbf{q}}(t) + \mathbf{c}\mathbf{q}(t) = \mathbf{Q}_0 e^{i\omega_e t}.$$

The meanings of the various individual terms in this equation will now be discussed separately bearing in mind that the axes employed are equilibrium axes. The 6 x 1 column matrices $\mathbf{Q}_0$ and $\mathbf{q}(t)$ represent amplitude of input and output respectively:

$$\mathbf{Q}_0 = \begin{bmatrix} X \\ Y \\ Z \\ K \\ M \\ N \end{bmatrix}; \quad \mathbf{q}(t) = \begin{bmatrix} x(t) \\ y(t) \\ z(t) \\ \phi(t) \\ \theta(t) \\ \psi(t) \end{bmatrix}.$$

Here

$X, Y, Z$ = amplitudes of surge, sway and heave forces
$K, M, N$ = amplitudes of roll, pitch and yaw moments

and

$x(t), y(t), z(t)$ = surge, sway and heave motions in the directions $C^*x^*, C^*y^*, C^*z^*$
$\phi(t), \theta(t), \psi(t)$ = variations of roll, pitch and yaw orientations with respect to the equilibrium axes.

The origin of the body axis in the ship is usually taken at the point of trisection of the midship section, longitudinal plane of symmetry and the waterplane. In general, this origin does not coincide with the centre of mass. However, for simplification, we shall assume that the ship, with lateral symmetry, has a centre of mass coincident with the origin of the body axes at the point of trisection. In this event the symmetric 6 x 6 generalised mass matrix **m** is

$$\mathbf{m} = \begin{bmatrix} m & 0 & 0 & 0 & 0 & 0 \\ 0 & m & 0 & 0 & 0 & 0 \\ 0 & 0 & m & 0 & 0 & 0 \\ 0 & 0 & 0 & I_{xx} & 0 & -I_{xz} \\ 0 & 0 & 0 & 0 & I_{yy} & 0 \\ 0 & 0 & 0 & -I_{xz} & 0 & I_{zz} \end{bmatrix}$$

where $m$ is the mass of the ship, $I_{xx}$, $I_{yy}$ and $I_{zz}$ are the principal moments of inertia and $I_{xz}$ is a product of inertia that is zero if the ship has fore- and aft-symmetry.

The elements of the 6 x 6 matrix **a** are a combination of hydrodynamic forces which are in phase with the acceleration of the ship. By analogy with the mass matrix, **a** is referred to as the 'added mass matrix' while (**m** + **a**) is the virtual mass matrix. Physically it may be described very loosely as the quantity of fluid in the vicinity of the ship which is accelerated with the ship. On the other hand, the elements of the 6 x 6 damping matrix **b** are a combination of hydrodynamic forces which are in phase with the velocity of the ship and are associated with energy dissipation. In an ideal fluid, energy can only be dissipated by the generation and propagation of waves in the free surface.

For a laterally symmetric ship, the added mass and damping matrices have the forms

$$\mathbf{a} = \begin{bmatrix} X_{\ddot{x}} & 0 & X_{\ddot{z}} & 0 & X_{\ddot{\theta}} & 0 \\ 0 & Y_{\ddot{y}} & 0 & Y_{\ddot{\phi}} & 0 & Y_{\ddot{\psi}} \\ Z_{\ddot{x}} & 0 & Z_{\ddot{z}} & 0 & Z_{\ddot{\theta}} & 0 \\ 0 & K_{\ddot{y}} & 0 & K_{\ddot{\phi}} & 0 & K_{\ddot{\psi}} \\ M_{\ddot{x}} & 0 & M_{\ddot{z}} & 0 & M_{\ddot{\theta}} & 0 \\ 0 & N_{\ddot{y}} & 0 & N_{\ddot{\phi}} & 0 & N_{\ddot{\psi}} \end{bmatrix}; \mathbf{b} = \begin{bmatrix} X_{\dot{x}} & 0 & X_{\dot{z}} & 0 & X_{\dot{\theta}} & 0 \\ 0 & Y_{\dot{y}} & 0 & Y_{\dot{\phi}} & 0 & Y_{\dot{\psi}} \\ Z_{\dot{x}} & 0 & Z_{\dot{z}} & 0 & Z_{\dot{\theta}} & 0 \\ 0 & K_{\dot{y}} & 0 & K_{\dot{\phi}} & 0 & K_{\dot{\psi}} \\ M_{\dot{x}} & 0 & M_{\dot{z}} & 0 & M_{\dot{\theta}} & 0 \\ 0 & N_{\dot{y}} & 0 & N_{\dot{\phi}} & 0 & N_{\dot{\psi}} \end{bmatrix}$$

where, for example $K_{\ddot{\phi}}$, $K_{\ddot{\psi}}$ are the added hydrodynamic roll moments associated with the roll angular acceleration $\ddot{\phi}(t)$ and with angular yaw acceleration $\ddot{\psi}(t)$ respectively. The quantities $Z_{\dot{z}}$ and $Z_{\dot{\theta}}$

are the hydrodynamic damping forces associated with the heave velocity $\dot{z}(t)$ and with pitch angular velocity $\dot{\theta}(t)$, respectively.

Once again it is prudent to notice the contrast between this theory, entailing the use of equilibrium axes, and that of directional stability and control in which body axes are employed. Here the symbols $X_{\ddot{x}}$, $K_{\dot{y}}$, . . . represent a convenient linear representation of fluid actions and they are *not* hydrodynamic 'derivatives'. Notice in particular that, quite apart from the way they are defined, they have the wrong sign for derivatives.

Finally the 6 x 6 restoring matrix **c** accounts for the hydrostatic contributions, although there may be some small additional dynamic effects if the centre of mass and origin of the body axes do not coincide. There then exist restoring moments associated with the centre of mass of the ship in an angular displacement of the ship. For a harmonically oscillating ship, it is seen that the total hydrodynamic force in phase with the displacement is $(-\omega_e{}^2\mathbf{a} + \mathbf{c})$. In effect, there therefore exists some arbitrariness in the definitions of the matrices **a** and **c**. On the assumption that **c** is due to hydrostatic effects only, however, the definitions become unique. For the surface ship, the restoring matrix then has the form

$$\mathbf{c} = \begin{bmatrix} 0 & 0 & 0 & 0 & 0 & 0 \\ 0 & 0 & 0 & 0 & 0 & 0 \\ 0 & 0 & Z_z & 0 & Z_\theta & 0 \\ 0 & 0 & 0 & K_\phi & 0 & 0 \\ 0 & 0 & M_z & 0 & M_\theta & 0 \\ 0 & 0 & 0 & 0 & 0 & 0 \end{bmatrix}$$

where, for example, $Z_z$ is the heave restoring force associated with a unit heave displacement, $z(t)$, etc.

One outcome of the use of equilibrium axes, instead of body axes, is removal of the explicit symbol $\overline{U}$ from the equations of motion. The quantity $\overline{U}$ *does* appear in equations used in the analysis of directional stability and control.

In general, the matrices **a**, **b** and **c** are supposed constant with respect to time but, on the assumption that only the response that is proportional to $e^{i\omega_e t}$ is of interest, some elements are dependent on the frequency of wave encounter $\omega_e$. Substituting the foregoing matrices into the initial differential equation yields two sets of coupled equations. One set involves the symmetric motions of surge, heave, pitch and the other involves the antisymmetric motions of roll, sway and yaw. By assuming that the surge motion is an order smaller than the other symmetric motions, a case can be made for studying only the two coupled differential equations involving the pitch and heave motions. In effect, the disturbances of the laterally symmetric ship have been divided into two categories:

a) equations of small symmetric departures from a steady reference motion
b) equations of small antisymmetric departures from a steady reference motion.

We shall now consider these cases.

### 11.2.1 Heave and pitch motions

From the previous section, the equations describing the coupled motions for heave and pitch are seen to be of the form

$$\begin{bmatrix} m + Z_{\ddot{z}} & Z_{\ddot{\theta}} \\ M_{\ddot{z}} & I_{yy} + M_{\ddot{\theta}} \end{bmatrix} \begin{bmatrix} \ddot{z}(t) \\ \ddot{\theta}(t) \end{bmatrix} + \begin{bmatrix} Z_{\dot{z}} & Z_{\dot{\theta}} \\ M_{\dot{z}} & M_{\dot{\theta}} \end{bmatrix} \begin{bmatrix} \dot{z}(t) \\ \dot{\theta}(t) \end{bmatrix}$$

$$+ \begin{bmatrix} Z_z & Z_\theta \\ M_z & M_\theta \end{bmatrix} \begin{bmatrix} z(t) \\ \theta(t) \end{bmatrix} = \begin{bmatrix} Z \\ M \end{bmatrix} e^{i\omega_e t}.$$

The various coefficients $Z_{\ddot{z}}$, $Z_{\dot{\theta}}$, $M_{\ddot{z}}$ etc., may be derived from experiment or by analytical means. Table IX shows the coefficients as derived from the modified strip theory of Gerritsma and Beukelman[10] and from the velocity potential approach of Vugts[13], and Salvesen, *et al.* [14]. In this table, $m(x)$ is the two dimensional sectional added mass per unit length and $N(x)$ is the two dimensional damping coefficient per unit length for heave while $B(x)$ is the sectional beam at the waterline. The position coordinate $x^A$, added mass $m^A$ and damping coefficient $N^A$ are associated with the ship's aftermost section. (We shall examine the origins of these quantities later on, merely noting at this stage that they reflect the geometry of the ship).

Gerritsma and Beukelman[10] invoked the strip theory hypothesis (the ship's beam and draught being much smaller than the length) at the commencement of their analysis whilst Salvesen *et al.* [14] introduced this assumption after the initial setting up of the mathematical model as a means of simplifying the subsequent analysis. The theories differ as to the effect of forward speed but are in

**Table IX**

| Coefficient | Modified strip theory of Gerritsma and Beukelman[10] | Theory of Vugts[13] and Salvesen *et al.*[14] |
|---|---|---|
| $Z_{\ddot{z}}$ | $\int m(x)dx$ | $\int m(x)dx - \bar{U}\omega_e^{-2} N^A$ |
| $Z_{\dot{z}}$ | $\int N(x)dx + \bar{U}m^A$ | $\int N(x)dx + \bar{U}m^A$ |
| $Z_z$ | $\rho g\int b(x)dx$ | same |
| $Z_{\ddot{\theta}}$ | $-\int xm(x)dx - \bar{U}\omega_e^{-2}\int N(x)dx - \bar{U}^2\omega_e^{-2} m^A$ | $-\int xm(x)dx - \bar{U}\omega_e^{-2}\int N(x)dx + \bar{U}\omega_e^{-2} x^A N^A - \bar{U}^2\omega_e^{-2} m^A$ |
| $Z_{\dot{\theta}}$ | $-\int xN(x)dx + \bar{U}\int m(x)dx - \bar{U}x^A m^A$ | $-\int xN(x)dx + \bar{U}\int m(x)dx - \bar{U}x^A m^A - \bar{U}^2\omega_e^{-2} N^A$ |
| $Z_\theta$ | $-\rho g\int xB(x)dx$ | same |
| $M_{\ddot{z}}$ | $-\int xm(x)dx$ | $-\int xm(x)dx + \bar{U}\omega_e^{-2}\int N(x)dx + \bar{U}\omega_e^{-2} x^A N^A$ |
| $M_{\dot{z}}$ | $-\int xN(x)dx - \bar{U}x^A m^A - \bar{U}\int m(x)dx$ | same |
| $M_z$ | $-\rho g\int xB(x)dx$ | same |
| $M_{\ddot{\theta}}$ | $\int x^2 m(x)dx + \bar{U}\omega_e^{-2}\int xN(x)dx + \bar{U}^2\omega_e^{-2}\int m(x)dx$ $+ \bar{U}\omega_e^{-2} x^A m^A$ | $\int x^2 m(x)dx + \bar{U}^2\omega_e^{-2}\int m(x)dx - \bar{U}\omega_e^{-2}(x^A)^2 N^A$ $+ \bar{U}^2\omega_e^{-2} x^A m^A$ |
| $M_{\dot{\theta}}$ | $\int x^2 N(x)dx + \bar{U}(x^A)^2 m^A$ | $\int x^2 N(x)dx + \bar{U}^2\omega_e^{-2}\int N(x)dx + \bar{U}(x^A)^2 m^A + \bar{U}^2\omega_e^{-2} x^A N^A$ |
| $M_\theta$ | $\rho g\int x^2 B(x)dx$ | same |
| $Z$ | $\zeta_0\int e^{ikx} e^{-kT_*}[\rho gB(x) - \omega\{\omega_e m(x) - iN(x)\}]dx$ $+ i\omega\zeta_0\bar{U}m^A e^{ikx^A} e^{-kT_*^A}$ | same $+ \zeta_0\bar{U}\omega_e^{-1} e^{ikx^A} e^{-kT_*^A}\omega$ |
| $M$ | $-\zeta_0\int e^{ikx} e^{-kT_*} x[\rho gB(x) - \omega\{\omega_e m(x) - iN(x)$ $- i\bar{U}m(x)\}]dx - i\zeta_0\omega\bar{U}x^A m^A e^{ikx^A} e^{-kT_*^A}$ | same $- \zeta_0\int e^{ikx} e^{-kT_*^A}\bar{U}\omega_e^{-1}\omega N(x)dx$ $- \omega\zeta_0\bar{U}\omega_e^{-1} x^A N^A e^{ikx^A} e^{-kT_*^A}$ |

*Note:* $\zeta_0$ = Amplitude of wave;
all integrals are over the length of the ship.

agreement when the ship has zero ahead speed. A slender ship with pointed ends has $m^A = 0 = N^A$, however, and both theories satisfy the Timman and Newman[15] relationship in the sense that the coefficients $Z_{\dot{\theta}}$ and $M_{\dot{z}}$ have the same forward speed dependence but are of opposite sign. The retention of terms involving $m^A$ and $N^A$ accounts for possible transom stern effects.

Wave excitation force and moment in head waves ($\chi = 180°$) agree closely between the theories, the additional terms in the third column of Table IX derived by Salvesen *et al.* [14] (which also differ slightly from those of Vugts' theory) having very little effect on the determined motions. In the derivation of the wave distrubance terms, it was assumed in the theories concerned that the pressure distribution on the ship is the pressure of the undisturbed incident wave evaluated at the ship's hull. This is commonly known as the 'Froude-Kriloff' hypothesis.

The quantity $T_*$ in the force and moment amplitudes in Tables IX and X is defined as

$$T_* = \frac{1}{k} \ln \left\{1 - \frac{2k}{B(x)} \int_{-T}^{0} y e^{kz} \, dz\right\}$$

where $T$ is the draught of the ship. A simplification to this expression is made by equating $T_*$ to the mean draught at each section, i.e.

$$T_* = \frac{\text{Area of cross-section}}{\text{sectional waterline beam}}.$$

This approximation holds for sections with regular shaped hulls, but care must be taken when considering, for example, a bulbous bow, since otherwise errors may be introduced in the evaluated wave exciting force and moments.

### 11.2.2 Vertical shear forces and bending moments

The upward external load per unit length, $w(x)$, on the hull at a given time, is equal to the difference between the inertia force and the sum of the external forces. This latter force comprises the static restoring force, the wave exciting forces and the hydrodynamic force due to the ship's motion and the weight per unit length. Thus we have

$$w(x) = -\frac{\mu(x)}{g}(\ddot{z} - x\ddot{\theta}) + F(x) + \rho g A_x - \mu(x)$$

where $F(x)$ is the prevailing upward hydrodynamic force per unit length, $\mu(x)$ is the weight of the section per unit length and $A_x$ is the

**Table X**

| Coefficient | Theory of Vugts[13] and Salvesen *et al.*[14] |
|---|---|
| $Y_{\ddot{y}}$ | $\int m_s(x)dx - \bar{U}\bar{\omega}_e^{-2} N_s^A$ |
| $Y_{\dot{y}}$ | $\int N_s(x)dx + \bar{U}m_s^A$ |
| $Y_{\ddot{\phi}}$ | $\int m_{sr}(x)dx - \bar{U}\bar{\omega}_e^{-2} N_{sr}^A$ |
| $Y_{\dot{\phi}}$ | $\int N_{sr}(x)dx + \bar{U}m_{sr}^A$ |
| $Y_{\ddot{\psi}}$ | $\int x m_s(x)dx + \bar{U}\bar{\omega}_e^{-2} \int N_s(x)dx - \bar{U}\bar{\omega}_e^{-2} x^A N_s^A + \bar{U}^2 \bar{\omega}_e^{-2} m_s^A$ |
| $Y_{\dot{\psi}}$ | $\int x N_s(x)dx - \bar{U}\int m_s(x)dx + \bar{U}x^A m_s^A + \bar{U}^2 \bar{\omega}_e^{-2} N_s^A$ |
| $K_{\ddot{y}}$ | $Y_{\ddot{\phi}}$ |
| $K_{\dot{y}}$ | $Y_{\dot{\phi}}$ |
| $K_{\ddot{\phi}}$ | $\int m_r(x)dx - \bar{U}\bar{\omega}_e^{-2} N_r^A$ |
| $K_{\dot{\phi}}$ | $\int N_r(x)dx + \bar{U}m_r^A$ |
| $K_{\phi}$ | $\rho g \nabla \overline{GM}$ |
| $K_{\ddot{\psi}}$ | $\int x m_{sr}(x)dx + \bar{U}\bar{\omega}_e^{-2} \int N_{sr}(x)dx - \bar{U}\bar{\omega}_e^{-2} x^A N_{sr}^A + \bar{U}^2 \bar{\omega}_e^{-2} m_{sr}^A$ |
| $K_{\dot{\psi}}$ | $\int x N_{sr}(x)dx - \bar{U}\int m_{sr}(x)dx + Ux^A m_{sr}^A + \bar{U}^2 \bar{\omega}_e^{-2} x^A N_{sr}^A$ |
| $N_{\ddot{y}}$ | $\int x m_s(x)dx - \bar{U}\bar{\omega}_e^{-2} \int N_s(x)dx - \bar{U}\bar{\omega}_e^{-2} x^A N_s^A$ |
| $N_{\dot{y}}$ | $\int x N_s(x)dx + \bar{U}\int m_s(x)dx + \bar{U}x^A m_s^A$ |
| $N_{\ddot{\phi}}$ | $\int x m_{sr}(x)dx - \bar{U}\bar{\omega}_e^{-2} \int N_{sr}(x)dx - \bar{U}\bar{\omega}_e^{-2} x^A N_{sr}^A$ |
| $N_{\dot{\phi}}$ | $\int x N_{sr}(x)dx + \bar{U}\int m_{sr}(x)dx + \bar{U}x^A m_{sr}^A$ |
| $N_{\ddot{\psi}}$ | $\int x^2 m_s(x)dx + \bar{U}^2 \bar{\omega}_e^{-2} \int m_s(x)dx - \bar{U}\bar{\omega}_e^{-2} (x^A)^2 N_s^A$ |
| | $+ \bar{U}^2 \bar{\omega}_e^{-2} x^A m_s^A$ |
| $N_{\dot{\psi}}$ | $\int x^2 N_s(x)dx + \bar{U}^2 \bar{\omega}_e^{-2} \int N_s(s)dx + \bar{U}(x^A)^2 m_s^A + \bar{U}^2 \bar{\omega}_e^{-2} x^A N_s^A$ |
| $(Y, K, N)$ | $-i\zeta_0 kg \sin\chi(Y', K', N')$ |
| $Y'$ | $\int e^{-ikx\cos\chi_e - kT_*}[\rho A_x + \omega\bar{\omega}_e^{-2}\{\omega_e m_s(x) - iN_s(x)\}]dx$ |
| | $- \bar{U}\bar{\omega}_e^{-2}(\omega_e m_s^A - iN_s^A)e^{-ikx^A\cos\chi_e - kT_*^A}$ |
| $K'$ | $\int e^{-ikx\cos\chi_e - kT_*}\left[\rho\left\{\frac{B^3(x)}{12} - A_x\overline{C^*B}\right\}\right.$ |
| | $\left. + \omega\bar{\omega}_e^{-2}\{\omega_e m_{sr}(x) - iN_{sr}(x)\}\right] dx$ |
| | $- \bar{U}\bar{\omega}_e^{-2}(\omega_e m_{sr}^A - iN_{sr}^A)e^{-ikx^A\cos\chi_e - kT_*^A}$ |
| $N'$ | $\int e^{-ikx\cos\chi_e - kT_*}[x[\rho A_x + \omega\bar{\omega}_e^{-2}\{\omega_e m_s(x) - iN_s(x)\}]$ |
| | $+ i\bar{U}\bar{\omega}_e^{-1}\{\omega_e m_s(x) - iN_s(x)\}]dx$ |
| | $- \bar{U}\bar{\omega}_e^{-2}(\omega_e m_s^A - iN_s^A)x^A e^{-ikx^A\cos\chi_e - kT_*^A}$ |

*Note:* All integrals are over the length of the ship and $\zeta_0 k$ is the amplitude of the wave slope.

immersed cross-sectional area in still water. The last two terms of this equation are associated with the still water condition so that the dynamic part of the loading is given by

$$w(x) = -\frac{\mu(x)}{g}(\ddot{z} - x\ddot{\theta}) + F(x).$$

The time-dependent shear force, $Q_s$, and bending moment, $M_b$, at a specified cross section $x'$ are given by the contribution of the time-dependent part of the load acting on the portion of the hull forward of the considered section (see Figure 62). That is

$$Q_s(x') = \int_{x'}^{\text{bows}} \left\{ \frac{\mu(x)}{g} (\ddot{z} - x\ddot{\theta}) - F(x) \right\} dx$$

and

$$M_b(x') = \int_{x'}^{\text{bows}} (x - x') \left\{ \frac{\mu(x)}{g} (\ddot{z} - x\ddot{\theta}) - F(x) \right\} dx$$

where the time dependent contribution of the hydrodynamic force over the portion of the hull concerned is

$$\int_{x'}^{\text{bows}} F(x)dx = Ze^{i\omega_e t} - Z_{\ddot{z}}\ddot{z} - Z_{\dot{z}}\dot{z} - Z_z z - Z_{\ddot{\theta}}\ddot{\theta} - Z_{\dot{\theta}}\dot{\theta} - Z_\theta \theta$$

and the coefficient $Z_{\dot{z}}$, for example, is

$$Z_{\dot{z}} = \int_{x'}^{\text{bows}} N(x)dx + \bar{U}m(x').$$

The other coefficients are similarly redefined where now the integrals are over the portion of the hull forward of the considered section. Due to change in the lower limit of the integration the terms $m^A$, $N^A$, $x^A$, $T_*^A$ in Table IX are replaced by $m(x')$, $N(x')$, $x'$ and $T_*(x')$ respectively.

Knowing the wave disturbance and the ship's responses, the dynamic shear and wave bending moment may be determined. This

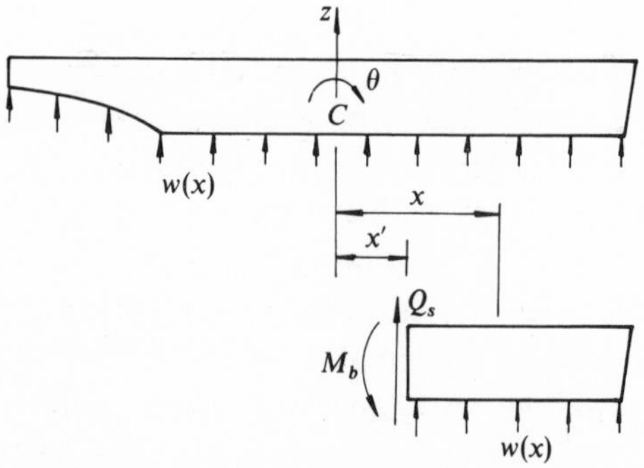


**Figure 62.** The shearing force $Q_s$ and bending moment $M_b$ due to wave action.

approach to ship strength calculations, while very common, nevertheless rests on the basic assumption that the ship may be supposed rigid. A more rigorous approach[16] is to investigate modal contributions to the shear force and bending moment.

### 11.2.3 Antisymmetric motions

From section 11.2, it is seen that the equations describing the sway, roll and yaw motions may be expressed in the matrix form

$$\begin{bmatrix} m + Y_{\ddot{y}} & Y_{\ddot{\phi}} & Y_{\ddot{\psi}} \\ K_{\ddot{y}} & I_{xx} + K_{\ddot{\phi}} & -I_{xz} + K_{\ddot{\psi}} \\ N_{\ddot{y}} & -I_{xz} + N_{\ddot{\phi}} & I_{zz} + N_{\ddot{\psi}} \end{bmatrix} \begin{bmatrix} \ddot{y}(t) \\ \ddot{\phi}(t) \\ \ddot{\psi}(t) \end{bmatrix}$$

$$+ \begin{bmatrix} Y_{\dot{y}} & Y_{\dot{\phi}} & Y_{\dot{\psi}} \\ K_{\dot{y}} & K_{\dot{\phi}} & K_{\dot{\psi}} \\ N_{\dot{y}} & N_{\dot{\phi}} & N_{\dot{\psi}} \end{bmatrix} \begin{bmatrix} \dot{y}(t) \\ \dot{\phi}(t) \\ \dot{\psi}(t) \end{bmatrix} + \begin{bmatrix} 0 & 0 & 0 \\ 0 & K_{\phi} & 0 \\ 0 & 0 & 0 \end{bmatrix} \begin{bmatrix} y(t) \\ \phi(t) \\ \psi(t) \end{bmatrix} = \begin{bmatrix} Y \\ K \\ N \end{bmatrix} e^{i\omega_e t}.$$

Table X shows the various coefficients as derived by the strip theory of Vugts, and Salvesen *et al.* The variables $m_s(x)$, $N_s(x)$ are the two dimensional sectional added mass and damping per unit length in sway; $m_r(x)$, $N_r(x)$ are the sectional added mass and damping per unit length in roll; $m_{sr}(x)$, $N_{sr}(x)$ are the two dimensional added mass and damping coefficients per unit length due to cross coupling between the sway and roll motions; $x^A$, $m_s{}^A$, $m_r{}^A$ etc., refer to the coefficients associated with the aftermost section of the ship; $\nabla$ is the displaced volume of the ship, $\overline{GM}$ is the metacentric height; $A_x$ is the immersed cross sectional area and $\overline{C^*B}$ is the distance of the centre of buoyancy from the waterplane in still water.

Since the theories are formulated with a velocity potential, no effect of viscosity is included. But a comparison between theory and experiment shows that the roll-damping coefficient, $K_{\dot{\phi}}$, is significantly affected by the viscosity. In order to calculate the amplitude of roll displacement with reasonable accuracy the coefficient $K_{\dot{\phi}}$ is sometimes modified to include such viscous influences.

With a slender ship for which $m_s^A = 0 = N_s^A = m_{sr}^A = N_{sr}^A$, the forward-speed terms in the damping coefficients make those coefficients satisfy the following Timman-Newman[15] relationships:

$$Y_{\dot{\phi}} = K_{\dot{y}}; \qquad Y_{\dot{\psi}} = -N_{\dot{y}}; \qquad K_{\dot{\psi}} = -N_{\dot{\phi}}.$$

The strip theory assumption is again invoked in the determination of the hydrodynamic force and moments. The wave disturbing effects in the two theories are slightly different. The Vugts expressions[13] (as given in Table X) are dependent on $(\omega^2/\omega_e)$, whereas in Salvesen *et al.*[14] we have instead a dependence on $\omega$ only. Vugts suggests that the additional term $(\omega/\omega_e)$ is especially important for the higher forward speeds and for wave directions within a sector of, say, 45° from the ship's track.

Although the three motions are coupled, many writers have sought to investigate rolling motions in isolation. Froude[17] established the rolling theory of ships in beam seas and postulated that the centre of mass of the ship lying broadside-on to waves, moves like a particle of water, in a circle of diameter equal to the wave height. A roll moment was created by the buoyancy force acting in a direction normal to the wave surface and the component of the weight force in a parallel direction passing through the centre of mass. The roll moment was affected more by the wave slope than the wave height. The hydrodynamic force and moments as given in Table X confirms Froude's findings in that they are dependent on the wave slope amplitude $(\zeta_0 k)$, in contrast to the disturbing effects of symmetric waves, which were functions of the wave amplitude $\zeta_0$ only.

## 11.3 The ship motion receptances

The reader will now see that the way ahead is clear (in theory at least) to compute a matrix of output mean square spectral densities, given the necessary input information. To achieve this end, however, it is necessary first to determine a receptance matrix, for it is this which relates the output to the input.

From section 11.2 we see that the ship's motion is governed by the matrix equation

$$(\mathbf{m} + \mathbf{a})\ddot{\mathbf{q}}(t) + \mathbf{b}\dot{\mathbf{q}}(t) + \mathbf{c}\mathbf{q}(t) = \mathbf{Q}_0 e^{i\omega_e t}$$

where $\mathbf{q}(t)$, $\mathbf{Q}_0$ are 2 x 1 matrices for the symmetric motions and 3 x 1 matrices for the antisymmetric motions, the other matrices having been previously defined. To illustrate how the receptances may be derived, we shall consider the simpler problem of symmetric motions, although the method we shall adopt holds good for the antisymmetric motions as well.

If the procedure discussed in section 10.1.1 is extended suitably, it is found that the receptance matrix for the heave and pitch motions is given by

$$\mathbf{H}(\omega_e) = [(\mathbf{c} - \omega_e^2\mathbf{a}) + i\omega_e\mathbf{b}]^{-1}.$$

This may be written in the form

$$\mathbf{H}(\omega_e) = \begin{bmatrix} H_{zZ} & H_{zM} \\ H_{\theta Z} & H_{\theta M} \end{bmatrix}$$

where, to take one element by way of example,

$$H_{\theta Z}(\omega_e) = \frac{M_z - \omega_e{}^2 M_{\ddot{z}} + i\omega_e M_{\dot{z}}}{|\mathbf{\Delta}|}$$

and $|\mathbf{\Delta}|$ is the determinant of $\mathbf{H}(\omega_e)$. This particular receptance gives the amplitude and phase of the pitch motion for a heave force of unit amplitude and frequency $\omega_e$.

Since the motions of pitching and heaving are coupled and the damping is by no means negligible, the receptances have significant values over a wide range of encounter frequencies. They are not sharply peaked in the vicinity of the natural frequency of heave and pitch motions.

As we have seen previously, the heave excitation $Z(t)$ and pitch moment $M(t)$ depend on the wave height, wave length or wave number and on the wave heading relative to the ship. The heave force is considerable, and effective on all courses, but the pitch moment is small in beam waves and large in head or following waves. The ratio (wave length)/(ship length) becomes important in determining the motion of the ship. For example, regardless of their steepness, short waves have little effect on pitching and heaving motions when coming from ahead or astern. However, the same waves could produce a considerable heave force as well as a large rolling moment if coming from abeam. In general, these symmetric forcing functions may be expressed in terms of the wave input

$$\zeta(\mathbf{r}, t) = \zeta_0 e^{i(\mathbf{k}.\mathbf{r} - \omega_e t)}.$$

The input becomes

$$\begin{bmatrix} Z(t) \\ M(t) \end{bmatrix} = \begin{bmatrix} Z(\mathbf{k}, \omega_e) \\ M(\mathbf{k}, \omega_e) \end{bmatrix} \zeta(\mathbf{r}, t)$$

where $Z(\mathbf{k}, \omega_e)$ and $M(\mathbf{k}, \omega_e)$ are functions of the wave properties, the frequency of encounter and the hydrodynamic and geometric properties of the ship. Table IX illustrates the theoretical forms of these latter functions for head seas, but the expressions are more complicated when the ship is at an arbitrary heading with respect to the waves.

The responses in heave and pitch are given by

$$\begin{bmatrix} z(t) \\ \theta(t) \end{bmatrix} = \mathbf{H}(\omega_e)\begin{bmatrix} Z \\ M \end{bmatrix} e^{i\omega_e t} = \mathbf{H}(\omega_e)\begin{bmatrix} Z(\mathbf{k}, \omega_e) \\ M(\mathbf{k}, \omega_e) \end{bmatrix} \zeta(\mathbf{r}, t).$$

This may be written in the alternative form

$$\begin{bmatrix} z(t) \\ \theta(t) \end{bmatrix} = \mathbf{H}(\mathbf{k}, \omega_e)\zeta(\mathbf{r}, t) = \mathbf{H}(\mathbf{k}, \omega_e)\zeta_0{}^{i(\mathbf{k}.\mathbf{r} - \omega_e t)}$$

where

$$\mathbf{H}(\mathbf{k}, \omega_e) = \begin{bmatrix} H_{z\zeta}(\mathbf{k}, \omega_e) \\ H_{\theta\zeta}(\mathbf{k}, \omega_e) \end{bmatrix} = \begin{bmatrix} H_{zZ}(\omega_e)Z(\mathbf{k}, \omega_e) + H_{zM}(\omega_e)M(\mathbf{k}, \omega_e) \\ H_{\theta Z}(\omega_e)Z(\mathbf{k}, \omega_e) + H_{\theta M}(\omega_e)M(\mathbf{k}, \omega_e) \end{bmatrix}.$$

The 'ship motion receptances' $H_{z\zeta}(\mathbf{k}, \omega_e)$ and $H_{\theta\zeta}(\mathbf{k}, \omega_e)$ give the outputs in heave and pitch for a sinusoidal wave input $\zeta$ of unit amplitude, wave number $\mathbf{k}$, and absolute frequency $\omega$. In deep water waves the wave encounter frequency

$$\omega_e = \omega - \frac{\bar{U}\omega^2}{g}\cos\chi$$

so that it is possible to determine $\mathbf{H}(\mathbf{k}, \omega_e)$ in terms of $\mathbf{k}$, $\omega$, $\mu$ and $\chi$.

In head waves $\chi = \pi$, and in following waves $\chi = 0$. The wave encounter frequency then reduces to

$$\omega_e = \omega \pm \frac{\bar{U}\omega^2}{g}$$

where the positive and negative signs apply to head and following wave conditions respectively.

By way of illustration, let us assume that $T_0$ $(= 2\pi/\omega_0)$ is the natural period of either the uncoupled heave or pitch motion. The wave encounter frequency at $\omega = \omega_0$ may be expressed as

$$\omega_e = \omega_0 \pm \frac{\bar{U}}{g}\omega_0^2.$$

Since $T_0 = 2\pi/\omega_0$ and $k = \omega_0^2/g$, or $\omega_0^2 = 2\pi g/\lambda$, the wave encounter frequency may be expressed in the form

$$T_0 \sqrt{\frac{g}{L}} = \frac{(\omega_e/\omega_0)(\lambda/L)}{\sqrt{\left(\frac{\lambda}{2\pi L}\right)} \pm F_n}$$

where

$$F_n = \frac{\bar{U}}{\sqrt{gL}}$$

is the Froude number.

Since little motion will result from a force or moment applied at a frequency greatly removed from the natural frequency of motion, it is to be expected that the greatest motion occurs in the frequency range

$$0{\cdot}75 < \omega_e/\omega_0 < 1{\cdot}25.$$

This is only a rough and ready guide and certainly not a necessary and sufficient condition for motions of large amplitude.

In the same way, the wavelengths encountered by the ship may be expected to cause significant heave and pitch moments when

$$0{\cdot}8 < \lambda/L < \lambda_{max}/L.$$

However, for long, shallow waves where $\lambda$ is large, the force and moments are small giving low amplitudes of motion. The upper limit ($\lambda_{max}/L$) gives the maximum wavelength which proves troublesome.

Figure 63 shows the operation zones for a ship travelling in head or following seas. The majority of ships usually operate in the 'sub-critical' and 'critical zones'. For a ship operating in the critical zone, a decrease or an increase (if possible) in speed results in a decrease in the ship's parasitic motion.

In the linear ship-wave system considered so far, the heave and pitch motions are the two outputs, whilst the input is taken either as the heave excitation and pitch moment or as the wave disturbance. The input and output are of course related by the appropriate receptances. In the same way other outputs may be considered – e.g. shear force or bending moment, which are also related to the wave disturbance input by the required receptances.

For the antisymmetric motions, the slope of the wave disturbance, $\zeta'$ $(= \zeta_0 k)$, is a better measure of input to the linear ship-wave system. The method of determining the required receptances between the sway, roll, yaw motions and the wave slope is similar to

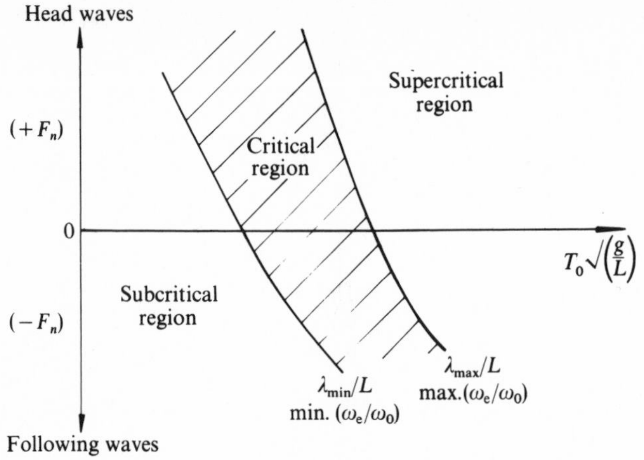


**Figure 63.** Operational zones for ship motions in head and following waves. Ships normally operate in the critical and sub-critical regions. (The sign in front of the Froude number $F_n$ is the sign to be used in the equation for $T_0\sqrt{(g/L)}$.)

that already described. As an illustration, we shall consider the simple case in which the roll motion is assumed decoupled from the other antisymmetric motions.

From section 11.2 we note that the equation describing the roll motion is

$$(I_{xx} + K_{\ddot{\phi}})\ddot{\phi}(t) + K_{\dot{\phi}}\dot{\phi}(t) + K_{\phi}\phi(t) = Ke^{i\omega_e t}.$$

This can be expressed in the form

$$\ddot{\phi}(t) + 2k_{\dot{\phi}}\omega_{\phi}\dot{\phi}(t) + \omega_{\phi}^2\phi(t) = \omega_{\phi}^2 \frac{K}{K_{\phi}} e^{i\omega_e t} = \omega_{\phi}^2 K_1 e^{i\omega_e t}$$

where the natural frequency of roll $\omega_{\phi}$, and the damping factor $k_{\dot{\phi}}$, are given by

$$\omega_{\phi} = \sqrt{\left(\frac{K_{\phi}}{I_{xx} + K_{\ddot{\phi}}}\right)}, \qquad k_{\dot{\phi}} = \frac{K_{\dot{\phi}}}{2\sqrt{\{K_{\phi}(I_{xx} + K_{\ddot{\phi}})\}}}.$$

For a sinusoidal roll motion $\phi(t) = \phi_0 e^{i\omega_e t}$, the receptance

$$H_{\phi K_1}(\omega_e) = \frac{\omega_{\phi}^2}{(\omega_{\phi}^2 - \omega_e^2) + 2ik_{\phi}\omega_{\phi}\omega_e}$$

gives the amplitude and phase of the roll motion for a rolling moment of unit amplitude and frequency $\omega_e$. The square of the modulus of this receptance or the roll response amplitude operator is

$$|H_{\phi K_1}(\omega_e)|^2 = \frac{\omega_{\phi}^4}{\{(\omega_{\phi}^2 - \omega_e^2)^2 + 4k_{\phi}^2\omega_{\phi}^2\omega_e^2\}}$$

and it is a function of $\omega_\phi$, $k_{\dot\phi}$, $\omega_e$. The rolling motion is given by

$$\phi(t) = \frac{\omega_\phi^2 K_1 e^{i(\omega_e t - \beta)}}{\sqrt{\{(\omega_\phi^2 - \omega_e^2)^2 + 4k_\phi^2 \omega_\phi^2 \omega_e^2\}}}$$

where the phase angle between the roll displacement and the roll moment, which is a function of the wave slope as shown in Table X, is

$$\beta = \tan^{-1}\left\{\frac{2k_\phi(\omega_e/\omega_\phi)}{1 - (\omega_e/\omega_\phi)^2}\right\}.$$

When $0 < \omega_e/\omega_\phi \leqslant 1$, $0 < \beta < \pi/2$, the ship rolls with the waves as shown in Figure 64a. When $1 < \omega_e/\omega_\phi < \infty$, $\pi/2 < \beta < \pi$, the ship rolls into the waves as shown in Figure 64b.

For small values of the damping factor $k_{\dot\phi}$ the modulus of the receptance $|H_{\phi K_1}(\omega_e)|$ has very large values for a sinusoidal wave moment excitation applied near the natural frequency of roll motion. Even if the excitation covers a large range of frequencies as in an irregular seaway, the roll response has a dominant value near the natural frequency, $\omega_\phi$. It is seen from Figure 65 that an increase in the damping factor flattens the peak of the receptance, resulting in a significant reduction of roll motion.

A stabilization unit in a ship may be either ‘passive’ or ‘active’. The function of such a system is either to create applied rolling moments that are opposed to that causing a rolling motion, or to increase the damping factor of the rolling motion.

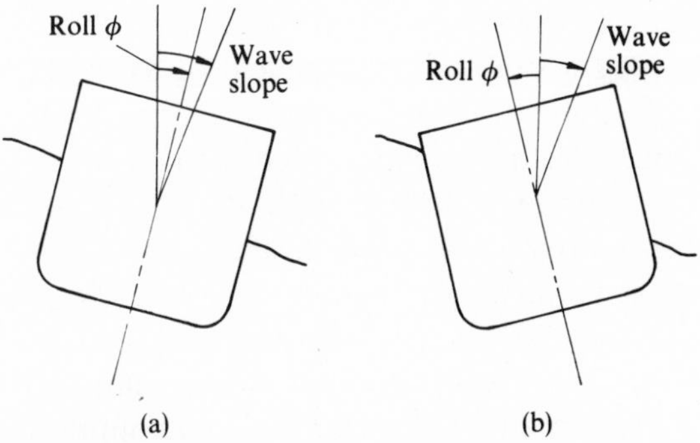


**Figure 64.** Rolling motion of a ship when

(a) $0 < \dfrac{\omega_e}{\omega_\phi} \leqslant 1$ and the ship rolls with the waves, and

(b) $1 < \dfrac{\omega_e}{\omega_\phi} \leqslant \infty$ and the ship rolls against the waves.

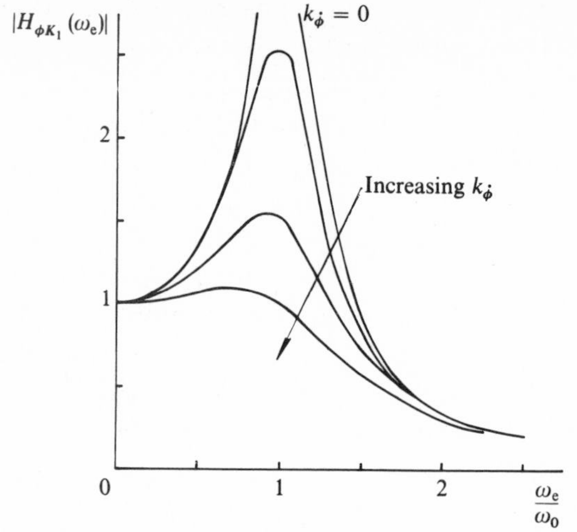


**Figure 65.** Variation of the modulus of the roll receptance with damping coefficient.

The principal passive systems are bilge keels, fixed fins, passive tank systems and passive moving weight systems. These require neither power sources nor special control devices. Bilge keels are the commonest device in this category; they are carefully aligned with the flow around the hull in calm water, so as to minimise the resistance to motion ahead. This resistance is partially offset by the reduction in roll amplitude (which may be as much as 40 per cent).

The principal active systems are active fins, active tank systems, active moving weights and gyroscopes. These use power to achieve a reduction of roll motion and usually depend on gyroscopes for the purposes of control. The gyro system senses the rolling motion of the ship and activates the stabilizing system. For example, external fins may be caused to deflect in such a direction as to produce a moment opposing the roll. The fins may be sited at the turn of the bilge, but they can be withdrawn into the hull when not in use so as to minimise additional resistance. Such systems provide no stabilization when the ship is stopped, or moving at very slow speed. At high speed, reductions of 90 per cent are possible in regular waves encountered at the natural roll frequency of the ship; but in practice a reduction of the order of 75 per cent is more common in normal sea states for a ship already fitted with bilge keels.

## 11.4 The two dimensional sectional parameter

The ideal linearised theory for calculating the coefficients $Z_{\ddot{z}}$, $Y_{\ddot{y}}$, $K_{\ddot{\phi}}$

etc. would make allowances for three dimensional flows satisfying boundary conditions on the hull while the ship is moving ahead and simultaneously performing the appropriate parasitic motion (of heave, pitch, sway . . .) at a prescribed frequency. Unfortunately such a theory does not exist and, as may be seen from the previous sections, the hydrodynamic coefficients are all expressed in terms of their respective two dimensional sectional added mass and damping parameters per unit length, i.e. $m(x)$, and $N(x)$ etc.

The theoretical calculations of these parameters are usually determined by using either the Lewis[18] form method or the Tasai[19, 20] and Porter[21] close fit mapping method or the Frank[22] close-fit source distribution. All three methods assume linear wave theory and neglect viscous effects, so that the velocity potential of a cylinder oscillating in otherwise calm water in either the heave, sway or roll mode may be determined. Knowing this potential, the sectional parameters can be obtained by integrating the pressures given by the Bernoulli equation.

In the Lewis form method the geometrical shape of the section is mathematically represented by a Lewis form which has the same beam, draught and sectional area as the given ship section but not necessarily the same shape as the section. The method is quite accurate for many common ship sections but breaks down at sections with large bulbs or small sectional area.

The close-fit method developed separately by Tasai and Porter conformally maps the ship section into a circle by applying a mapping function with as many coefficients as necessary to obtain the desired close-fit accuracy. In the Frank source distribution method, the shape of the given section is represented by a given number of offset points with straight line segments joining the points. The velocity potential is obtained by distributing pulsating source singularities with constant strength over each of the straight segments. Figure 66 compares some experimental results of Vugts[23] and the theoretical predictions derived from these three methods. (Note that the added masses, the damping coefficients and the frequency $\omega_e$ have all been made dimensionless as signified by the 'overbar'; this has been accomplished as follows:

$$\bar{Y}_{\ddot{y}} = \frac{Y_{\ddot{y}}}{\rho BT}, \quad \bar{Y}_{\dot{y}} = \frac{Y_{\dot{y}}}{\rho T\sqrt{2gB}}, \quad \bar{K}_{\ddot{\phi}} = \frac{\bar{K}_{\ddot{\phi}}}{B^3 T},$$

$$\bar{K}_{\dot{\phi}} = \frac{K_{\dot{\phi}}}{\rho B^2 T\sqrt{2gB}}, \quad \bar{\omega}_e = \omega\sqrt{B/2g},$$

where $\rho$ is the density of water.)

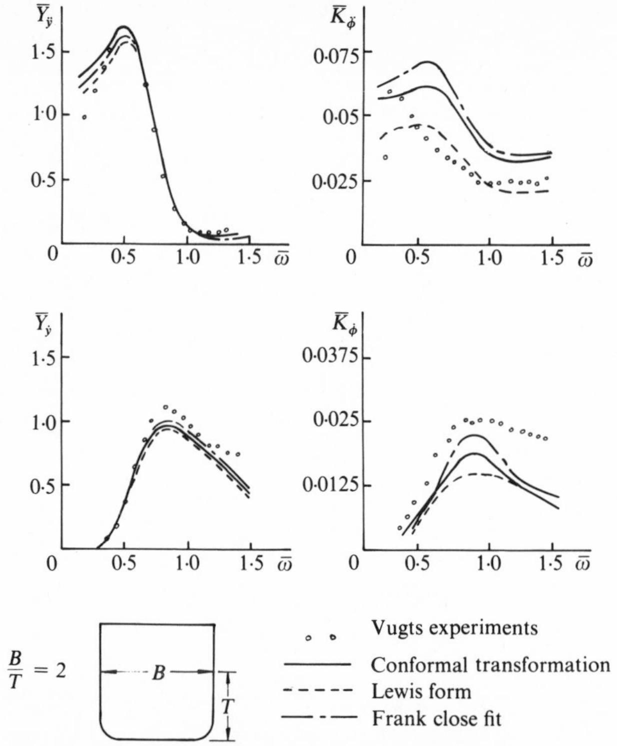


**Figure 66.** Comparison of the theoretical and measured non-dimensional added mass and damping coefficients for sway and roll of the sectional shape shown.

### 11.4.1 Added mass

Although we refer to 'added mass' there is no suggestion that mass is really 'added' in the sense that an identifiable body of water is 'entrained by' the hull. From Tables IX and X we see that it is the added masses or inertias per unit length that are needed and are coupled with the appropriate in-phase motion of the ship. They are physically associated with a stationary wave system created by the motion of the ship. For example, if the oscillatory frequency is high, as in ship vibration, the surface disturbance is negligible. At the low frequencies induced by the waves, the added mass values change with frequency so that information derived from potential flow considerations in deeply submerged forms no longer strictly applies, and needs to be modified.

Among the simplest examples of the theoretical determination of the added mass parameters are those for a circular two dimensional cylinder and those for an ellipsoid. Consider an infinitely long circular cylinder of diameter B moving in a direction perpendicular to its axis in an infinite, incompressible, inviscid liquid of density $\rho$ which is at rest at infinity. Let the velocity of the cylinder at any instant be $u$, so that the kinetic energy per unit length is $mu^2/2$. As the cylinder moves, it causes particles of liquid to move out of its way and round to its 'rear'. The liquid thus acquires kinetic energy[24] and the value at any instant per unit length of cylinder is $\rho\pi B^2 u^2/8$. The total kinetic energy of the cylinder and liquid per unit length is thus $(m + m')u^2/2$ where $m' = \rho\pi B^2/4$. In this case $m'$ is equal to the mass of liquid displaced by unit length of cylinder and this, generally speaking, is far from negligible. If an external force $F$ per unit length acts on the cylinder in the direction of $u$, then

$$Fu = \frac{d}{dt}\left(\frac{mu^2}{2} + \frac{m'u^2}{2}\right)$$

so that

$$F = (m + m')\dot{u}$$

assuming that there is no change in potential energy. The quantity $(m + m')$ is the virtual mass and $m'$ is the added mass, both per unit length.

This result for a moving cylinder is central to the discussion of added mass. Its derivation by consideration of kinetic energy is convenient, but it may also be found by the use of time-dependent potential functions[24, 25]. Being readily available to the analyst, the potential flow around an accelerating cylinder provides a convenient basis of comparison. Let

$$C = \frac{\text{added mass per unit length of other body in two dimensional flow}}{\text{added mass per unit lingth of cylinder having the appropriate 'comparable' size}}.$$

For a ship floating at a free surface there will be two coefficients:

$$C_V = \frac{m'_V}{\pi\rho B^2/8} \qquad \text{and} \qquad C_H = \frac{m'_H}{\pi\rho T^2/2}$$

for vertical (i.e. symmetric) motion and for horizontal (i.e. antisymmetric) motion respectively. Here $B$ is the beam and $T$ is the draught of the ship's section.

Since the added moment of inertia per unit length of a circular cylinder rotating in an inviscid fluid is nil, the added moment of inertia $J'$ of a hull is therefore compared arbitrarily with the quantity $\pi\rho T^4$. Thus

$$C_T = \frac{J'}{\pi\rho T^4}.$$

Actual $C$ values for different cross sectional forms have been tabulated by Korvin-Kroukovsky[5].

The three dimensional ellipsoid is also amenable to a theoretical analysis, and some results for the added mass parameters are shown in Figure 67 in which $\rho$ is the density of water.

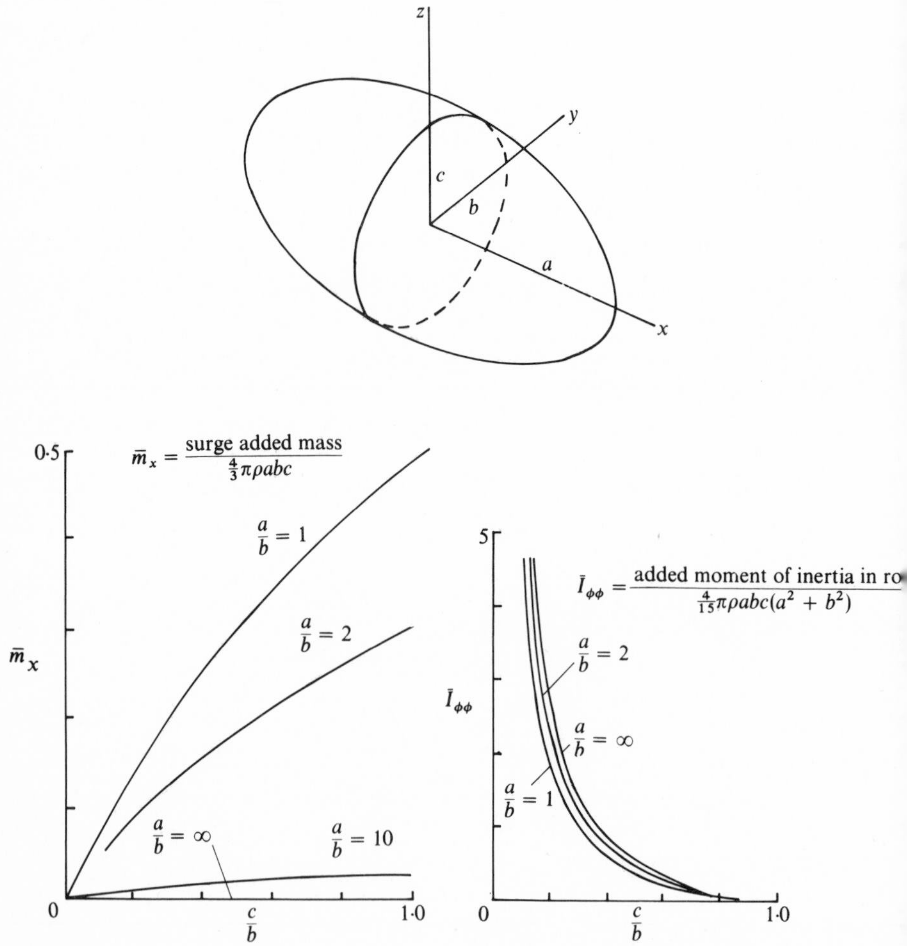

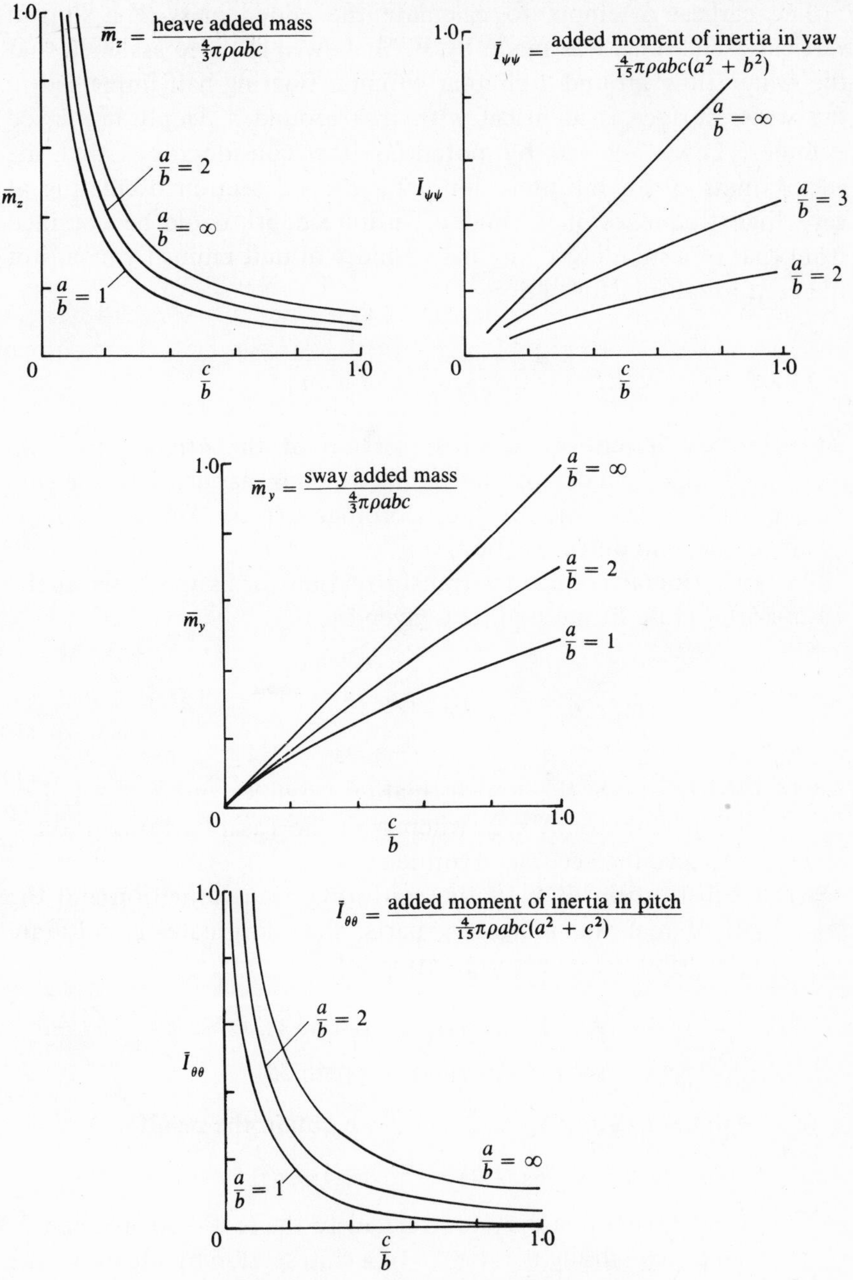


**Figure 67.** Added masses and inertias of an ellipsoid with major semi-axis of length *a* and minor semi-axes of lengths *b* and *c*.

The earliest attempts to calculate the added mass of a ship in vertical oscillations was made by F. M. Lewis[18]. He assumed that the water flow around a circular cylinder floating half immersed in the water surface is identical with that around a deeply immersed cylinder. Lewis showed by potential flow considerations that the added mass of an infinitely long ship-shaped section oscillating at very high frequency in a fluid of infinite depth could be obtained from that of a submerged circular cylinder of unit radius by means of the conformal transformation

$$\kappa = \xi + i\nu = \eta + \frac{a_1}{\eta} + \frac{a_3}{\eta^3}$$

where $\eta = e^{i\beta}$ describes the cross section of the original half immersed circular cylinder of unit radius which is mapped into the ship shaped section described by the coordinates $(\xi, \nu)$. The coefficients $a_1$ and $a_3$ depend on the section.

A more general conformal transformation forms the basis of the Tasai-Porter close fit method. It is given by

$$\kappa = \sum_{p=0}^{N} a_{2p-1}\, \eta^{-(2p-1)}$$

where the $a_{2p-1}$ are the transformation variables and $a_{-1} = 1$. For the Lewis[18] method $N = 2$, whereas in the Tasai-Porter method $N$ is chosen to give the accuracy required.

After substitution of $\eta$ in the conformal transformation and the separation of real and imaginary parts, the coordinates $\xi$ and $\nu$ are given for the simple Lewis formulation by

$$\xi = (1 + a_1)\cos\beta + a_3 \cos 3\beta,$$
$$\nu = (1 - a_1)\sin\beta - a_3 \sin 3\beta.$$

When $\beta = 0$, $\xi = B/2$ and $\beta = \pi/2$, $\nu = T$ we obtain the results

$$B/2 = 1 + a_1 + a_3, \qquad T = 1 - a_1 + a_3.$$

The coefficients $a_1$ and $a_3$ are related to the sectional area coefficient $\sigma_s$ and beam-draught ratio $\lambda_s$ of a ship section by means of the equations

$$\sigma_s = \frac{A_x}{BT} = \frac{\pi(1 - a_1^2 - 3a_3^2)}{4\{(1 + a_3)^2 - a_1^2\}}, \qquad \lambda_s = \frac{B}{T} = \frac{2(1 + a_1 + a_3)}{1 - a_1 + a_3}$$

where $A_x$ is the cross-sectional area of the ship section given by

$$A_x = 2 \int_0^{\pi/2} \xi dv$$

$$= 2 \int_0^{\pi/2} \{(1 + a_1)\cos\beta + a_3 \cos 3\beta\}\{(1 - a_1)\cos\beta - 3a_3 \cos 3\beta\} d\beta$$

$$= \frac{\pi}{2}(1 - a_1^2 - 3a_3^2).$$

Landweber and Macagno[26] have shown that the added mass of a symmetrical hull sectional shape in vertical oscillatory motion is given by

$$m'_V = \frac{\pi\rho}{2}\left\{1 + 2a_1 + \sum_{p=1}^{N}(2p - 1)a_{2p-1}^2\right\}$$

and for horizontal oscillatory motion

$$m'_H = \frac{\pi\rho}{2}\left\{1 - 2a_1 + \sum_{p=1}^{N}(2p - 1)a_{2p-1}^2\right\}$$

where for Lewis forms $N = 2$. Thus it follows that the coefficients $C$ are given as

$$C_V = \frac{m'_V}{\pi\rho B^2/8} = \frac{(1 + a_1)^2 + 3a_3^2}{(1 + a_1 + a_3)^2}$$

and

$$C_H = \frac{m'_H}{\pi\rho T^2/2} = \frac{(1 - a_1)^2 + 3a_3^2}{(1 - a_1 + a_3)^2}$$

where $B$ and $T$ are previously determined.

Other conformal transformations have been used, but the more transformation variables employed the more complicated becomes the analysis – although the transformation does become more accurate.

The previous analysis is only valid for oscillatory motion of high frequency. For a ship in waves, the frequency of oscillation is much decreased. There is now generated, together with the stationary wave disturbance, a progressive wave resulting in a loss of energy by radiation from the ship due to the interaction of the ship and wave disturbance. To allow for this, Ursell[27] introduced a frequency dependent correcting factor $k_4$ which modified the $C_V$ factor to $k_4 C_V$.

### 11.4.2 The damping coefficient

In the same way as the added mass coefficients are influenced by the frequency of oscillation in waves, so are the damping coefficients. From Tables IX and X, we see that the damping coefficients are associated with terms that are dependent on the velocity or angular velocity of the ship's motion. Physically the damping coefficients are associated with a travelling wave system which is set by the motion of the ship and is responsible for the energy dissipated from the ship-wave system.

By considering the rate of energy loss when a ship performs steady vertical oscillations, an expression may be found for the theoretical damping coefficients of a ship section. It is $\rho g^2 \bar{A}^2/\omega_e^2$ where $\bar{A}$ is the 'wave height ratio' given by

$$\bar{A} = \frac{\text{Amplitude of travelling wave}}{\text{Amplitude of body motion}}.$$

The functions $k_4$, $C_V$ and $\bar{A}$ for various ship cross sections are illustrated by E. V. Lewis[28].

## 11.5 Wave receptance

Given the elevation of a wave train at some position $A$ on the free surface, it is sometimes required to know the wave elevation at some other position $B$ not necessarily on the surface, as shown in Figure 68. Consider a regular sinusoidal wave of number $k$ and frequency $\omega$ travelling in deep water and having a wave elevation

$$\zeta(X, Z, t) = \zeta_0 e^{+kZ} e^{i(kX - \omega t)}$$

where the exponential term $e^{+kZ}$ is a correction factor accounting for the attenuation of pressure fluctuations with depth, as discussed in section 7.1.2.

The wave elevation at position A on the free surface ($X = X_1$, $Z = 0$) is given by

$$\zeta(X_1, 0, t) = \zeta_0 e^{i(kX_1 - \omega t)}$$

whilst at position B ($X = X_2, Z = -Z_0$) we have

$$\zeta(X_2, -Z_0, t) = \zeta_0 e^{-kZ_0} e^{i(kX_2 - \omega t)}.$$

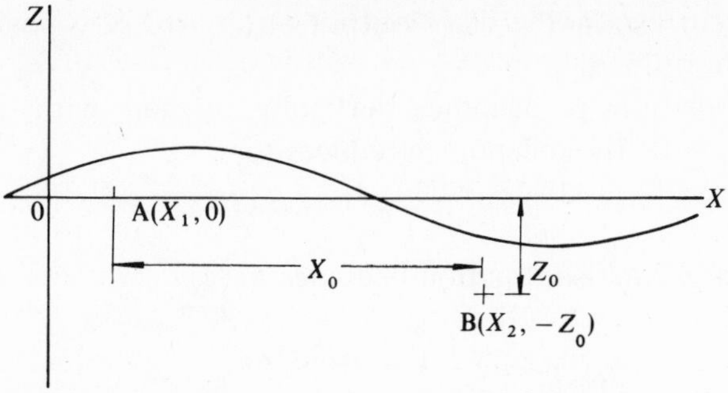


**Figure 68.** Definition of points $A(X_1, O)$ and $B(X_2, -Z_0)$ in a sinusoidal wave.

If the position of $B$ is a distance $X_0$ from $A$ such that $X_2 = X_0 + X_1$ then it follows that

$$\zeta(X_2, -Z_0, t) = e^{-kZ_0} e^{ikX_0} \zeta(X_1, 0, t)$$

which indicates that the observation at $B$ displays a phase difference of $(kX_0)$ with respect to the original observation at $A$.

For deep water waves, $k = \omega^2/g$, so that

$$\zeta(X_2, -Z_0, t) = e^{-(\omega^2/g)(Z_0 - iX_0)} \zeta(X_1, 0, t).$$

On comparing this result with others given in section 10.1.1, we see that it may be interpreted as a wave input $\zeta(X_1, 0, t)$ having an output $\zeta(X_2, -Z_0, t)$ related to it by the receptance

$$H(\omega) = e^{-(\omega^2/g)(Z_0 - iX_0)}.$$

This has been shown by Breslin *et al.* [29] to have a Fourier transform given by the wave impulse function

$$h_\zeta(t) = \frac{1}{2\pi} \int_{-\infty}^{\infty} H(\omega) e^{i\omega t} d\omega = \text{Re} \frac{1}{\pi} \int_0^{\infty} e^{-(\alpha\omega^2/g)} e^{i\omega t} d\omega$$

which, on integration, becomes

$$h_\zeta(t) = \text{Re}\left[\sqrt{\left(\frac{g}{\pi\alpha}\right)} e^{-gt^2/4\alpha} \left\{0{\cdot}5 + \text{erf}\, it \sqrt{\left(\frac{g}{2\alpha}\right)}\right\}\right]$$

where Re denotes the real part of the expression and

$$\alpha = Z_0 - iX_0.$$

Since the wave impulse function is a characteristic of the linear wave system, it follows from section 10.1.1 that for any known wave input the corresponding wave output at an arbitrary position $(X, Z)$ may be determined.

When position $B$ becomes vertically aligned with $A$ (so that $X_0 = 0, Z_0 \neq 0$) the receptance reduces to

$$H(\omega) = e^{-(\omega^2/g)Z_0}$$

and the wave impulse function becomes

$$h_\zeta(t) = \mathrm{Re}\,\frac{1}{\pi}\int_0^\infty e^{-(\omega^2/g)Z_0}\cos\omega t\, d\omega$$

$$= \frac{1}{2}\sqrt{\left(\frac{g}{\pi Z_0}\right)}\, e^{-gt^2/4Z_0}.$$

## 11.6 Transient wave testing

A ship's receptances $H(\omega)$ may either be calculated, as described previously in this chapter, or measured in a series of model tests in regular waves over a required range of wave lengths. A considerable reduction in testing time would result if these tests in regular waves could be replaced by a test with a single wave. In effect, the various regular waves must be replaced by a single wave disturbance which encompasses the required range of wave lengths. Such a disturbance is known as a 'transient wave' and was utilised by Davis and Zarnich[30] to determine the receptances for the ship's heave and pitch motions.

A transient wave is produced by generating a wave train consisting of individual deep water waves, each having velocity

$$c = \frac{\lambda}{T} = \frac{\omega}{k} = \frac{g}{\omega}.$$

The waves of low frequencies travel faster than those of higher frequencies. By the skilful generation of waves in a towing tank, it is possible to arrange for the fast-moving waves of low frequency to catch up the slow-moving waves of higher frequency and to coalesce at a determined position and time. This produces a single large wave having components extending over the range of wave lengths required, and may be thought of as having the shape of an approximate unit impulse function. A model ship at this predetermined position in the tank will now be excited into motion by a unit impulsive wave input.

The theory given in the previous section can now be applied, with $A$ the point of wave generation and $B$ the position of the impulse in the free surface ($Z = 0$). Thus we have

$$\zeta(X_2, t) = H(\omega) \,.\, \zeta(X_1, t)$$

which has a Fourier transform given by

$$\zeta(X_2, \omega) = H(\omega) \,.\, \zeta(X_1, \omega).$$

But at position $X_2$, where the wave impulse occurs,

$$\zeta(X_2, t) = \delta(t)$$

so that

$$\zeta(X_2, \omega) = \frac{1}{2\pi} \int_{-\infty}^{\infty} \zeta(X_2, t) e^{-i\omega t} dt = \frac{1}{2\pi}$$

as we saw in section 6.2. At the position of wave generations, we now have

$$\zeta(X_1, \omega) = \frac{1}{2\pi} H^{-1}(\omega) = \frac{1}{2\pi} e^{-i(\omega^2/g)X_0}$$

which has an inverse Fourier transform given by

$$\zeta(X_1, t) = \int_{-\infty}^{\infty} \zeta(X_1, \omega) e^{i\omega t} d\omega = \frac{1}{2\pi} \int_{-\infty}^{\infty} e^{-i\{(\omega^2 X_0/g) - \omega t\}} d\omega$$

$$= \frac{1}{\pi} \int_0^{\infty} \cos\left(\frac{\omega^2 X_0}{g} - \omega t\right) d\omega.$$

Davis and Zarnich[30] have shown that the integration of this expression gives the initial wave at position $A$ as

$$\zeta(X_1, t) = \sqrt{\frac{g}{2\pi X_0}} \left[\cos\left(\frac{gt^2}{4X_0}\right)\left\{\tfrac{1}{2} + C\left(t\sqrt{\frac{g}{2\pi X_0}}\right)\right\}\right.$$

$$\left. + \sin\left(\frac{gt^2}{4X_0}\right)\left\{\tfrac{1}{2} + S\left(t\sqrt{\frac{g}{2\pi X_0}}\right)\right\}\right]$$

where $C(x)$ and $S(x)$ are Fresnel cosine and sine integrals defined by

$$C(x) = \int_0^x \cos(\pi z^2/2)\,dz$$

$$S(x) = \int_0^x \sin(\pi z^2/2)\,dz$$

with the property that

$$\lim_{x\to\infty} C(x) = \tfrac{1}{2} = \lim_{x\to\infty} S(x).$$

For large values of $\sqrt{g/2\pi X_0}$, i.e. $X_0 \to 0$, the previous expression reduces to

$$\zeta(X_1, t) = \sqrt{\frac{g}{\pi X_0}}\cos\left(\frac{gt^2}{4X_0} - \frac{\pi}{4}\right).$$

Figure 69 shows in non-dimensional form and time scale reversed the initial wave profile which will yield coalescence of the waves to form a unit impulse at a distance $X_0$ away from the position of generation. Thus a transient wave observed in a towing tank which initially has a very high frequency that linearly decreases to zero with constant amplitude, as shown in Figure 69, produces for a brief instant a very large wave at a given position in the tank having the shape of a unit impulse. Figures 70a and b show a comparison of the heave and pitch responses on a stationary model obtained from theory[31] and from regular wave and transient wave experiments.

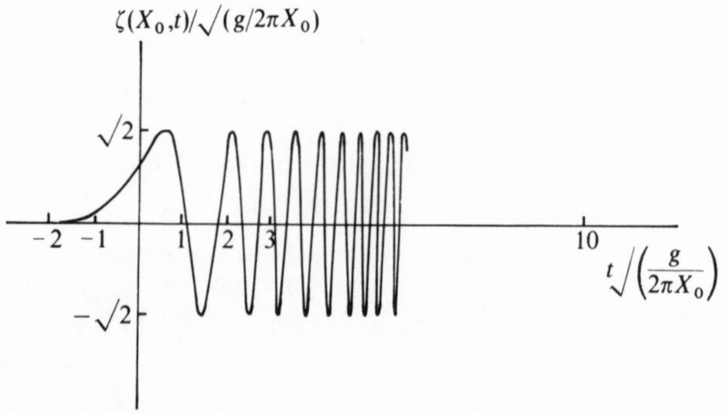


**Figure 69.** Initial wave profile (in non-dimensional form and with time reversed) that will produce an impulsive wave at distance $X_0$ from the point of generation.

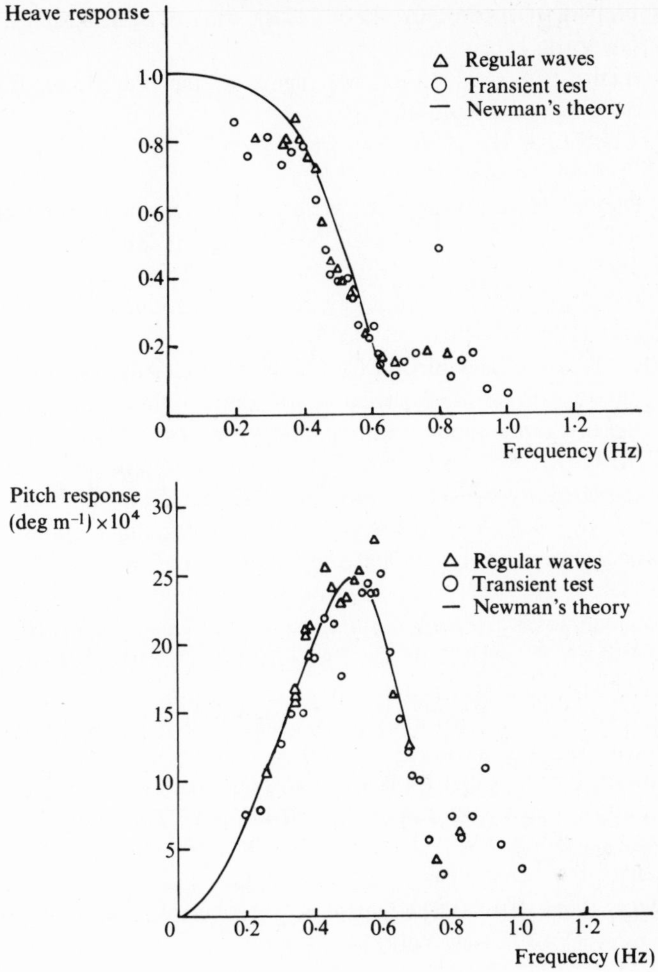


**Figure 70.** Comparison between theoretical and experimental results of a ship's motion in (a) heave, and (b) pitch.

## References


[1] NEWMAN, J. N. and TUCK, E. O., 1964, 'Current progress in the slender body theory of ship motions'. *Fifth Symposium on Naval Hydrodynamics*, Bergen, Norway, pp. 129–166.

[2] NEWMAN, J. N., 1961, 'A linearised theory for the motion of a thin ship in regular waves', *J. Ship Res.*, **5**, 34–55.

[3] KIM, W. D., 1963, 'On the force oscillations of shallow draft ships', *J. Ship Res.*, **7**, 7–18.

[4] VOSSERS, G., 1962, 'Some applications of the slender body theory in ship hydrodynamics', *Dissertation, Tech. Univ. of Delft*, pub. no. 214.

[5] KORVIN-KROUKOVSKY, B. V., 1961, *Theory of Seakeeping, SNAME,* New York.

[6] KRILOFF, A., 1898, 'A general theory of the oscillations of a ship on waves'. *Trans. I.N.A.,* **40**, 135.

[7] MITCHELL, J. H., 1898, 'The wave resistance of a ship'. *Philosophical Magazine, London,* **45**, 106–123.

[8] St. DENIS, M. and PIERSON, W. J., 1953, 'On the motions of ships in confused seas'. *Trans. SNAME*, New York, **61**, 280–332.

[9] PETERS, A. S. and STOKER, J. J., 1957, 'The motion of a ship as a floating rigid body in a seaway'. *Communications in Pure and Applied Math.,* **10**, 399–490.

[10] GERRITSMA, J. and BEUKELMAN, W., 1967, 'Analysis of the modified strip theory for the calculation of ship motions and wave bending moments'. *International Shipbuilding Prog.,* **14**, 319–337.

[11] BEUKELMAN, W., 1970, 'Pitch and heave characteristics of a destroyer'. *International Shipbuilding Prog.,* **17**, 235–252.

[12] TASAI, F., 1967, 'On the swaying, yawing and rolling motions of ships in oblique waves'. *International Shipbuilding Prog.,* **14**, 216–228.

[13] VUGTS, J. H., 1971, 'The hydrodynamic forces and ship motions in oblique waves', *Netherlands Ship Research Centre*, Report No. 150S.

[14] SALVESEN, N., TUCK, E. O., FALTINSEN, O., 1970, Ship motions and sea loads, *SNAME,* **78**, 250–287.

[15] TIMMAN, R. and NEWMAN, J. N., 1962, 'The coupled damping coefficients of a symmetric ship', *J. Ship Res.,* **5**, 1–7.

[16] BISHOP, R. E. D. and EATOCK TAYLOR, R., 1973, 'On wave-induced stress in a ship executing symmetric motions'. *Phil. Trans. Roy. Soc. of London,* **275**, 1–32.

[17] FROUDE, W., 1861, 'On the rolling of ships', *Trans. I.N.A.,* **11**, 180–229.

[18] LEWIS, F. M., 1929, 'The inertia of water surrounding a vibrating ship'. *Trans. SNAME*, New York, **27**, 1–20.

[19] TASAI, F., 1961, 'Hydrodynamic force and moment produced by swaying and rolling oscillation of cylinder on the free surface'. *Research Institute of Applied Mechanics*, Kyushu University, Hakozaki, Fukuoka, Japan.

[20] TASAI, F., 1970, 'On the damping force and added mass of ships heaving and pitching'. *Research Institute of Applied Mechanics*, Kyushu University, Hakozaki, Fukuoka, Japan.

[21] PORTER, W., 1960, 'Pressure distributions, added mass and damping coefficients for cylinder oscillating in a free surface'. *University of California Engineering Publications*, Series no. 82–16.

[22] FRANK, W., 1967, 'Oscillations of cylinders in or below the free surface of deep fluids', *NSRDC. Washington D.C.*, Report no. 2375.

[23] VUGTS, J. H., 1969, 'The coupled roll-sway-yaw performance in oblique waves', 12th International Towing Tank Conference, Rome.

[24] MILNE THOMPSON, L. M., 1962. *Theoretical Hydrodynamics*, Macmillan, London.

[25] LAMB, H., 1945, *Hydrodynamics*, C.U.P. Cambridge.

[26] LANDWEBER, L. and MACAGNO, M. C., 1957, 'Added mass of two dimensional forms oscillating in a free surface'. *J. Ship Res.*, **1**, 20–29.

[27] URSELL, F., 1949, 'On the heaving motions of a circular cylinder on the surface of a fluid'. *Quarterly J. Mechanics and Applied Maths.*, **2**, 335–353.

[28] LEWIS, E. V., 1967, 'The motions of ships in waves'. *Principles of Naval Architecture*, edited J. P. Comstock, *SNAME*, New York, 607–715.

[29] BRESLIN, J. P., SAVITSKY, D. and TSAKONAS, S., 1964, 'Deterministic evaluation of motions of marine craft in irregular seas', *Fifth symposium on Naval Hydrodynamics, Bergen*, 461–505.

[30] DAVIS, M. C. and ZARNICH, E. E., 1964, 'Testing ship models in transient waves'. *Fifth symposium on Naval Hydrodynamics, Bergen*, 507–543.

[31] NEWMAN, J. N., 1964, 'A slender body theory for ship oscillations in waves'. *J. Fluid Mechanics,* **18**, 602–608.

# 12 Motion in a seaway

Anything like a frontal attack on the general problem of motion in a random seaway is likely to be very complicated and hence quite unsatisfactory for our present purpose. For simplicity, we shall therefore assume that the only motions under study are the symmetric motions of heave $z(t)$ and pitch $\theta(t)$ and that these motions are not coupled with surge. The problem of antisymmetric motions may be tackled in a similar way to that described in this chapter, but the exciting forces and moments are then functions of wave slope instead of wave height.

Initially we shall consider a ship travelling in a long-crested seaway described mathematically by a one-dimensional sea spectrum $S_{\zeta\zeta}(\omega)$. The ship-wave system thus has a single input from the waves and we shall examine two outputs. Later we shall widen the discussion by relaxing the restriction to a *long*-crested seaway.

## 12.1 Heave and pitch in a long crested sea

It was shown in section 10.3 that if $\mathbf{S}_{\mathbf{qq}}(\omega)$ and $\mathbf{S}_{\mathbf{QQ}}(\omega)$ are the matrices of mean square spectral densities of output and input respectively and $\mathbf{H}(\omega)$ is the receptance matrix, then

$$\mathbf{S}_{\mathbf{qq}}(\omega) = \mathbf{H}^*(\omega) \,.\, \mathbf{S}_{\mathbf{QQ}}(\omega) \,.\, \mathbf{H}^T(\omega)$$

or

$$\begin{bmatrix} S_{zz}(\omega) & S_{z\theta}(\omega) \\ S_{\theta z}(\omega) & S_{\theta\theta}(\omega) \end{bmatrix} = \begin{bmatrix} H^*_{z\zeta}(\omega) \\ H^*_{\theta\zeta}(\omega) \end{bmatrix} S_{\zeta\zeta}(\omega)[H_{z\zeta}(\omega) \quad H_{\theta\zeta}(\omega)].$$

That is to say the ship receptance must be employed since we wish to relate $z(t)$ and $\theta(t)$ to the wave elevation $\zeta(t)$ rather than to the force and moment, $Z(t)$ and $M(t)$ respectively. It is convenient to refer, not to the mathematical double-sided spectra $S(\omega)$ but rather to the physical single-sided spectra $\Phi(\omega)$; thus we have, alternatively,

$$\begin{bmatrix} \Phi_{zz}(\omega) & \Phi_{z\theta}(\omega) \\ \Phi_{\theta z}(\omega) & \Phi_{\theta\theta}(\omega) \end{bmatrix} = \begin{bmatrix} H^*_{z\zeta}(\omega) \\ H^*_{\theta\zeta}(\omega) \end{bmatrix} \Phi_{\zeta\zeta}(\omega) \; [H_{z\zeta}(\omega) \quad H_{\theta\zeta}(\omega)].$$

The ship receptances are here defined in terms of the absolute frequency $\omega$. In fact, however, the moving ship responds to a seaway with the encounter frequency $\omega_e$. Now the energy of the waves is the same whether it is expressed in terms of $\omega$ or $\omega_e$ so that

$$E[\zeta^2(\mathbf{r}, t)] = \int_0^\infty \Phi_{\zeta\zeta}(\omega)d\omega = \int_0^\infty \Phi_{\zeta\zeta}(\omega_e)d\omega_e$$

where

$$\Phi_{\zeta\zeta}(\omega_e) = \frac{\Phi_{\zeta\zeta}(\omega)}{|d\omega_e/d\omega|}$$

(see section 9.2). The modulus of $d\omega_e/d\omega$ is used because $\Phi_{\zeta\zeta}(\omega_e)$ must always be positive (c.f. section 6). In a long-crested sea, with $\mu$ a constant, the frequency of encounter $\omega_e$ is given by

$$\omega_e = \omega - \frac{\omega^2}{g}\bar{U} \cos \chi$$

so that

$$\Phi_{\zeta\zeta}(\omega_e) = \frac{\Phi_{\zeta\zeta}(\omega)}{\left|1 - \dfrac{2\omega\bar{U}}{g} \cos \chi\right|}.$$

Given the wave spectrum $\Phi_{\zeta\zeta}(\omega)$, it is thus possible to find the 'wave encounter spectrum', $\Phi_{\zeta\zeta}(\omega_e)$. In general the effect of the transformation is to produce a flattened spectrum extending over a wider range of frequencies as the ship speed increases. This is illustrated in Figure 71. The ship receptances $H_{z\zeta}(\omega)$ and $H_{\theta\zeta}(\omega)$ may be replaced by $H_{z\zeta}(\omega_e)$ and $H_{\theta\zeta}(\omega_e)$ so that the output-input relations become

$$\Phi_{zz}(\omega_e) = |H_{z\zeta}(\omega_e)|^2\Phi_{\zeta\zeta}(\omega_e),$$
$$\Phi_{z\theta}(\omega_e) = H^*_{z\zeta}(\omega_e)H_{\theta\zeta}(\omega_e)\Phi_{\zeta\zeta}(\omega_e),$$
$$\Phi_{\theta z}(\omega_e) = H_{z\zeta}(\omega_e)H^*_{\theta\zeta}(\omega_e)\Phi_{\zeta\zeta}(\omega_e),$$
$$\Phi_{\theta\theta}(\omega_e) = |H_{\theta\zeta}(\omega_e)|^2\Phi_{\zeta\zeta}(\omega_e),$$

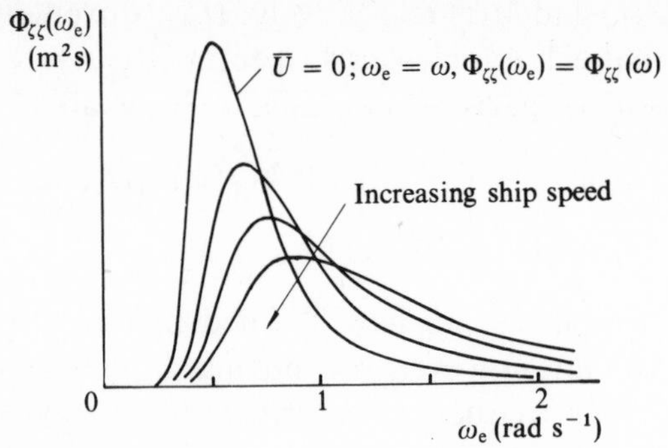


**Figure 71.** Variation of a wave encounter spectrum with ship speed.

where the receptances

$$H_{z\zeta}(\omega_e) = H_{z\zeta}(\omega); \qquad H_{\theta\zeta}(\omega_e) = H_{\theta\zeta}(\omega)$$

for a given relationship between the frequencies $\omega$ and $\omega_e$.

The one-sided mean square spectral densities $\Phi_{zz}(\omega_e)$, $\Phi_{\zeta\zeta}(\omega_e)$, etc. have an important physical interpretation. They may be thought of as relating to records taken on board ship. The above relationships between these quantities are important, because they permit useful calculations to be performed. Consider some of the possibilities with a ship in head seas:

(a) If $\Phi_{\zeta\zeta}(\omega_e)$ and $|H_{z\zeta}(\omega_e)|^2$ are known, $\Phi_{zz}(\omega_e)$ can be determined. That is, if $|H_{z\zeta}(\omega_e)|^2$ has been found by calculation or model tests, the spectral density of heave motion can be found for any one-dimensional seaway whose spectrum is given.

(b) If $\Phi_{\zeta\zeta}(\omega_e)$ and $\Phi_{zz}(\omega_e)$ are known, $|H_{z\zeta}(\omega_e)|^2$ may be calculated. This has found application in experimentation with models, because $|H_{z\zeta}(\omega_e)|^2$ can be determined in one run in a known irregular wave system. This is to obviate the need for many runs, each in a regular wave system of different wave length.

(c) Suppose that $\Phi_{zz}(\omega_e)$ and $|H_{z\zeta}(\omega_e)|^2$ are known. This is the case when measurements are made of response on board a ship whose modified receptance is known. The spectral density $\Phi_{\zeta\zeta}(\omega_e)$ can be calculated in theory, but this method of obtaining the sea spectrum is full of difficulties because of spreading of the sea waves.

Further, the ship receptances $H_{z\zeta}(\omega_e)$ and $H_{\theta\zeta}(\omega_e)$ can be calculated if the cross spectral densities of input and output are known. Thus, following the theory of section 10.3 we have

$$\Phi_{\zeta z}(\omega_e) = H_{z\zeta}(\omega_e)\Phi_{\zeta\zeta}(\omega_e),$$
$$\Phi_{\zeta\theta}(\omega_e) = H_{\theta\zeta}(\omega_e)\Phi_{\zeta\zeta}(\omega_e).$$

The phase relationship between the responses and the waves can be found as shown in section 10.2.1. The magnitudes of the receptances and the phase relationships between the responses and the waves are given by

$$|H_{z\zeta}(\omega_e)| = \frac{|\Phi_{\zeta z}(\omega_e)|}{\Phi_{\zeta\zeta}(\omega_e)}$$

with phase relation

$$\epsilon_{\zeta z}(\omega_e) = \tan^{-1}\left[\frac{Q_{\zeta z}(\omega_e)}{C_{\zeta z}(\omega_e)}\right]$$

and

$$|H_{\theta\zeta}(\omega_e)| = \frac{|\Phi_{\zeta\theta}(\omega_e)|}{\Phi_{\zeta\zeta}(\omega_e)}$$

with phase relation

$$\epsilon_{\zeta\theta}(\omega_e) = \tan^{-1}\left[\frac{Q_{\zeta\theta}(\omega_e)}{C_{\zeta\theta}(\omega_e)}\right].$$

The phase relationship between the heave and pitch random processes is given by

$$\epsilon_{z\theta}(\omega_e) = \epsilon_{\zeta z}(\omega_e) - \epsilon_{\zeta\theta}(\omega_e)$$
$$= \tan^{-1}\left[\frac{Q_{\zeta z}(\omega_e)C_{\zeta\theta}(\omega_e) - Q_{\zeta\theta}(\omega_e)C_{\zeta z}(\omega_e)}{C_{\zeta z}(\omega_e)C_{\zeta\theta}(\omega_e) + Q_{\zeta\theta}(\omega_e)Q_{\zeta z}(\omega_e)}\right].$$

The theory given in section 6.4 shows that the coherence functions for the wave input and ship's motions have the form

$$\gamma_{z\zeta}(\omega_e) = \frac{|\Phi_{z\zeta}(\omega_e)|^2}{\Phi_{\zeta\zeta}(\omega_e)\Phi_{zz}(\omega_e)}.$$

Figure 72 illustrates typical coherence functions as determined by Dalzell and Yamanouchi[1].

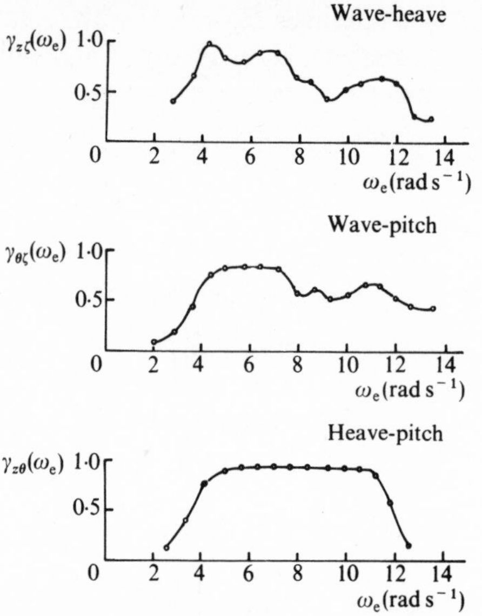


**Figure 72.** Typical coherence functions (after Dalzell and Yamanouchi).

## 12.2 Heave and pitch in a general seaway

Consider now the general three-dimensional sea spectrum $\Phi_{\zeta\zeta}(\mathbf{k}, \omega)$. If $H_{z\zeta}(\mathbf{k}, \omega)$ and $H_{\theta\zeta}(\mathbf{k}, \omega)$ are the ship receptances in the absolute frequency $\omega$, then the individual one-sided spectral density functions of the response motions are

$$\Phi_{zz}(\mathbf{k}, \omega) = |H_{z\zeta}(\mathbf{k}, \omega)|^2 \Phi_{\zeta\zeta}(\mathbf{k}, \omega),$$
$$\Phi_{z\theta}(\mathbf{k}, \omega) = H^*_{z\zeta}(\mathbf{k}, \omega) H_{\theta\zeta}(\mathbf{k}, \omega) \Phi_{\zeta\zeta}(\mathbf{k}, \omega),$$
$$\Phi_{\theta z}(\mathbf{k}, \omega) = H_{z\zeta}(\mathbf{k}, \omega) H^*_{\theta\zeta}(\mathbf{k}, \omega) \Phi_{\zeta\zeta}(\mathbf{k}, \omega),$$
$$\Phi_{\theta\theta}(\mathbf{k}, \omega) = |H_{\theta\zeta}(\mathbf{k}, \omega)|^2 \Phi_{\zeta\zeta}(\mathbf{k}, \omega).$$

These results are in the most general form for a stationary ship, but they may be simplified since the wave number components in deep water are related to the frequency and to the relative angle between ship and wave.

For any component wave we may write

$$k_x = k \cos \chi, \qquad k_y = k \sin \chi$$

where $\chi = \mu - \xi$ (see section 11.1). But $k = \omega^2/g$ in deep water so that $k_x$ and $k_y$ are functions of $\chi$ and $\omega$. Therefore each component wave is determined by its amplitude, relative direction of propagation and absolute frequency. Hence the wave spectral density is of the form $\Phi_{\zeta\zeta}(\chi, \omega)$ and, if the ship receptances are defined accordingly, we find that

$$\Phi_{zz}(\chi, \omega) = |H_{z\zeta}(\chi, \omega)|^2 \Phi_{\zeta\zeta}(\chi, \omega),$$
$$\Phi_{z\theta}(\chi, \omega) = H^*_{z\zeta}(\chi, \omega) H_{\theta\zeta}(\chi, \omega) \Phi_{\zeta\zeta}(\chi, \omega),$$
$$\Phi_{\theta z}(\chi, \omega) = H_{z\zeta}(\chi, \omega) H^*_{\theta\zeta}(\chi, \omega) \Phi_{\zeta\zeta}(\chi, \omega),$$
$$\Phi_{\theta\theta}(\chi, \omega) = |H_{\theta\zeta}(\chi, \omega)|^2 \Phi_{\zeta\zeta}(\chi, \omega).$$

Again the ship experiences excitation by waves, not at the absolute frequency $\omega$, but at the frequency of encounter $\omega_e$. Again a transformation is required. The energy of the waves is the same in both the $\omega$ domain and the $\omega_e$ domain, so

$$E[\zeta^2(\mathbf{r}, t)] = \int_0^\infty \int_0^\infty \int_0^\infty \Phi_{\zeta\zeta}(\mathbf{k}, \omega) d\mathbf{k}\, d\omega = \int_0^\infty \int_0^\infty \int_0^\infty \Phi_{\zeta\zeta}(\mathbf{k}, \omega_e) d\mathbf{k}\, d\omega_e$$

and so

$$\Phi_{\zeta\zeta}(\mathbf{k}, \omega) d\omega = \Phi_{\zeta\zeta}(\mathbf{k}, \omega_e) d\omega_e.$$

which reduces to

$$\Phi_{\zeta\zeta}(\chi, \omega) d\omega = \Phi_{\zeta\zeta}(\chi, \omega_e) d\omega_e.$$

With

$$\omega_e = \omega - \frac{\bar{U}\omega^2}{g} \cos \chi$$

it follows for the deep-water waves that

$$\frac{d\omega_e}{d\omega} = 1 - \frac{2\bar{U}\omega}{g} \cos \chi$$

so that

$$\Phi_{\zeta\zeta}(\chi, \omega_e) = \frac{\Phi_{\zeta\zeta}(\chi, \omega)}{\left| 1 - \dfrac{2\bar{U}\omega}{g} \cos \chi \right|}.$$

Hence, given $\Phi_{\zeta\zeta}(\chi, \omega)$ we can determine $\Phi_{\zeta\zeta}(\chi, \omega_e)$.

The ship receptance $H_{z\zeta}(\mathbf{k}, \omega)$ and $H_{\theta\zeta}(\mathbf{k}, \omega)$ can be redefined in terms of the variables $\chi$ and $\omega_e$ so that we find

$$\Phi_{zz}(\chi, \omega_e) = |H_{z\zeta}(\chi, \omega_e)|^2 \Phi_{\zeta\zeta}(\chi, \omega_e),$$
$$\Phi_{z\theta}(\chi, \omega_e) = H^*_{z\zeta}(\chi, \omega_c) H_{\theta\zeta}(\chi, \omega_e) \Phi_{\zeta\zeta}(\chi, \omega_e),$$
$$\Phi_{\theta z}(\chi, \omega_e) = H_{z\zeta}(\chi, \omega_e) H^*_{\theta\zeta}(\chi, \omega_e) \Phi_{\zeta\zeta}(\chi, \omega_c),$$
$$\Phi_{\theta\theta}(\chi, \omega_e) = |H_{\theta\zeta}(\chi, \omega_e)|^2 \Phi_{\zeta\zeta}(\chi, \omega_e).$$

Since these spectra are of the most general type that we shall discuss in this book, it will be convenient to recall their significance. It will be remembered that, for instance,

$$E[z^2(\mathbf{r}, t)] = \int_0^{\infty} \int_{\chi} \Phi_{zz}(\chi, \omega_e) d\chi \, d\omega_e$$

$$= \langle z^2(\mathbf{r}, t) \rangle$$

which denotes the mean square value of the heave motion of the ship measured at time $t$ and arbitrary position $\mathbf{r}$ in the seaway. Again, for the cross spectral density $\Phi_{z\theta}(\chi, \omega_e)$

$$E[z(\mathbf{r}, t)\theta(\mathbf{r}, t)] = \int_0^{\infty} \int_{\chi} \Phi_{z\theta}(\chi, \omega_e) d\chi \, d\omega_e = \langle z(\mathbf{r}, t)\theta(\mathbf{r}, t) \rangle.$$

This latter result refers to the mean product of the ergodic random processes $z(\mathbf{r}, t)$ and $\theta(\mathbf{r}, t)$ measured at the same time $t$ and position $\mathbf{r}$ in the seaway.

Notice that the one-dimensional response spectra may be obtained by integrating over all values of the angle $\chi$ of relative direction. For example, the one-dimensional heave spectrum is given by

$$\Phi_{zz}(\omega_e) = \int_{\chi} |H_{z\zeta}(\chi, \omega_e)|^2 \Phi_{\zeta\zeta}(\chi, \omega_e) d\chi$$

though it must be remembered that $\omega_e$ and $\chi$ are not always independent of each other and care must be taken in deciding the limits of integration, as we shall discuss in the following sections.

### 12.2.1 Relative motion

The vertical displacement upward at any position $x$ from the origin of the body axes fixed in the ship is

$$z(t) - x\theta(t)$$

where the heave and pitch responses are measured at the origin. The relative vertical motion, $z_r(t)$, between the ship and the wave elevation measured at $x$ is

$$z_r(t) = z(t) - x\theta(t) - \zeta(t).$$

The auto-correlation function of the relative motion is

$$
\begin{aligned}
R_{z_r z_r}(\tau) &= \langle z_r(t) z_r(t+\tau)\rangle \\
&= \langle\{z(t) - x\theta(t) - \zeta(t)\}\{z(t+\tau) - x\theta(t+\tau) - \zeta(t+\tau)\}\rangle \\
&= R_{zz}(\tau) + x^2 R_{\theta\theta}(\tau) + R_{\zeta\zeta}(\tau) - \{R_{\zeta z}(\tau) + R_{\zeta z}(-\tau)\} \\
&\qquad - x\{R_{\theta z}(\tau) + R_{\theta z}(-\tau)\} + x\{R_{\zeta\theta}(\tau) + R_{\zeta\theta}(-\tau)\}
\end{aligned}
$$

where results from section 5.4 have been used under the assumption that the ship and wave motion processes are ergodic. By transforming, and employing the theory in section 6.4, the spectral density function of the relative vertical motion may be found; it is

$$
\begin{aligned}
\Phi_{z_r z_r}(\omega) = \Phi_{zz}(\omega) + x^2 \Phi_{\theta\theta}(\omega) + \Phi_{\zeta\zeta}(\omega) - 2\Phi^{R}_{\zeta z}(\omega) \\
- 2x\Phi^{R}_{\theta z}(\omega) + 2x\Phi^{R}_{\zeta\theta}(\omega)
\end{aligned}
$$

where the superscript R refers to the real part. However, we see from the previous sections that the ship motions may be written in terms of the wave input so that this last result reduces to the form

$$\Phi_{z_r z_r}(\omega) = |H_{z_r\zeta}(x;\omega)|^2 \Phi_{\zeta\zeta}(\omega)$$

where

$$
\begin{aligned}
|H_{z_r\zeta}(x;\omega)|^2 = 1 + x^2|H_{\theta\zeta}(\omega)|^2 + |H_{z\zeta}(\omega)|^2 - 2H^{R}_{z\zeta}(\omega) \\
- 2x\{H^{*}_{\theta\zeta}(\omega)H_{z\zeta}(\omega)\}^{R} + 2xH^{R}_{\theta\zeta}(\omega).
\end{aligned}
$$

When $x = L/2$, the relative motion at the bows is obtained and the mean square relative motion is given by

$$\langle z_r^2(t)\rangle = \int_0^\infty \left|H_{z_r\zeta}\left(\frac{L}{2};\omega\right)\right|^2 \Phi_{\zeta\zeta}(\omega)d\omega.$$

The moving ship experiences excitation at the encounter frequency $\omega_e$ rather than the absolute frequency $\omega$, so that the relative motion spectrum becomes

$$\Phi_{z_r z_r}(\omega_e) = |H_{z_r\zeta}(x;\omega_e)|^2 \Phi_{\zeta\zeta}(\omega_e)$$

and from section 10.22 it follows that the relative velocity spectrum is

$$\Phi_{\dot z_r \dot z_r}(\omega_e) = \omega_e^2 |H_{z_r\zeta}(x;\omega_e)|^2 \Phi_{\zeta\zeta}(\omega_e).$$

Figure 73[2] illustrates the forms of relative motion, velocity and the response amplitude operator $|H_{z_r\zeta}(x; \omega_e)|^2$ at station 3 (0·1 times ship's length $L$ aft of the forward perpendicular) for a ship travelling with forward speed of 10 knots in a head sea of sea state 7.

## 12.3 Practical heaving and pitching processes

It has been mentioned that the seaways which are commonly observable may be described as 'long-crested' or 'short-crested'. It is timely to recall the difference.

In a long-crested irregular seaway no spreading occurs and $\mu$ is a constant. The wave crests are parallel to one another, all advancing in the same direction. They are irregular in the sense that the distances between crests vary and the amplitudes vary. In this case, the general expression for the frequency of encounter in deep water, namely

$$\omega_e = \omega - \frac{\bar{U}\omega^2}{g} \cos(\mu - \xi)$$

takes the particular form

$$\omega_e = \omega - \frac{\bar{U}\omega^2}{g} \cos \chi$$

where $\chi$ is a constant for a ship with a fixed heading.

By contrast, spreading does occur in a short-crested seaway. Component waves do not normally occur travelling in *all* directions with equal probabilities and there is normally a 'dominant' direction of propagation. Since $\mu$ is not now unique, the relative direction $\chi$ between the wave component and the ship heading is to be regarded, not as a constant but as a random variable. There is thus a fundamentally different interpretation to make when applying the last equation for $\omega_e$ to short crested seas.

Notice that, for both long and short crested seas

$$\frac{d\omega_e}{d\omega} = 1 - \frac{2\bar{U}\omega}{g} \cos \chi$$

giving a maximum value

$$\omega_e = \frac{g}{4\bar{U} \cos \chi}$$

when

$$\omega = \frac{g}{2\bar{U} \cos \chi}$$

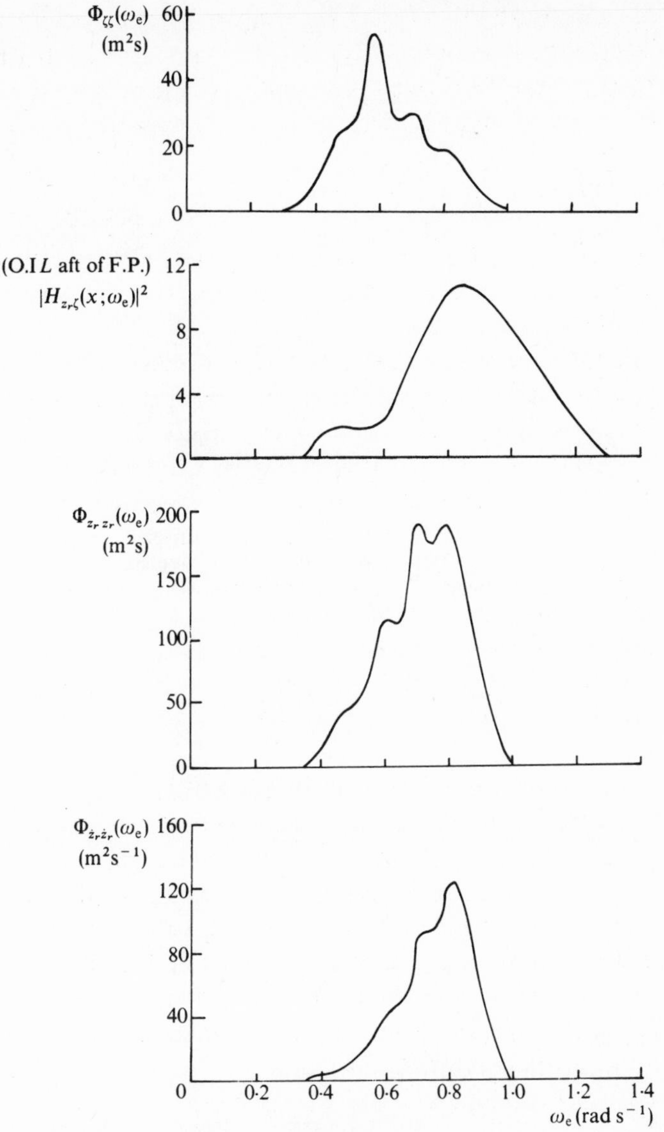


**Figure 73.** Encounter and relative motion spectra for a ship travelling at 10 knots in a head sea of state 7.

where $\chi$ is constant for the former case and is a random variable for the latter. The physical meaning of this result may be seen by thinking of a ship travelling with constant speed $\bar{U}$ in long-crested waves such that $\chi = 0$. As Figure 74 shows there is an algebraic maximum frequency of encounter with the stern when $\omega = g/2\bar{U}$, and when $\omega > g/\bar{U}$ the encounter is with the bows of the ship. This is

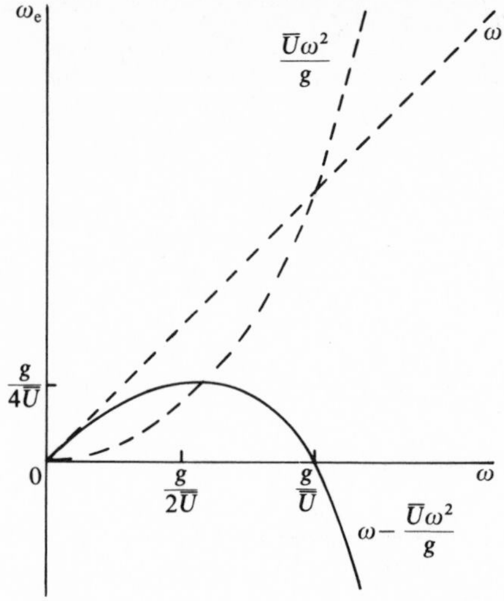


**Figure 74.** Variation of encounter frequency with frequency for a given ship speed $\bar{U}$ in head and following seas.

most easily visualised if it is remembered that the speed of propagation of a wave is

$$c = \frac{\omega}{k}$$

and that in deep water this becomes

$$c = \frac{\omega}{\omega^2/g} = \frac{g}{\omega}.$$

Wave encounter is with the stern if $\bar{U} < c$ and with the bow if $\bar{U} > c$.

### 12.3.1 Head seas

For waves approaching a ship from the bow,

$$\frac{\pi}{2} < \chi < \frac{3\pi}{2}$$

so that cos $\chi$ is negative and the frequency of encounter $\omega_e$ is an uniquely defined positive quantity whose value is greater than $\omega$, the absolute frequency of the waves.

A long crested sea will provide a single constant value of $\chi$ within the range mentioned. If $\chi = \pi$, the ship moves in a true head sea. A possible spectrum is then $\Phi_{\zeta\zeta}(\omega_e)$ as shown in Figure 75a and the modified pitching receptance is such that $|H_{\theta\zeta}(\omega_e)|^2$ is as indicated

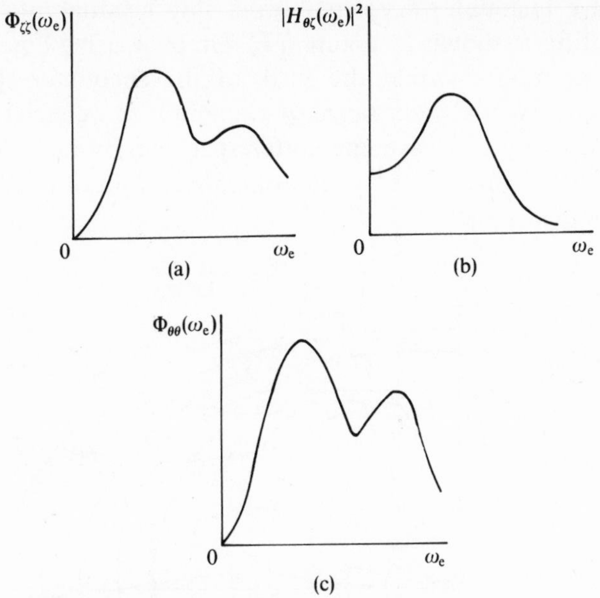


**Figure 75.** Possible forms of (a) a wave encounter spectrum, (b) a typical pitch response, and (c) a pitch encounter spectrum.

in Figure 75b. It has been shown by Conolly[3] that for a particular ship travelling at 19 knots in a head sea the pitching spectrum $\Phi_{\theta\theta}(\omega_e)$ was as shown sketched in Figure 75c.

For a short crested sea in deep water, two dimensional spectra must be used. Figure 76 shows a two dimensional sea spectrum for waves approaching the bows from all directions within 60° of the dominant direction.

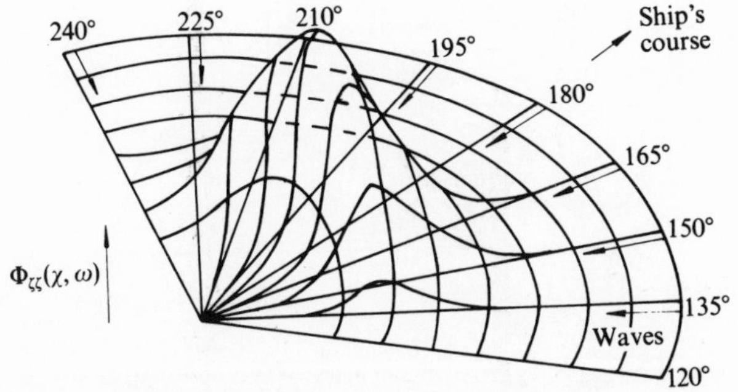


**Figure 76.** A typical directional wave spectrum for a ship travelling in the dominant direction of a short-crested head sea.

The wave encounter spectrum for a ship heading into the dominant direction is shown in Figure 77a. On comparing Figures 76 and 77a it is seen that, while the peak of the encounter spectrum is depressed, it covers a wider range of encounter frequencies $\omega_e$.

By multiplying the wave encounter spectrum by the required ship receptance of Figure 77b, the encounter spectrum of motion, Figure 77c, is obtained.

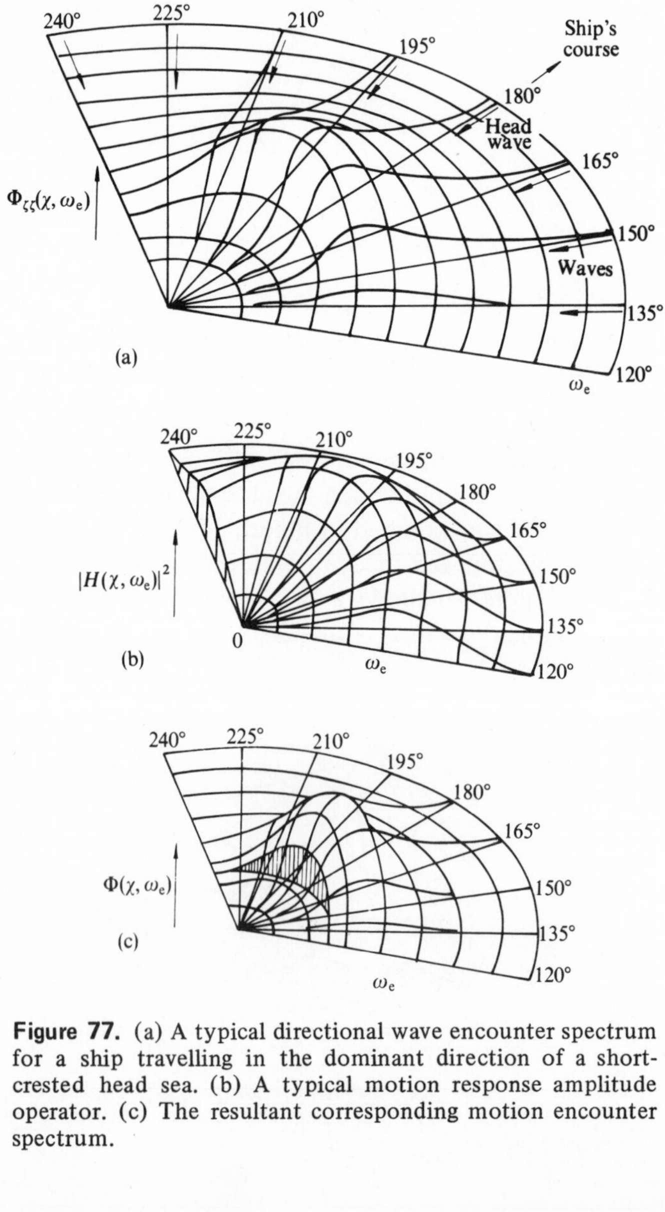


**Figure 77.** (a) A typical directional wave encounter spectrum for a ship travelling in the dominant direction of a short-crested head sea. (b) A typical motion response amplitude operator. (c) The resultant corresponding motion encounter spectrum.

The mean square of the pitching motion $\Phi_{\theta\theta}(\omega_e)$ observed on board ship at each frequency of encounter, contains contributions from all waves having relative directions ranging from

$$\frac{\pi}{2} < \chi < \frac{3\pi}{2}$$

as is suggested by the shaded area of Figure 77c. Thus the one-dimensional encounter spectrum of pitching motion is given by

$$\Phi_{\theta\theta}(\omega_e) = \int_{\chi} |H_{\theta\zeta}(\chi, \omega_e)|^2 \, \Phi_{\zeta\zeta}(\chi, \omega_e) d\chi$$

from which the statistical properties discussed in section 9.6 may be calculated.

### 12.3.2 Following seas

For waves of absolute frequency $\omega$ approaching the ship from astern

$$0 < \chi < \frac{\pi}{2} \text{ and } \frac{3\pi}{2} < \chi < 2\pi$$

so that $\cos\chi$ is positive. The encounter frequency over the stern is

$$\omega_e = \omega - \frac{\bar{U}\omega^2}{g} \cos\chi$$

which becomes negative if

$$\frac{\bar{U}\omega}{g} \cos\chi > 1.$$

As we have seen (Figure 74) this is because the ship may outstrip the wave.

Following St. Denis and Pierson[4] we may redefine the encounter frequency as

$$\omega_e = \omega \left(1 - \frac{\bar{U}\omega}{g} \cos\chi\right)$$

for $0 \leqslant \omega \leqslant \dfrac{g}{\bar{U}\cos\chi}$, and

$$\omega_e = -\,\omega \left(1 - \frac{\bar{U}\omega}{g} \cos\chi\right)$$

for $\dfrac{g}{\bar{U}\cos\chi} < \omega \leqslant \infty$. The variation of $\omega_e$ with $\omega$, when it is defined in this way, is as shown in Figure 78. For waves in deep water.

$$c = \omega/k = g/\omega.$$

The axis of $\omega$ can usefully be divided into three ranges:

I: $$\omega < \frac{g}{2\bar{U}\cos\chi}$$

so that

$$\bar{U}\cos\chi < \frac{c}{2}$$

II: $$\frac{g}{2\bar{U}\cos\chi} < \omega < \frac{g}{\bar{U}\cos\chi}$$

so that

$$\frac{c}{2} < \bar{U}\cos\chi < c$$

III: $$\frac{g}{\bar{U}\cos\chi} < \omega < \infty$$

so that

$$c < \bar{U}\cos\chi < \infty.$$

For each value of $\omega$ there is a unique corresponding value of $\omega_e$. The converse is not true, however, since for $0 \leqslant \omega_e \leqslant g/4\bar{U}\cos\chi$, there are three values of $\omega$ which give the same encounter frequency.

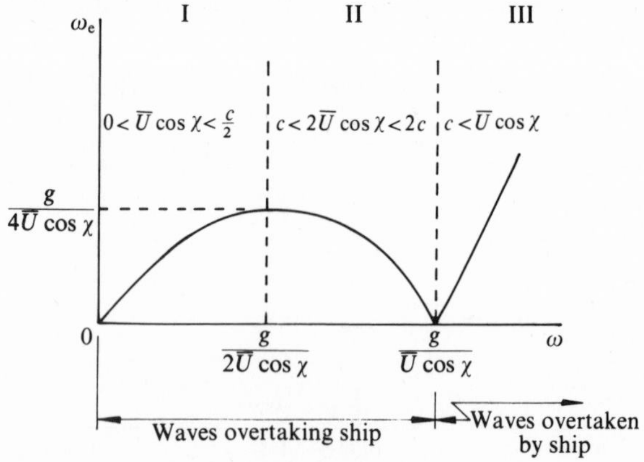


**Figure 78.** Variation of encounter frequency $\omega_e$ with wave frequency $\omega$ for a ship in a following sea showing three possible régimes.

Consider a long-crested sea having the spectral density $\Phi_{\zeta\zeta}(\omega)$ shown in Figure 79a. The encounter spectral density is

$$\Phi_{\zeta\zeta}(\omega_e) = \frac{\Phi_{\zeta\zeta}(\omega)}{\left|1 - \frac{2\bar{U}}{g}\cos\chi\right|}$$

where $\chi$ is now a constant (since the waves are long-crested) and where the modulus of the expression in the denominator is now used in view of the new definition of $\omega_e$. The curve of $\Phi_{\zeta\zeta}(\omega_e)$ has the form shown in Figure 79b. The curve is replotted in a different form in Figure 79c.

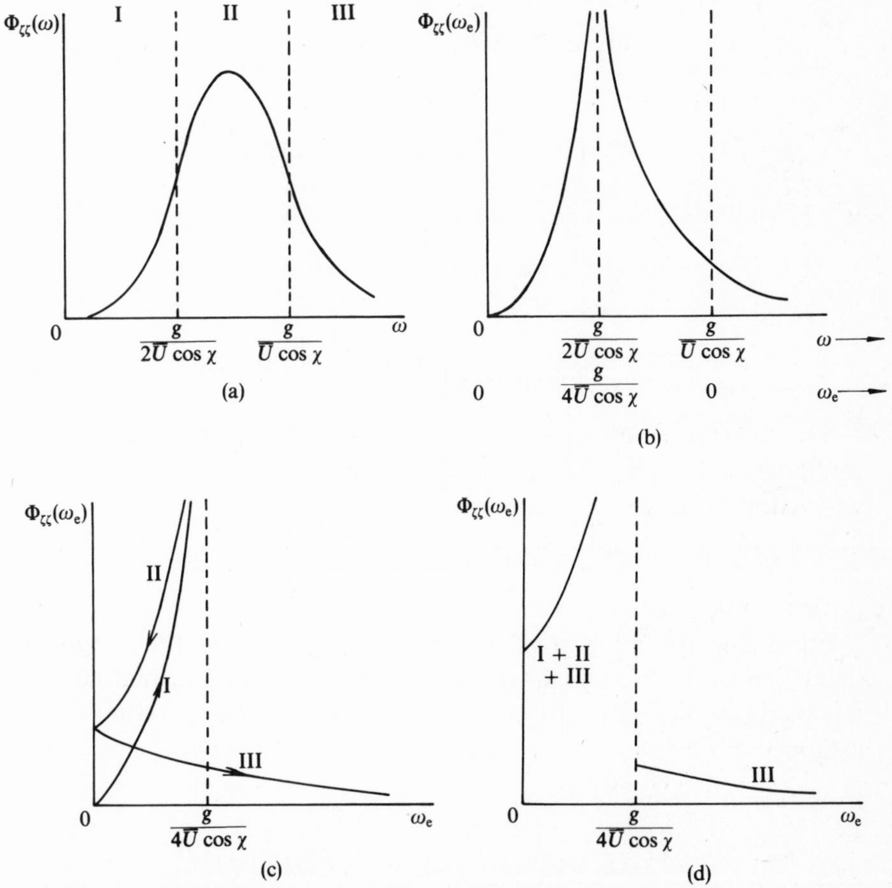


**Figure 79.** Typical forms of (a) wave spectrum, (b) wave encounter spectrum in a following sea, (c) the contributions to the wave encounter spectrum, and (d) the total wave encounter spectrum for the three regions I, II, III shown in figure 78.

The contributions from the three regions of $\omega_e$ are additive in the range $0 \leqslant \omega_e \leqslant g/4\bar{U} \cos \chi$. The spectrum that would be calculated from a wave record taken in a moving ship would therefore be of the form shown in Figure 79d. Although it is possible to derive the encounter spectrum $\Phi_{\zeta\zeta}(\omega_e)$ from the sea spectrum $\Phi_{\zeta\zeta}(\omega)$, the converse is not true in a following sea because we are not able to apportion $\Phi_{\zeta\zeta}(\omega_e)$ between the three regions in the range $0 \leqslant \omega_e \leqslant g/4\bar{U} \cos \chi$.

In order to obtain the motion spectra in the encounter frequency it is necessary to employ Figure 79d together with the ship receptances appropriate to the three regions. Thus, for $\Phi_{zz}(\omega_e)$ we have:

from region I

$$\Phi_{zz}^{I}(\omega_e) = |H_{z\zeta}^{I}(\omega_e)|^2 \Phi_{\zeta\zeta}^{I}(\omega_e);$$

from region II

$$\Phi_{zz}^{II}(\omega_e) = |H_{z\zeta}^{II}(\omega_e)|^2 \Phi_{\zeta\zeta}^{II}(\omega_e);$$

from region III

$$\Phi_{zz}^{III}(\omega_e) = |H_{z\zeta}^{III}(\omega_e)|^2 \Phi_{\zeta\zeta}^{III}(\omega_e);$$

such that

$$\Phi_{zz}(\omega_e) = \Phi_{zz}^{I}(\omega_e) + \Phi_{zz}^{II}(\omega_e) + \Phi_{zz}^{III}(\omega_e).$$

Notice that this produces an infinite value or delta function at the frequency $\omega_e = g/4\bar{U} \cos \chi$, indicating that the initial random process has a dominant periodic component of this frequency, (c.f. section 6.2.1) but that the area beneath the curve of $\Phi_{zz}(\omega_e)$ is finite, so that the statistical properties of the motion discussed in section 9.6 can be determined.

For a ship travelling in a following *short*-crested sea, the foregoing theory must be modified, since $\chi$ is no longer a constant, but a random variable. Thus, to take the heave encounter spectrum as an example:

from region I

$$\Phi_{zz}^{I}(\chi, \omega_e) = |H_{z\zeta}^{I}(\chi, \omega_e)|^2 \Phi_{\zeta\zeta}^{I}(\chi, \omega_e);$$

from region II

$$\Phi_{zz}^{II}(\chi, \omega_e) = |H_{z\zeta}^{II}(\chi, \omega_e)|^2 \Phi_{\zeta\zeta}^{II}(\chi, \omega_e);$$

from region III

$$\Phi_{zz}^{III}(\chi, \omega_e) = |H_{z\zeta}^{III}(\chi, \omega_e)|^2 \Phi_{\zeta\zeta}^{III}(\chi, \omega_e);$$

such that

$$\Phi_{zz}(\chi, \omega_e) = \Phi_{zz}^{I}(\chi, \omega_e) + \Phi_{zz}^{II}(\chi, \omega_e) + \Phi_{zz}^{III}(\chi, \omega_e).$$

The one-dimensional heave encounter spectrum $\Phi_{zz}(\omega_e)$ observed on board ship at any value of $\omega_e$ contains a contribution from waves in the range $0 < \chi < \pi/2$ and $3\pi/2 < \chi < 2\pi$. It is given by

$$\Phi_{zz}(\omega_e) = \int_{\chi} \{\Phi_{zz}^{I}(\chi, \omega_e) + \Phi_{zz}^{II}(\chi, \omega_e) + \Phi_{zz}^{III}(\chi, \omega_e)\} \, d\chi$$

but the limits of this integration require some thought because $\omega_e$ and $\chi$ are related to each other.

From Figure 78 we see that the encounter frequency $\omega_e$ and angular variable $\chi$ satisfy the relationship

$$0 \leqslant \omega_e \leqslant \frac{g}{4\bar{U} \cos \chi}$$

in regions I, II and III, implying that $\cos \chi$ is a positive quantity i.e. $0 < \chi < \pi/2$ and $3\pi/2 < \chi < 2\pi$. The additional requirement in region III is given by

$$\frac{g}{4\bar{U} \cos \chi} < \omega_e \leqslant \infty.$$

These relationships dictate the limits of integration, as in the following special cases.

(i) $\omega_e < g/4\bar{U}$. In regions I, II and III

$$\omega_e \leqslant \frac{g}{4\bar{U} \cos \chi}$$

so that

$$\cos \chi \leqslant \frac{g}{4\bar{U}\omega_e}.$$

Hence

$$\frac{g}{4\bar{U}\omega_e} > 1 = \max (\cos \chi),$$

implying that the relative angle between the ship and wave components lies in the range

$$0 < \chi < \frac{\pi}{2} \quad \text{and} \quad \frac{3\pi}{2} < \chi < 2\pi.$$

The additional requirement of region III is not satisfied, since the minimum value of $\omega_e$ occurs when $\cos\chi$ is a maximum i.e. $\cos\chi = 1$ and $\omega_e = g/4\bar{U}$. Thus the one-dimensional heave encounter spectrum is given by

$$\Phi_{zz}(\omega_e) = \left\{ \int_0^{\pi/2} + \int_{3\pi/2}^{2\pi} \Phi_{zz}^{\mathrm{I}}(\chi, \omega_e) + \Phi_{zz}^{\mathrm{II}}(\chi, \omega_e) + \Phi_{zz}^{\mathrm{III}}(\chi, \omega_e) \right\} d\chi$$

where the term in { } is integrated separately between the stated limits.

(ii) $\omega_e > g/4\bar{U}$. In regions I, II and III we have,

$$\cos\chi \leqslant \frac{g}{4\bar{U}\omega_e} < 1$$

indicating that the angular variable $\chi$ lies in the range

$$\cos^{-1}\left(\frac{g}{4\bar{U}\omega_e}\right) \leqslant \chi \leqslant \frac{\pi}{2} \quad \text{and} \quad \frac{3\pi}{2} \leqslant \chi < 2\pi - \cos^{-1}\left(\frac{g}{4\bar{U}\omega_e}\right).$$

Further, in region III

$$\frac{g}{4\bar{U}\omega_e} < \cos\chi$$

so that

$$1 > \cos\chi > \frac{g}{4\bar{U}\omega_e}.$$

That is

$$0 < \chi < \cos^{-1}\left(\frac{g}{4\bar{U}\omega_e}\right) \quad \text{and} \quad 2\pi - \cos^{-1}\left(\frac{g}{4\bar{U}\omega_e}\right) < \chi < 2\pi.$$

The one-dimensional heave encounter spectrum is therefore defined by

$$\Phi_{zz}(\omega_e) = \left\{ \int_{\cos^{-1}(g/4\bar{U}\omega_e)}^{\pi/2} + \int_{3\pi/2}^{2\pi - \cos^{-1}(g/4\bar{U}\omega_e)} \right.$$

$$\left. \Phi_{zz}^{\mathrm{I}}(\chi, \omega_e) + \Phi_{zz}^{\mathrm{II}}(\chi, \omega_e) \right\} d\chi + \left\{ \int_0^{\pi/2} + \int_{3\pi/2}^{2\pi} \Phi_{zz}^{\mathrm{III}}(\chi, \omega_e) \right\} d\chi.$$

whence

$$\Phi_{zz}(\omega_e) = \left\{ \int_{\cos^{-1}(g/4\bar{U}\omega_e)}^{\pi/2} + \int_{3\pi/2}^{2\pi - \cos^{-1}(g/4\bar{U}\omega_e)} |H_{z\zeta}^{\mathrm{I}}(\chi, \omega_e)|^2 \Phi_{\zeta\zeta}^{\mathrm{I}}(\chi, \omega_e) \right.$$

$$\left. + |H_{z\zeta}^{\mathrm{II}}(\chi, \omega_e)|^2 \Phi_{\zeta\zeta}^{\mathrm{II}}(\chi, \omega_e) \right\} d\chi$$

$$+ \left\{ \int_0^{\pi/2} + \int_{3\pi/2}^{2\pi} |H_{z\zeta}^{\mathrm{III}}(\chi, \omega_e)|^2 \Phi_{\zeta\zeta}^{\mathrm{III}}(\chi, \omega_e) \right\} d\chi$$

from which the statistical properties as described in section 9.6 may be determined.

### 12.3.3 Non-dimensional form of statistical properties

In general the exact form of the spectrum – for example $\Phi_{zz}(\omega_e)$ – is not required, since we mainly describe the motion in terms of the moments

$$m_n = \int_0^\infty \omega_e^n \Phi_{zz}(\omega_e) d\omega_e.$$

Evaluation of this integral is difficult, however, because the spectrum has an infinite value at $\omega_e = g/4\bar{U} \cos \chi$ in a long-crested following sea as we show in Figure 79.

To overcome problems involved with this integration, a non-dimensional approach may be adopted using the wave-length rather than the frequency as variable. Thus for a ship of length $L$, the encounter frequency may be expressed in the form

$$\omega_e = \sqrt{\frac{g}{L}} \left\{ \sqrt{\left(\frac{2\pi L}{\lambda}\right)} - F_n \frac{2\pi L}{\lambda} \cos \chi \right\}$$

where the Froude number $F_n = \bar{U}/\sqrt{gL}$. The relationship for deep water waves $k = \omega^2/g$ may be written in the non dimensional form

$$\frac{L}{\lambda} = \frac{\omega^2}{2\pi} \left(\frac{L}{g}\right)$$

so that

$$d\left(\frac{L}{\lambda}\right) = \frac{\omega}{\pi} \left(\frac{L}{g}\right) d\omega.$$

But the theory of section 9.2 shows that

$$\Phi_{zz}\left(\frac{L}{\lambda}\right)d\left(\frac{L}{\lambda}\right) = \Phi_{zz}(\omega)d\omega$$

and hence

$$\Phi_{zz}\left(\frac{L}{\lambda}\right) = \frac{\Phi_{\zeta\zeta}(\omega)}{\left|\dfrac{d(L/\lambda)}{d\omega}\right|} = \frac{\Phi_{zz}(\omega)}{\omega}\,\frac{\pi g}{L}.$$

Furthermore

$$\Phi_{zz}(\omega_e)d\omega_e = \Phi_{zz}\left(\frac{L}{\lambda}\right)d\left(\frac{L}{\lambda}\right).$$

Now the moments $m_n$ are given by

$$m_n = \int_0^\infty \omega_e^n \Phi_{zz}(\omega_e)d\omega_e = \int_0^\infty \omega_e^n |H_{z\zeta}(\omega_e)|^2 \Phi_{\zeta\zeta}(\omega_e)d\omega_e$$

$$= \int_0^{(L/\lambda)_{max}} \frac{\pi g}{L}\,\omega_e^n\,\frac{\Phi_{\zeta\zeta}(\omega)}{\omega}\left|H_{z\zeta}\left(\frac{L}{\lambda}\right)\right|^2 d\left(\frac{L}{\lambda}\right)$$

where the limit $(L/\lambda)_{max}$ (= 4 say) has been selected as representing the wavelength at which ship motions are very small.

For the Pierson-Moskowitz spectrum described in section 9.5,

$$\Phi_{\zeta\zeta}(\omega) = \frac{A}{\omega^5}\,e^{-B/\omega^4}.$$

Thus

$$\frac{\Phi_{\zeta\zeta}(\omega)}{\omega} = \frac{A}{\omega^6}\,e^{-B/\omega^4} = \frac{AL^3}{8\pi^3 g^3}\left(\frac{\lambda}{L}\right)^3 e^{-(BL^2/4\pi^2g^2)(\lambda/L)^2}$$

and the moment integral becomes

$$m_n = \int_0^{(L/\lambda)_{max}} \frac{\pi g}{L}\,\frac{AL^3}{8\pi^3 g^3}\left(\frac{g}{L}\right)^{n/2}\left\{\left(\frac{2\pi L}{\lambda}\right)^{1/2} - F_n\,\frac{2\pi L}{\lambda}\cos\chi\right\}^n \times$$

$$\cdot\left(\frac{\lambda}{L}\right)^3 e^{-(BL^2/4\pi^2g^2)(\lambda/L)^2}\left|H_{z\zeta}\left(\frac{L}{\lambda}\right)\right|^2 d\left(\frac{L}{\lambda}\right).$$

This integral now presents no special computational difficulties.

The form of $|H_{z\zeta}(L/\lambda)|^2$ must be quoted in non-dimensional form. For linear motions it is convenient to use (motion/wave height) while for angular motions (motion/wave slope) is a convenient parameter. In this latter case, to obtain the correct dimensional quantities, the wave spectrum must be multiplied by $(\omega^4/g^2)(180/\pi)^2$ to give a resulting mean square motion value in (degrees)$^2$.

The moment of the Pierson-Moskowitz spectrum in the non-dimensional form reduces to

$$m_n = \int_0^{(L/\lambda)_{max}} \frac{AL^2}{8\pi g^2}\left(\frac{g}{L}\right)^{n/2}\left\{\left(\frac{2\pi L}{\lambda}\right)^2 - F_n\frac{2\pi L}{\lambda}\cos\chi\right\}^n\left(\frac{\lambda}{L}\right)^3 \times$$

$$e^{-(BL^2/4\pi^2 g^2)(\lambda/L)^2}\,|H_{z\zeta}(L/\lambda)|^2\,d\left(\frac{L}{\lambda}\right).$$

For the International Towing Tank Conference spectrum described in section 9.5, in which $A = 8{\cdot}1 \times 10^{-3}g^2$ and $B = 3{\cdot}11\ h_{1/3}^{-2}$ (in metric units), $BL^2$ is a function of $(L/h_{1/3})$. Thus given $|H_{z\zeta}(L/\lambda)|^2$ for a particular ship form, the effect of change of ship length is immediately known for a range of values of $(L/h_{1/3})$.

## References

[1] DALZELL, J. F. and YAMANOUCHI, Y., 1958, 'Analysis of model test results in irregular head seas to determine motion amplitudes and phase relationships to waves'. *Stevens Institute of Technology*. Report no. 708. Hoboken, New Jersey.

[2] OCHI, M. K., 1964, 'Prediction of occurrence and severity of ship slamming at sea'. *Fifth Symposium of Naval Hydrodynamics, Bergen*, pp. 545–596.

[3] CONOLLY, J. E., 1963, 'Ship motions in irregular waves', in *Sea going Qualities of Ships*, N.P.L., H.M.S.O., London, pp. 55–88.

[4] St. DENIS, M., and PIERSON, W. J., 1953, 'On the motions of ships in confused seas', *Trans. SNAME*, **61**, 280–357.

# 13 Excessive motions and design factors

The combination of heave, pitch and roll motions produces a vertical motion at any point along the length of the hull. This motion is usually less at the stern than at the bow where large vertical velocities and accelerations may occur. Further, at forward positions the relative vertical motion between the ship and sea may be excessive, leading to the shipping of green water and to slamming. These events are most severe in head seas and increase with increased sea state. For the crew, they make working conditions difficult and increase the possible risk of injury. Damage to the ship may also occur and the captain is obliged either to reduce speed or change heading to ensure the safety of his vessel.

What the ship's speed ahead shall be is usually decided by the captain and his decision will depend on his experience. Unfortunately he may demand a speed that his ship cannot maintain with safety. It thus becomes desirable that criteria for ship operation be devised. This chapter contains some methods used in formulating criteria.

## 13.1 Narrow band motion process

From the theory of sections 9.7 and 9.7.1 it was shown that the peak or envelope random process of the wave motion may be assumed to be a narrow band random process with band width parameter $\epsilon = 0$. Since this same wave motion excites the ship, it follows that the random process $X(t)$ describing the peak distribution of the ship's motion random process is also a narrow band process which may be described by a Rayleigh probability density function. That is

$$f_X(x) = \frac{x}{m_X}\, e^{-x^2/2m_X} \qquad \text{for } x > 0$$

and the probability of the process $X$ at time $t$ exceeding the value $x$ is shown in section 9.7 to be

$$P[X > x] = e^{-x^2/2m_X}$$

or

$$\frac{x^2}{m_X} = -\,4{\cdot}605 \log_{10} P[X > x].$$

Figure 46 illustrates this relationship.

In the above, the moments $m_X$ have the values

$$m_X = \int_0^\infty \Phi_{XX}(\omega)d\omega = m_0$$

if the random ship's motion is a displacement;

$$m_X = \int_0^\infty \omega^2 \Phi_{XX}(\omega)d\omega = m_2$$

if the random ship's motion is a velocity;

$$m_X = \int_0^\infty \omega^4 \Phi_{XX}(\omega)d\omega = m_4$$

if the random ship's motion is an acceleration.

For the purpose of illustration in this chapter, it is assumed desirable that the motion under consideration shall exceed the value $x$ no more than 1 in every 20 oscillations so that $P[X > x] = 0{\cdot}05$. Hence, by the foregoing result

$$x = 2{\cdot}45\sqrt{m_X}.$$

A probability of 0·05 implies that the ship's operation is not seriously impaired, but that if this value is increased then the damage to the ship and or injury to the crew may result.

## 13.2 Absolute and relative motion

In a long crested head sea, the dominant motions are pitch and heave. The non-dimensional plots in Figures 80–85[1] illustrate the

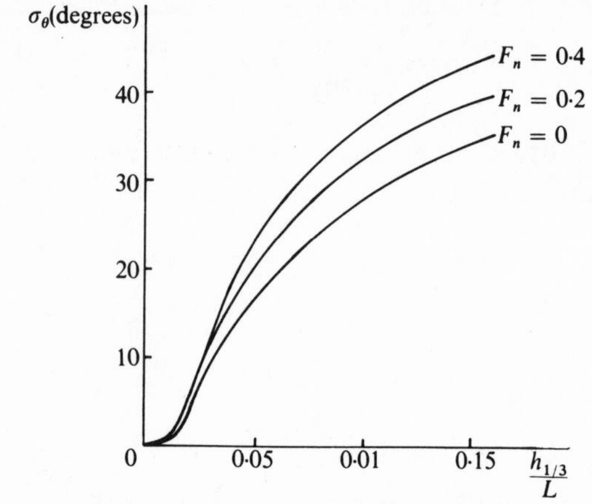


**Figure 80.** Non-dimensional variation with significant wave height of root mean square value of the pitching motion of a ship in a head sea.

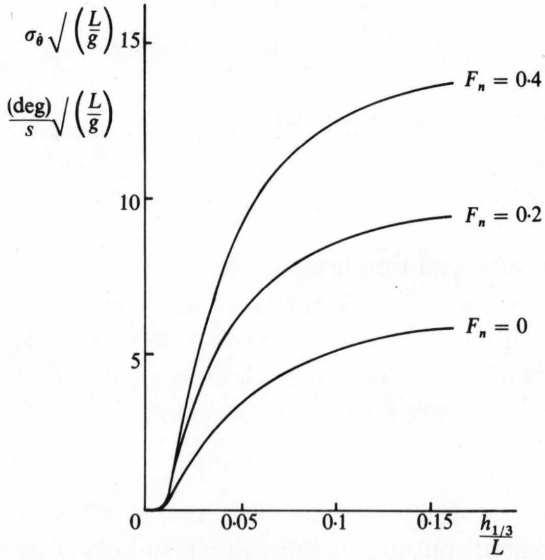


**Figure 81.** Non-dimensional variation with significant wave height of root mean square value of the pitching velocity of a ship in a head sea.

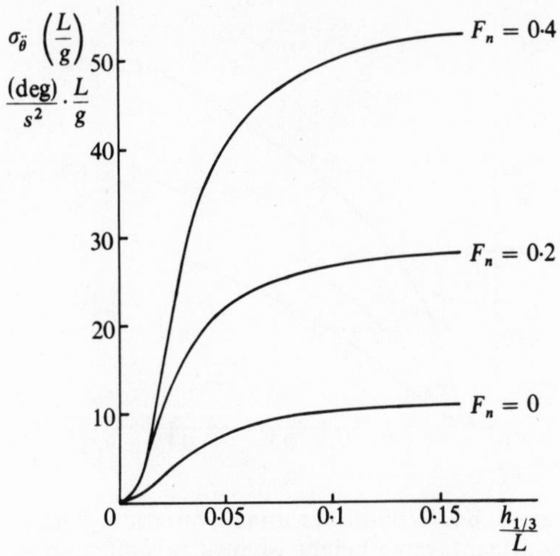


**Figure 82.** Non-dimensional variation with significant wave height of root mean square value of the pitching acceleration of a ship in a head sea.

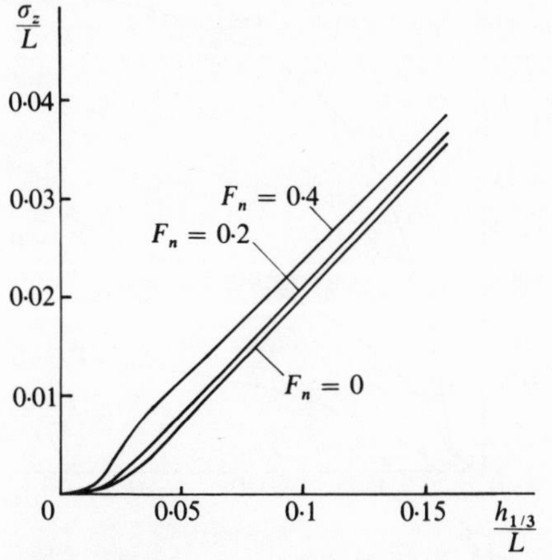


**Figure 83.** Non-dimensional variation with significant wave height of root mean square value of the heave motion of a ship in a head sea.

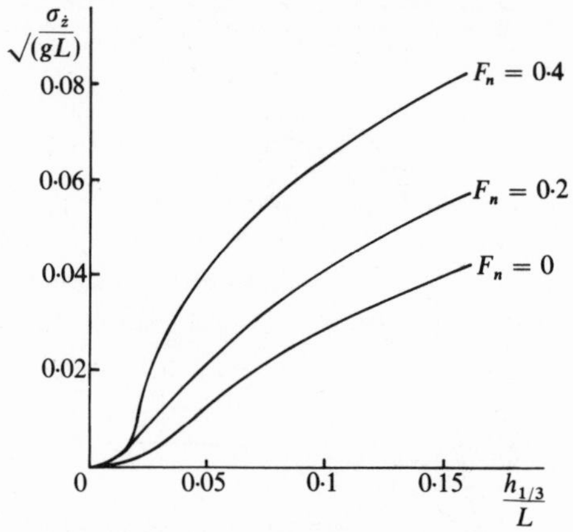


**Figure 84.** Non-dimensional variation with significant wave height of root mean square value of the heave velocity of a ship in a head sea.

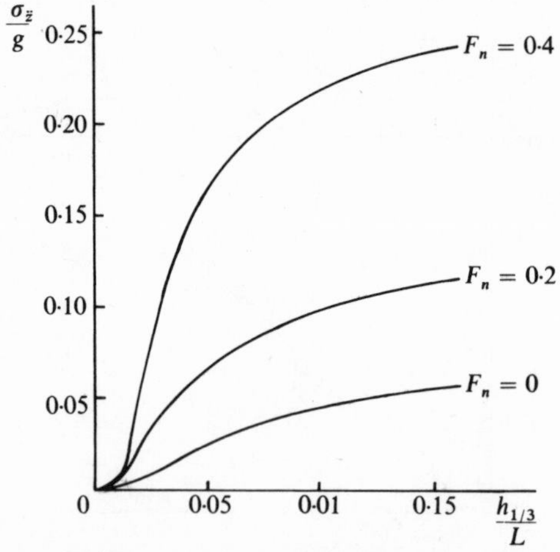


**Figure 85.** Non-dimensional variation with significant wave height of root mean square value of the heave acceleration of a ship in a head sea.

effects of forward speed and sea state on the heave and pitch displacements, velocities and accelerations measured at the centre of mass of ships with similar hull characteristics. For example, such theoretical curves may be obtained for naval ships with fine forms, cargo vessels with fuller forms, tankers etc.

From the theory of section 12.2.1, a combination of the heave and pitch motions gives the absolute vertical motion at different positions along the ship's hull. Figures 86–88 illustrates the effect of forward speed and sea state in the vertical motions at station 3 (i.e. 0·1 $L$ aft of the forward perpendicular). By accounting for the wave profile, the relative motions between the ship and wave surface may be calculated at different stations along the ship. Figures 89–91 show how the relative bow motion in head seas is influenced by the forward speed and sea state.

It is seen from all these figures that the motions (especially those of velocity and acceleration), are dependent on the ship's forward velocity $\bar{U}$. This is because the frequency of wave encounter in head seas is

$$\omega_e = \omega + \frac{\bar{U}\omega^2}{g}$$

and this increases with the ship's forward velocity.

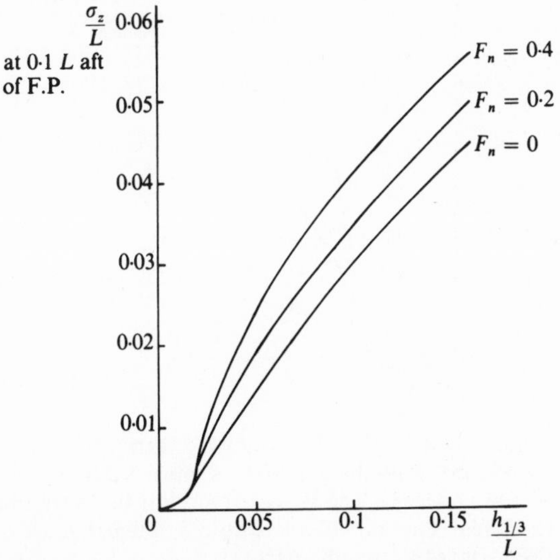


**Figure 86.** Non-dimensional variation with significant wave height of root mean square value of the vertical motion of a point in a ship in a head sea. The curves relate to a point 0·1 L aft of the forward perpendicular.

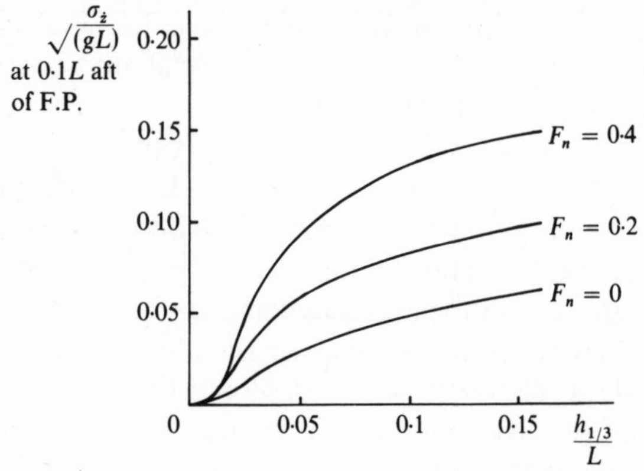


**Figure 87.** Non-dimensional variation with significant wave height of root mean square value of the vertical velocity of a point in a ship in a head sea. The curves relate to a point 0.1 L aft of the forward perpendicular.

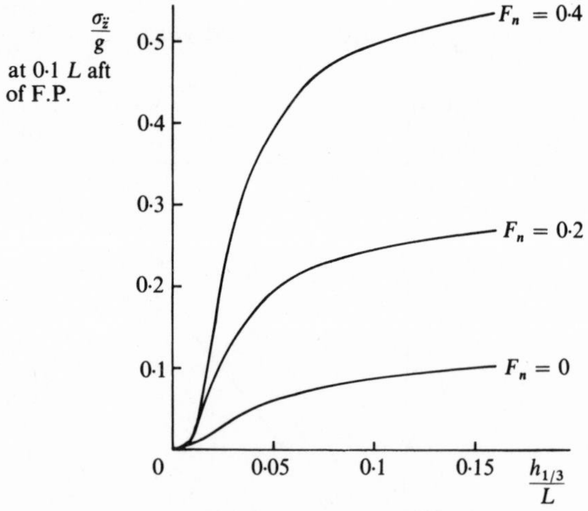


**Figure 88.** Non-dimensional variation with significant wave height of root mean square value of the vertical acceleration of a point in a ship in a head sea. The curves relate to a point 0.1 L aft of the forward perpendicular.

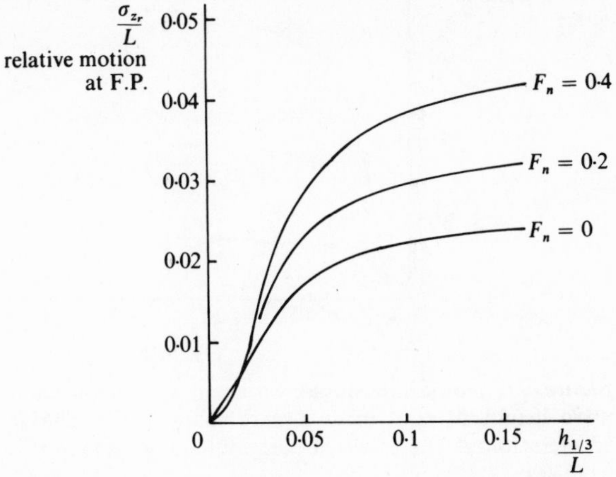


**Figure 89.** Non-dimensional variation with significant wave height of root mean square value of the relative motion at the forward perpendicular of a ship in a head sea.

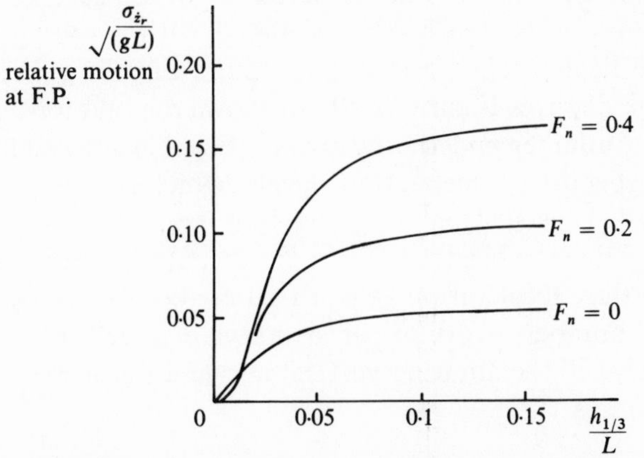


**Figure 90.** Non-dimensional variation with significant wave height of root mean square value of the relative velocity at the forward perpendicular of a ship in a head sea.

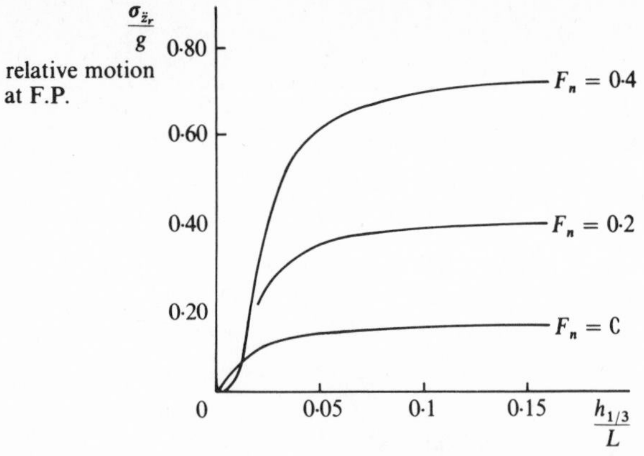


**Figure 91.** Non-dimensional variation with significant wave height of root mean square value of the relative acceleration at the forward perpendicular of a ship in a head sea.

Figures 80–88 show that the heave, pitch and vertical displacement at station 3 are least affected by forward speed. But the relative bow displacement and all velocity and acceleration curves show pronounced increases with forward speed.

An acceptable limiting value of vertical acceleration which the efficiency of the ship is not impaired is somtimes taken as $\pm 0{\cdot}6\,g$ measured at station 3. This is associated with a root mean square value of vertical acceleration at this station of $0{\cdot}25\,g$. Cross plotting from Figure 88 gives Figure 92 which shows the relationship between the Froude number and the ratio (significant wave height/length of ship) for specified acceleration levels which are expected to be exceeded in 1 oscillation in 20. As an example, a ship of length 100 m travelling in a head sea of 5 m significant wave height reaches the limiting vertical acceleration value of $\pm 0{\cdot}6\,g$ once in 20 oscillations at a Froude number of 0·25 or a forward speed of 7·83 m $s^{-1}$. Alternatively, if the limiting vertical acceleration is $\pm 0{\cdot}3\,g$ then the speed is 3·5 m $s^{-1}$.

## 13.3 Wetness

Deck wetness due to the shipping of green water occurs when the relative displacement between the ship and sea surface exceeds the local freeboard. This together with spray blown by the wind restricts the activity and ability of the crew to see, as well as causing deck damage in severe seas.

For a particular ship, Figure 89 shows that the variation of relative displacement at the bow increases with increased sea state and ship's forward velocity. From section 13.1 we see that the mean square value of the relative displacement, $m_0$ (i.e. the area under the curve of the mean square spectral density of relative displacement) and the freeboard $H$ at the bow are given by

$$H = 2{\cdot}45\sqrt{m_0}.$$

By multiplying the ordinate values of Figure 89 by 2·45 we obtain Figure 93. This shows the relationship between the freeboard, ship's forward velocity and the sea state in which 1 in 20 oscillations will result in deck wetness. For example, a ship 100 m long having a freeboard at the bow of 7 m travelling in a unidirectional long-crested head sea of significant wave height 6 m has a maximum speed limit of 9·4 m s$^{-1}$ (or $F_n = 0{\cdot}3$) before the probabilistic criterion is attained.

### 13.3.1 Expected number of deck wettings

Esimating occurrences of deck wetting per unit time is the same as determining the expected number of downcrossings per unit time at some prescribed level. This was discussed in section 9.6.1, and we see that the expected number of occurrences of deck wetting per unit time may be expressed as

$$N_-^H = \frac{1}{2\pi}\sqrt{\frac{m_2}{m_0}}\, e^{-H^2/2m_0}.$$

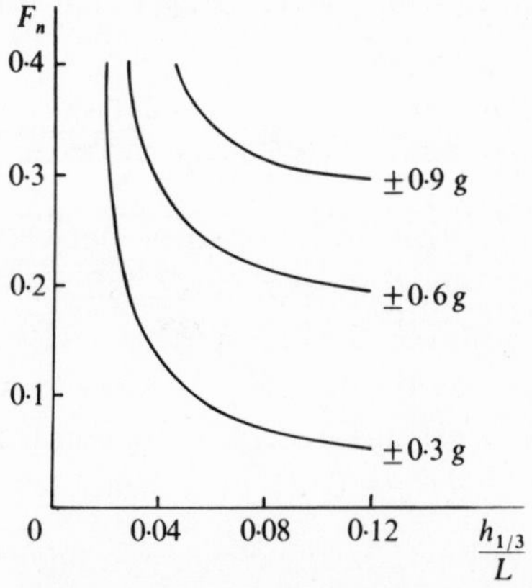


**Figure 92.** Limiting values of Froude number to be imposed in order to restrict vertical acceleration at a point 0·1 L aft of the forward perpendicular to various levels.

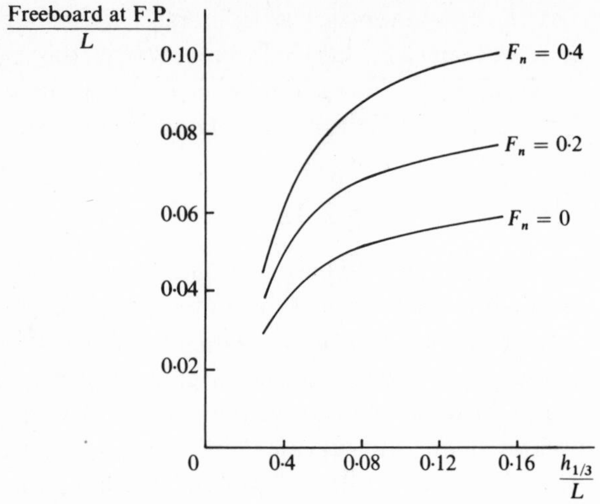


**Figure 93.** Limits placed on the dimensionless freeboard at the forward perpendicular by deck wetness for various Froude numbers.

The theory of section 9.7.1 shows that this result may be written as

$$N_{-}^{H} = \frac{1}{2\pi}\sqrt{\frac{m_2}{m_0}}\, P[X > H]$$

where $P[X > H]$ denotes the probability of the relative displacement exceeding the freeboard and $m_2$ is the mean square value of the relative velocity of the ship's foredeck and the wave. Alternatively, in order to utilise the information of Figures 89 and 90 the expected number of occurrences of deck wetting per hour ($3600\, N_{-}^{H}$) may be written as

$$3600\, N_{-}^{H}\sqrt{\frac{L}{g}} = \frac{3600}{2\pi}\frac{L}{\sqrt{m_0}}\sqrt{\frac{m_2}{Lg}}\, e^{-0{\cdot}5(H/L)^2(L^2/m_0)}.$$

Figure 94 illustrates this relationship between the number of occurrences of deck wetting per hour, ship's forward speed and significant wave height for a ship having a (freeboard/ship length) ratio of 0·065. For example, if this ship is of length 100 m it experiences 39 deck wettings in an hour when travelling at a forward velocity of 6·3 m s$^{-1}$ ($F_n = 0{\cdot}2$) in a head sea of significant wave height 8 m.

## 13.4 Slamming[2]

When a ship proceeds at certain speeds in rough seas, slamming takes place. That is to say, the hull bottom (over perhaps the forward one third of the ship's length) sustains large forces resulting from impact with the sea surface. The slams occur randomly and are most severe in head seas, although stern slamming may occur when the ship travels slowly in following seas. We shall consider the problem of head seas only.

A prediction of slamming may be quoted in terms of cycles of wave encounter or of time and the conditions associated with slamming are usually that

a) the bow emerges
b) a certain magnitude of relative velocity between ship and wave measured at the bow (a 'threshold' velocity) is exceeded

Two other conditions are sometimes cited[3], *viz.*,

c) there is a critical angle between keel line and water surface at the instant of impact
d) there is an unfavourable phase between the bow and wave motion.

Conditions (a) and (b) are usually considered necessary and sufficient for slamming to occur, whereas (c) and (d) are not thought to be so critical. For the sake of simplicity (c) and (d) will be ignored here.

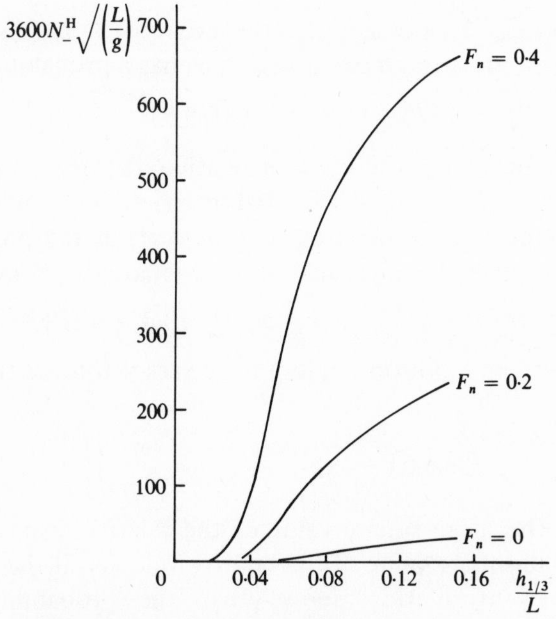


**Figure 94.** Expected number of occurrences of deck wetting per hour for various Froude numbers.

Slamming will only occur when conditions (a) and (b) are both met together. It is possible to have no slam despite bow emergence provided that the threshold velocity is below its critical value at hull entry. It has been found experimentally[4] that for a cargo ship of length 160 m the relative speed for which slamming first appeared was approximately $3{\cdot}7\ \mathrm{m\ s^{-1}}$. This magnitude of threshold velocity cannot be used universally, since ships of different lengths and geometric forms have different threshold velocity values. Ochi originally stated that the threshold velocity $v_*$ was given by

$$v_* = 0{\cdot}093\sqrt{Lg}$$

but this does not hold for all experimental evidence[5].

By contrast with the phenomena discussed in sections 13.2 and 13.3, where the theoretical model depended on one variable (either acceleration or displacement), the theoretical model for slamming must be modified so as to include two parameters, i.e. displacement and a threshold velocity.

### 13.4.1 Probability of slamming

Consider the events, assumed statistically independent,

$A$ denoting the relative displacement at the bow
$B$ denoting the relative velocity at the bow.

From section 2.2.1, we see that the event $(A \cap B)$, in which the independent events $A$ and $B$ occur together, has a probability given by

$$P[A \cap B] = P[A]P[B]$$

A slam occurs when the random relative displacement process $A$ exceeds the forefoot $T$, and the random relative velocity process $B$ exceeds the threshold velocity $v_*$ at time $t$. Thus the probability of the occurrence of a slam per cycle of wave encounter is given by

$$P[\text{slam}] = P[(A > T) \cap (B > v_*)] = P[A > T]P[B > v_*].$$

For narrow band motion random processes it follows from section 13.1 that,

$$P[\text{slam}] = \exp\left[-\left(\frac{T^2}{2m_0} + \frac{v_*^2}{2m_2}\right)\right]$$

where $m_0$ is the mean square value of the relative displacement and $m_2$ is the mean square value of the relative velocity at the bow.

The above information shows that the probability of bow emergence is given by

$$P[A > T] = e^{-T^2/2m_0}$$

or

$$T^2/m_0 = -4{\cdot}605 \log_{10} P[A > T].$$

For the probability of the relative bow motion to exceed the draught once in 20 oscillations, i.e. $P[A > T] = 0{\cdot}05$, the previous result may be written as

$$\frac{T}{L} = \frac{2{\cdot}45\sqrt{m_0}}{L}.$$

Thus from this equation we have a direct relationship between the draught and the mean square value of the relative displacement at the bow. By an appeal to Figure 89, we may equate the draught to the sea state and the ship's forward velocity. This relationship is shown in Figure 93 (replace $H/L$ by $T/L$).

In the same way, the probability of exceeding the threshold velocity is given by

$$\frac{v_*^2}{m_2} = -4{\cdot}605 \log_{10} P[B > v_*]$$

and for

$$P[B > v_*] = 0{\cdot}05$$

we have

$$\frac{v_*}{\sqrt{Lg}} = 2{\cdot}45\sqrt{\frac{m_2}{Lg}}$$

so that we have a relationship between threshold velocity, the mean square value of the relative velocity at the bow and so, by the use of Figure 90, the sea state and ship's forward velocity. This relationship is shown in Figure 95.

As an example, a ship of length 100 m with Froude number 0·2 travels in a seaway of 8 m significant wave height. For 1 in 20 oscillations of bow emergence the draught has a minimum height of 6·4 m and the threshold velocity is 8·3 m $s^{-1}$. Slamming occurs once in 400 oscillations.

### 13.4.2 Expected number of slams

The expected number of occurrences of slamming per unit time may be identified with the expected number of downcrossings per unit time at a given level with a governing condition on the velocity at the

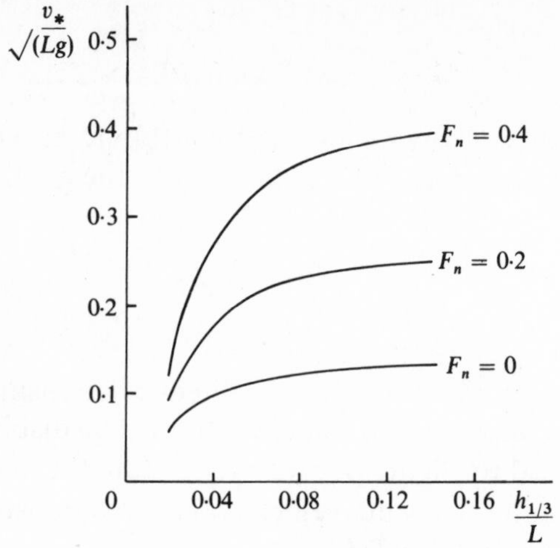


**Figure 95.** Limits placed on the dimensionless threshold velocity by slamming for various Froude numbers.

point of cross over. The underlying theory was derived in section 9.6.2 from which we see that the expected number of occurrences of slamming per unit time is

$$N_-^S = \frac{1}{2\pi}\sqrt{\frac{m_2}{m_0}}\exp\left[-\frac{1}{2}\left(\frac{T^2}{m_0}+\frac{v_*^2}{m_2}\right)\right].$$

According to section 13.4.1, this result may be written as

$$N_-^S = \frac{1}{2\pi}\sqrt{\frac{m_2}{m_0}}P[\text{slam}] = \frac{1}{2\pi}\sqrt{\frac{m_2}{m_0}}P[A > T]\,P[B > v_*].$$

Alternatively, in order to utilise the information of Figure 96 the expected number of slams per hour (3600 $N_-^S$) may be written as

$$3600\,N_-^S\sqrt{\frac{L}{g}} = \frac{3600}{2\pi}\frac{L}{\sqrt{m_0}}\sqrt{\frac{m_2}{Lg}}\exp\left[-0·5\left\{\left(\frac{T}{L}\right)^2\left(\frac{L}{m_0}\right)^2 + \left(\frac{v_*^2}{Lg}\right)\left(\sqrt{\frac{Lg}{m_2}}\right)^2\right\}\right].$$

Figure 96 illustrates the relationship between the expected number of occurrences of slams in an hour, the ship's forward speed and the significant wave height for a ship having a (draught/ship's length) ratio of 0·04 and $v_* = 0·093\sqrt{Lg}$. For example, if this ship is of

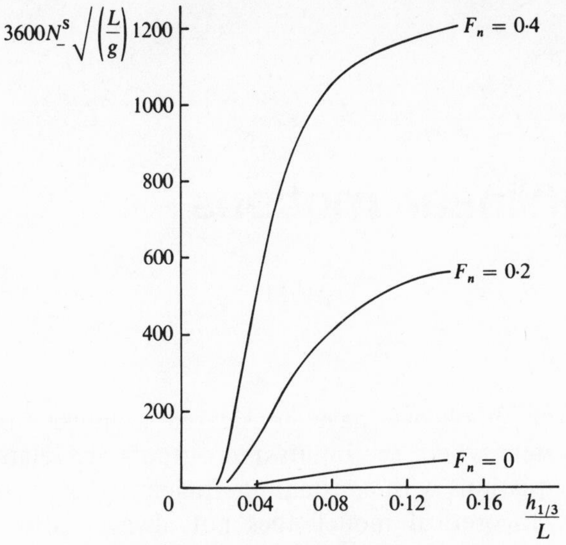


**Figure 96.** Expected number of occurrences of slamming per hour for various Froude numbers.

length 100 m it experiences 123 slams in an hour at a forward velocity of 6·3 m s$^{-1}$ or $F_n = 0·2$ in a head sea of significant wave height of 8 m.

## References


[1] CONOLLY, J. E., 1974, 'Standards of good seakeeping for destroyers and frigates in head seas'. *Proc. Internat. Symp. on the Dynamics of Marine Vehicles and Structures in Waves*, I. Mech. E., London, Ed. R. E. D. Bishop and W. G. Price.

[2] OCHI, M. K., 1964, 'Prediction of occurrence and severity of ship slamming at sea'. *Fifth Symposium of Naval Hydrodynamics, Bergen*, pp. 545–596.

[3] TICK, L. J., 1958, 'Certain probabilities associated with bow submergence and ship slamming in irregular seas'. *J. Ship Res.*, **2**, 30–37.

[4] OCHI, M. K., 1964, 'Extreme behaviour of a ship in rough seas – slamming and shipping of green water'. Trans. SNAME, **72**, 143–202.

[5] AERTSSEN, G., 1966, 'Service performance and seakeeping trials on nv 'Jordeans' ' *TRINA*, **108**, 305–343.

# 14 Nonlinear motions

So far in this book we have considered a linear time-invariant ship-wave system where the inputs and outputs are related by linear differential equations with constant coefficients. Unfortunately, such an idealised theoretical model does not always suffice. Accurate analysis provides many instances of non-linear behaviour.

Significant departures from linearity arise from free-surface conditions, wave added resistance, viscous effects, flow separation, severe motions, geometric properties of the hull, loadings, etc. These nonlinearities must be reflected in the mathematical model and we shall, for simple illustration, assume that they are adequately covered in the differential equation

$$\ddot{q}(t) + r\{\dot{q}(t), q(t)\} = Q(t)$$

governing motion of a system with only one degree of freedom. Here $r$ is a nonlinear function of the displacement $q(t)$ and velocity $\dot{q}(t)$ and $Q(t)$ is the random input.

For normal ship motions, rolling is probably the most obviously nonlinear. It is also the motion that can most realistically be treated in isolation. To show how nonlinearities arise we shall consider the changes in the roll restoring moment when the ship is wall-sided, as shown in Figure 97. The section has length $\delta L$ and beam $B$, $G$ is the centre of gravity, $B_0$ the centre of buoyancy, and $M$ is the metacentre (or more strictly the 'pro-metacentre', since $\phi$ is finite).

Due to a finite roll rotation $\phi$ of the section shown in Figure 97, the volume transferred in the element $(a_1 a_2 a_3)$ is $(B^2 \tan\phi\ \delta L/8)$ which has a moment in a direction parallel to $(a_1 a_2)$ of $(B^3 \tan\phi\ \delta L/12)$. Thus the horizontal displacement of the centre of buoyance for the whole ship, whose displacement is $\nabla$, is

$$B_0 B_1 = \frac{1}{\nabla}\int \frac{B^3}{12} \tan\phi\ dL = \frac{I}{\nabla} \tan\phi = B_0 M \tan\phi$$

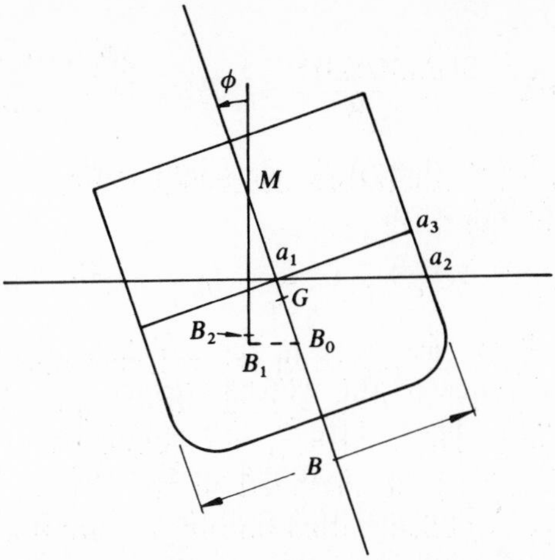


**Figure 97.** Variation in the position of the centre of buoyancy for a wall-sided ship in rolling motion.

where $I$ is the second moment of area about the ship's longitudinal axis. Similarly, the vertical displacement of the centre of buoyancy is

$$B_1 B_2 = \frac{1}{\nabla} \int \frac{B^3}{24} \tan^2\phi \; dL = \frac{I}{2\nabla} \tan^2\phi = \frac{B_0 M}{2} \tan^2\phi$$

so that the righting moment arm becomes

$$GZ = B_0 B_1 \cos\phi + B_1 B_2 \sin\phi - B_0 G \sin\phi$$

$$= \sin\phi\left(GM + \frac{B_0 M}{2} \tan^2\phi\right).$$

For the linear analysis to which table X refers, the righting arm was approximated to $GZ = GM \sin\phi \doteqdot GM\phi$. Such an expression is exact only for ships whose sides are formed by circular arcs with centres lying along the longitudinal axis. Froude[1] approximated $GZ$ to $GM\phi$ and showed that the simplification is valid for ships with tumble-home sides.

If the trigonometric functions are expanded in powers of $\phi$ for a wall-sided ship, it is found that

$$GZ = GM \sum_{n=1,2} d_{2n-1} \phi^{2n-1}$$

where

$$d_1 = 1, d_3 = \frac{3(B_0M/GM) - 1}{6}, d_5 = \frac{30(B_0M/GM) + 1}{120}.$$

etc.

The roll motion is therefore described approximately by a nonlinear equation of the form

$$\ddot{\phi}(t) + 2k_{\dot{\phi}}\omega_\phi\dot{\phi}(t) + \omega_\phi^2 \sum_{n=1,2} d_{2n-1}\phi^{2n-1}(t) = K(t).$$

Nonlinearities may also arise in the roll damping moment due to the frictional resistance of the wetted surface and the generation of eddies by the hull. These effects are associated with the angular velocity term and may be included in the differential equation if necessary. Having established this nonlinear equation of motion, we now obtain approximate solutions. In doing so we shall employ a selection of the many techniques that are available.

## 14.1 The perturbation method

A perturbation method developed by Crandall[2] for random vibration of nonlinear time-invariant systems is suitable for equations in which the nonlinear terms in the differential equation are small. To use it we first write the equation of motion in the form

$$\ddot{q}(t) + 2k\omega_0\dot{q}(t) + \omega_0^2 q(t) + \delta r\{q(t), \dot{q}(t)\} = Q(t)$$

where $\omega_0$ is the natural frequency of the motion, $k$ is the damping factor and $\delta$ is a suitable parameter that is sufficiently small to enable this method to be used.

We shall assume that the solution of the nonlinear equation can be expressed as a series,

$$q(t) = q_0(t) + \delta q_1(t) + \delta^2 q_2(t) + \ldots$$

where the zero order approximation $q_0(t)$ represents the solution of the linear equation, $q_1(t)$ is the solution of the homogeneous differential equation including the first approximation to the nonlinearity, and so on. Substituting this series solution into the nonlinear equation and grouping terms of the same order of smallness, we obtain the recursive chain of linear equations:

$$\ddot{q}_0(t) + 2k\omega_0\dot{q}_0(t) + \omega_0^2 q_0(t) = Q(t),$$

$$\ddot{q}_1(t) + 2k\omega_0\dot{q}_1(t) + \omega_0^2 q_1(t) + r\{q_0(t), \dot{q}_0(t)\} = 0,$$

$$\ddot{q}_2(t) + 2k\omega_0\dot{q}_2(t) + \omega_0^2 q_2(t) + r\{q_0(t), \dot{q}_0(t), q_1(t), \dot{q}_1(t)\} = 0.$$

etc.

For a stationary excitation the solutions of these equations are

$$q_0(t) = \int_{-\infty}^{\infty} h(\tau)Q(t-\tau)d\tau,$$

$$q_1(t) = -\int_{-\infty}^{\infty} h(\tau)r\{q_0(t-\tau), \dot{q}_0(t-\tau)\}d\tau,$$

etc., where the linear impulse function of the system is

$$h(t) = \frac{\sin \omega_0\sqrt{(1-k^2)}t}{\omega_0\sqrt{(1-k^2)}} e^{-k\omega_0 t}.$$

From these solutions, the expectation of the motion is seen to be

$$E[q(t)] = E[q_0(t)] + \delta E[q_1(t)] + \ldots.$$

$$= \int_{-\infty}^{\infty} h(\tau)\{E[Q(t-\tau)] - \delta E[r\{q_0(t-\tau), \dot{q}_0(t-\tau)\}] + \ldots\}d\tau$$

since $E[\ ]$ operates on the time variable $t$ only. The number of terms in the right hand side of this equation depends on the order of approximation of $\delta$ required. The auto-correlation function of the motion is

$$E[q(t)q(t+\tau)] = E[q_0(t)q_0(t+\tau)] + \delta\{E[q_0(t)q_1(t+\tau)] + E[q_1(t)q_0(t+\tau)]\} + \ldots.$$

or

$$R_{qq}(\tau) = R_{q_0q_0}(\tau) + \delta\{R_{q_0q_1}(\tau) + R_{q_1q_0}(\tau)\} + \ldots.$$

where the auto-correlation function

$$R_{q_0q_0}(\tau) = E[q_0(t)q_0(t+\tau)]$$

$$= E[\int_{-\infty}^{\infty} h(\tau_1)Q(t-\tau_1)d\tau_1 \int_{-\infty}^{\infty} h(\tau_2)Q(t+\tau-\tau_2)d\tau_2]$$

$$= \int_{-\infty}^{\infty}\int_{-\infty}^{\infty} h(\tau_1)h(\tau_2)E[Q(t-\tau_1)Q(t+\tau-\tau_2)]d\tau_1\, d\tau_2$$

$$= \int_{-\infty}^{\infty}\int_{-\infty}^{\infty} h(\tau_1)h(\tau_2)R_{QQ}(\tau+\tau_1-\tau_2)d\tau_1 d\tau_2$$

since it is assumed that the input $Q(t)$ is a stationary function. Further, the theory of section 5.4 shows that the cross-correlation function satisfies the relationship

$$R_{q_1 q_0}(\tau) = R_{q_0 q_1}(-\tau)$$

where

$$R_{q_0 q_1}(\tau) = -\int_{-\infty}^{\infty}\int_{-\infty}^{\infty} h(\tau_1)h(\tau_2)E[Q(t-\tau_1)r\{q_0(t+\tau-\tau_2),$$

$$\dot{q}_0(t+\tau-\tau_2)\}]\,d\tau_1\,d\tau_2.$$

The spectral density of the response is found by Fourier transforming the auto-correlation function. It is

$$\Phi_{qq}(\omega) = \Phi_{q_0 q_0}(\omega) + \delta\{\Phi_{q_0 q_1}(\omega) + \Phi_{q_1 q_0}(\omega)\} + \ldots$$

But the theory of section 6.4 shows that $\Phi_{q_1 q_0}(\omega) = \Phi^{*}_{q_0 q_1}(\omega)$ so the spectral density function of the motion response becomes

$$\Phi_{qq}(\omega) = \Phi_{q_0 q_0}(\omega) + 2\delta\,\Phi^{R}_{q_0 q_1}(\omega) + \ldots .$$

where $\Phi^{R}_{q_0 q_1}(\omega)$ indicates the real part of the cross spectral function $\Phi_{q_0 q_1}(\omega)$.

Yamanouchi[3] employed this method when considering the effects of nonlinear damping in roll motion. However, as an example we shall consider the effect of a nonlinear restoring moment, restricting attention to wall-sided ships. The equation of motion is now

$$\ddot{q}(t) + 2k\omega_0\dot{q}(t) + \omega_0^2\{q(t) + \sum_{n=2,3} d_{2n-1}q^{2n-1}(t)\} = Q(t)$$

where it is assumed that $Q(t)$ is a stationary Gaussian random process and the $d$ parameters are properties of the hull, as indicated in the preceding section. The recursive chain of linear differential equations is

$$\ddot{q}_0(t) + 2k\omega_0\dot{q}_0(t) + \omega_0^2 q(t) = Q(t)$$

$$\ddot{q}_1(t) + 2k\omega_0\dot{q}_1(t) + \omega_0^2 q_1(t) = -\,\omega_0^2\sum_{n=2} d_{2n-1}q_0^{2n-1}(t),$$

etc. and these have solutions

$$q_0(t) = \int_{-\infty}^{\infty} h(\tau)Q(t-\tau)d\tau,$$

$$q_1(t) = -\,\omega_0^2\sum_{n=2,3} d_{2n-1}\int_{-\infty}^{\infty} h(\tau)q_0^{2n-1}(t-\tau)d\tau,$$

etc.

The auto-correlation function of the first approximation to the solution is

$$R_{q_0 q_0}(\tau) = \int_{-\infty}^{\infty}\int_{-\infty}^{\infty} h(\tau_1)h(\tau_2)R_{QQ}(\tau + \tau_1 - \tau_2)d\tau_1 d\tau_2$$

which, when it is Fourier transformed, can be expressed as

$$R_{q_0 q_0}(\tau) = \int_{0}^{\infty} |H(\omega)|^2 \Phi_{QQ}(\omega)e^{i\omega\tau}d\omega$$

or

$$\Phi_{q_0 q_0}(\omega) = |H(\omega)|^2 \Phi_{QQ}(\omega).$$

Here the receptance is

$$H(\omega) = \int_{-\infty}^{\infty} h(t)e^{-i\omega t}dt = (\omega_0^2 - \omega^2 + 2ik\omega_0\omega)^{-1}$$

and the response amplitude operator,

$$|H(\omega)|^2 = \{(\omega_0^2 - \omega^2)^2 + 4k^2\omega_0^2\omega^2\}^{-1}.$$

The cross correlation function of the first and second order approximations to the roll motion is given by

$$R_{q_0 q_1}(\tau) = \omega_0^2 \sum_{n=2}' d_{2n-1} \int_{-\infty}^{\infty}\int_{-\infty}^{\infty} h(\tau_1)h(\tau_2)$$
$$\times E[Q(t - \tau_1)q_0^{2n-1}(t + \tau - \tau_2)]d\tau_1\, d\tau_2.$$

Simplifications may be made to this expression for a Gaussian random input process, since the first order approximation to the motion, $q_0(t)$, is the response of the linear system, and is therefore also a Gaussian random process. For simplicity, let $E[Q(t)] = 0$ so that $E[q_0(t)] = 0$ and hence the results of section 4.4.1 may be employed – whence

$$\begin{aligned} E[Q(t - \tau_1)q_0^{2n-1}(t + \tau - \tau_2)] &= E[Q(t - \tau_1 - \tau + \tau_2)q_0^{2n-1}(t)] \\ &= (2n - 1)E[Q(t - \tau - \tau_1) \\ &\qquad \times q_0(t - \tau_2)]E[q_0^{2n-2}(t)] \end{aligned}$$

where

$$\begin{aligned} E[q_0^{2n-2}(t)] &= \sigma_{q_0}^{2n-2}(2n - 3)(2n - 5)(2n - 7)\ldots 5.3.1 \\ &= \frac{(\sqrt{2}\sigma_{q_0})^{2n-2}}{\sqrt{\pi}}\, \Gamma\left(\frac{2n - 1}{2}\right) \end{aligned}$$

as shown in section 4.3.1, $\sigma_{q_0}^2$ being the mean square value of the linear response $q_0(t)$.

Since the linear response $q_0(t)$ is known, we have

$$E[Q(t-\tau-\tau_1)q_0(t-\tau_2)] = \int_{-\infty}^{\infty} h(\tau_3)E[Q(t-\tau-\tau_1) \times Q(t-\tau_2-\tau_3)]d\tau_3$$

$$= \int_{-\infty}^{\infty} h(\tau_3)R_{QQ}(\tau+\tau_1-\tau_2-\tau_3)d\tau_3$$

and the cross-correlation function therefore becomes

$$R_{q_0q_1}(\tau) = \frac{-\omega_0^2}{\sqrt{\pi}} \sum_{n=2,3} 2d_{2n-1}(\sqrt{2}\sigma_{q_0})^{2n-2}\,\Gamma\left(\frac{2n+1}{2}\right) \times \int_{-\infty}^{\infty}\int_{-\infty}^{\infty}\int_{-\infty}^{\infty} h(\tau_1)h(\tau_2)h(\tau_3)R_{QQ}(\tau+\tau_1-\tau_2-\tau_3)d\tau_1\, d\tau_2\, d\tau_3.$$

The cross correlation function $R_{q_1q_0}(\tau)$ is obtained by replacing $\tau$ by $-\tau$ in the previous equation. When it is Fourier transformed, this cross-correlation function becomes

$$R_{q_0q_1}(\tau) = \frac{-\omega_0^2}{\sqrt{\pi}} \sum_{n=2,3} 2d_{2n-1}(\sqrt{2}\sigma_{q_0})^{2n-2} \times \Gamma\left(\frac{2n+1}{2}\right) \int_0^{\infty} |H(\omega)|^2 H(\omega)\Phi_{QQ}(\omega)e^{i\omega\tau}\, d\omega$$

and the cross spectral density function is found to be

$$\Phi_{q_0q_1}(\omega) = \frac{-\omega_0^2}{\sqrt{\pi}} \sum_{n=2,3} 2d_{2n-1}(\sqrt{2}\sigma_{q_0})^{2n-2} \times \Gamma\left(\frac{2n+1}{2}\right) |H(\omega)|^2 H(\omega)\Phi_{QQ}(\omega)d\omega.$$

The total auto-correlation function for the roll response is

$$R_{qq}(\tau) = \int_0^{\infty} \{e^{i\omega\tau} - 2\omega_0^2 N_m H(\omega)\cos\omega\tau\}\, |H(\omega)|^2 \Phi_{QQ}(\omega)d\omega$$

to this first order of approximation.

The mean square spectral density function for rolling motion is

$$\Phi_{qq}(\omega) = \{1 - 2\omega_0^2 N_m H^R(\omega)\}\, |H(\omega)|^2 \Phi_{QQ}(\omega)$$

where

$$N_m = \sum_{n=2,3} \frac{2d_{2n-1}}{\sqrt{\pi}} (\sqrt{2}\sigma_{q_0})^{2n-2}\, \Gamma\left(\frac{2n+1}{2}\right).$$

If a similar form of nonlinearity occurred in the damping term such that

$$\ddot{q}(t) + 2k\omega_0\{\dot{q}(t) + \sum_{n=2,3} g_{2n-1}\, \dot{q}^{2n-1}(t)\} + \omega_0^2 q(t) = Q(t),$$

then the auto-correlation and spectral density functions of the roll response would be

$$R_{qq}(\tau) = \int_0^\infty \{e^{i\omega\tau} + 4k\omega_0(i\omega)N_d H(\omega)\cos\omega\tau\}\, |H(\omega)|^2 \Phi_{QQ}(\omega)d\omega$$

and

$$\Phi_{qq}(\omega) = [1 + 4k\omega_0 N_d\{i\omega H(\omega)\}^R]\,|H(\omega)|^2 \Phi_{QQ}(\omega).$$

In these expressions, $\{i\omega H(\omega)\}^R$ means the real part of $\{i\omega H(\omega)\}$ and

$$N_d = \sum_{n=2,3} \frac{2g_{2n-1}}{\sqrt{\pi}} (\sqrt{2}\sigma_{\dot{q}_0})^{2n-2}\, \Gamma\left(\frac{2n+1}{2}\right)$$

where the root mean square value of the linear velocity response $\dot{q}_0(t)$ is

$$\sigma_{\dot{q}_0} = \left\{\int_0^\infty \omega^2 |H(\omega)|^2\, \Phi_{QQ}(\omega)d\omega\right\}^{1/2}.$$

## 14.2 The functional method

In the nonlinear time-invariant system being considered, the motion response $q(t)$ is due to the random excitation process $Q(t)$. By varying $Q(t)$ we vary $q(t)$ so that the response may be described by the relationship $q\{Q(t)\}$; this is, in fact, a function of a function, or 'functional'. The original development of functionals was due to Volterra[4] who expressed the functional by the series

$$\begin{aligned} q\{Q(t)\} &= \int_{-\infty}^{\infty} h_0(\tau)Q(t-\tau)d\tau + \int_{-\infty}^{\infty}\int_{-\infty}^{\infty} h_1(\tau_1,\tau_2)Q(t-\tau_1)Q(t-\tau_2)d\tau_1\, d\tau_2 \\ &+ \int_{-\infty}^{\infty}\int_{-\infty}^{\infty}\int_{-\infty}^{\infty} h_2(\tau_1,\tau_2,\tau_3)Q(t-\tau_1)Q(t-\tau_2) \times Q(t-\tau_3)d\tau_1\, d\tau_2\, d\tau_3 + \dots \\ &= H_0\{Q(t)\} + H_1\{Q(t)\} + H_2\{Q(t)\} + \dots \end{aligned}$$

The functions $h(\tau)$ are called the 'Volterra kernels'. The first of these, $h_0(\tau)$, is the impulse response function of the linear time-invariant system; $h_1(\tau_1, \tau_2)$ is a two-dimensional function and is referred to as the quadrature kernel or Volterra kernel of the second order, and so on. The functional $H_n\{Q(t)\}$ is referred to as the 'nth degree Volterra functional' and it behaves as an operator; i.e. $H_n$ represents an integration of $Q(t)$.

By way of illustration we shall again consider the equation

$$\ddot{q}(t) + 2k\omega_0\dot{q}(t) + \omega_0^2\{q(t) + \sum_{n=2,3} d_{2n-1}q^{2n-1}(t)\} = Q(t).$$

On substituting the function representation into this equation and grouping the terms of the same order of $Q(t)$, we find the following recursive chain of equations:

$$\ddot{H}_0 + 2k\omega_0\dot{H}_0 + \omega_0^2 H_0 = Q(t)$$

$$\ddot{H}_1 + 2k\omega_0\dot{H}_1 + \omega_0^2 H_1 = 0$$

$$\ddot{H}_2 + 2k\omega_0\dot{H}_2 + \omega_0^2 H_2 = -d_3\omega_0^2 H_0^3$$

$$\ddot{H}_3 + 2k\omega_0\dot{H}_3 + \omega_0^2 H_3 = -3d_3\omega_0^2 H_0^2 H_1$$

$$\ddot{H}_4 + 2k\omega_0\dot{H}_4 + \omega_0^2 H_4 = -3d_3\omega_0^2(H_0^2 H_2 + H_0 H_1^2) - d_5\omega_0^2 H_0^5$$

etc.

The first equation yields the solution

$$H_0\{Q(t)\} = \int_{-\infty}^{\infty} h(\tau)Q(t-\tau)d\tau$$

where again $h(\tau)$ is the impulse response of the time-invariant linear system as defined in the previous section. The solution of the second equation is

$$H_1\{Q(t)\} = 0$$

which implies

$$H_3\{Q(t)\} = 0$$

and indeed that

$$H_{2n+1}\{Q(t)\} = 0$$

for all integer values of $n$.

The solution of the third equation is

$$H_2\{Q(t)\} = -d_3\omega_0^2 \int_{-\infty}^{\infty} h(\tau)[H_0\{Q(t)\}]^3\, d\tau$$

$$= -d_3\omega_0^2 \int_{-\infty}^{\infty}\int_{-\infty}^{\infty}\int_{-\infty}^{\infty}\int_{-\infty}^{\infty} h(\tau)h(\tau_1)h(\tau_2)h(\tau_3)$$

$$\times Q(t-\tau_1)Q(t-\tau_2)Q(t-\tau_3)d\tau_1\, d\tau_2\, d\tau_3\, d\tau$$

$$= -d_3\omega_0^2 \int_{-\infty}^{\infty}\int_{-\infty}^{\infty}\int_{-\infty}^{\infty} h(\tau_1, \tau_2, \tau_3)Q(t-\tau_2)$$

$$\times Q(t-\tau_2)Q(t-\tau_3)d\tau_1\, d\tau_2\, d\tau_3$$

where the third-order Volterra kernel is given by

$$h(\tau_1, \tau_2, \tau_3) = \int_{-\infty}^{\infty} h(\tau)h(\tau_1)h(\tau_2)h(\tau_3)d\tau$$

with the higher order kernels similarly defined.

On comparing the series solutions derived by the functional method and the perturbation method, it is seen that identical results are obtained for the motion response $q(t)$. That is,

$$q_0(t) = H_0\{Q(t)\}; \qquad q_1(t) = H_2\{Q(t)\},$$

etc.

In principle the perturbation or functional method may be applied to any nonlinear problem in which the nonlinear terms are small and for which the corresponding linear problem, obtained by ignoring the nonlinear terms, is solvable.

## 14.3 Equivalent linearisation method

The method of equivalent linearisation was first applied to problems of random oscillations by Caughey[5]. Kaplan[6] used the method when he considered the effect of nonlinear damping in roll motion. The equation

$$\ddot{q}(t) + r\{\dot{q}(t), q(t)\} = Q(t)$$

may be replaced by the 'equivalent' linear equation

$$\ddot{q}(t) + b_e\dot{q}(t) + c_e q(t) = Q(t)$$

where the error random process due to the linearisation is

$$e(t) = r(\dot{q}, q) - b_e\dot{q}(t) - c_e q(t).$$

For this error to be a minimum we must make the mean square value of the error $E[e^2(t)]$ as small as possible. This is accomplished when

$$\frac{\partial}{\partial b_e} E[e^2(t)] = 0 \quad \text{and} \quad \frac{\partial}{\partial c_e} E[e^2(t)] = 0.$$

After interchanging the order of differentiation and integration these conditions become

$$E[\dot{q}r(\dot{q}, q)] - b_e E[\dot{q}^2] - c_e E[q\dot{q}] = 0$$

$$E[qr(\dot{q}, q)] - b_e E[q\dot{q}] - c_e E[q^2] = 0.$$

Solving for the coefficients $b_e$ and $c_e$ we find

$$b_e = \{E[q^2]E[\dot{q}r(\dot{q}, q)] - E[q\dot{q}]E[qr(\dot{q}, q)]\} D^{-1}$$

$$c_e = \{E[\dot{q}^2]E[qr(\dot{q}, q)] - E[q\dot{q}]E[\dot{q}r(\dot{q}, q)]\} D^{-1}$$

where

$$D = E[q^2]E[\dot{q}^2] - (E[q\dot{q}])^2.$$

These are not explicit expressions for the coefficients, however, since the expectations themselves depend on the coefficients.

Since the input $Q(t)$ is assumed to be a stationary Gaussian random process with zero expectation, the approximate values of the displacement $q(t)$ and velocity $\dot{q}(t)$ determined from the equivalent linearised equation, are also Gaussian random processes which satisfy the condition $E[q(t)\dot{q}(t)] = 0$, as shown in section 5.4. The equivalent linearised constant relationships, therefore, reduce to

$$b_e = \frac{E[\dot{q}r(\dot{q}, q)]}{E[\dot{q}^2]} \quad \text{and} \quad c_e = \frac{E[qr(\dot{q}, q)]}{E[q^2]}.$$

With nonlinearities in the roll restoring moment such that

$$r(\dot{q}, q) = \omega_0^2 \sum_{n=1,2} d_{2n-1} q^{2n-1}(t),$$

the equivalent restoring moment is given by

$$c_e = \omega_0^2 \left\{1 + \sum_{n=2,3} \frac{d_{2n-1} E[q^{2n}]}{E[q^2]}\right\}.$$

But from section 4.3.1 it is seen that

$$\frac{E[q^{2n}]}{E[q^2]} = \frac{2^n}{\sqrt{\pi}} \sigma_q^{2n-2} \Gamma\left(\frac{2n+1}{2}\right)$$

so that the equivalent linearised restoring moment becomes

$$c_e = \omega_0^2(1 + N_m)$$

where

$$N_m = \sum_{n=2,3} d_{2n-1} \frac{2^n}{\sqrt{\pi}} \sigma_q^{2n-2} \Gamma\left(\frac{2n+1}{2}\right).$$

The spectral density function of the roll response is

$$\Phi_{qq}(\omega) = |H_e(\omega)|^2 \Phi_{QQ}(\omega)$$

where the square of the modulus of the receptance or response amplitude operator of the equivalent linear system is

$$\begin{aligned}|H_e(\omega)|^2 &= \{(c_e - \omega^2)^2 + 4k^2 \omega_0^2 \omega^2\}^{-1} \\ &= [\{\omega_0^2(1 + N_m) - \omega^2\}^2 + 4k^2 \omega_0^2 \omega^2]^{-1}.\end{aligned}$$

This latter has a first order approximation in the coefficient $N_m$ given by

$$|H_e(\omega)|^2 = |H(\omega)|^2 \ \{1 - 2N_m \ \omega_0^2 H^R(\omega)\}$$

where the response amplitude operator of this linear system is

$$|H(\omega)|^2 = [(\omega_0^2 - \omega^2)^2 + 4k^2 \omega_0^2 \omega^2]^{-1}$$

and $H^R(\omega)$ is the real part of the receptance of the linear system.

The spectral density of the roll response becomes

$$\Phi_{qq}(\omega) = \left\{1 - 2\omega_0^2 \sum_{n=2,3} d_{2n-1} \frac{2^n}{\sqrt{\pi}} \sigma_q^{2n-2} \Gamma\left(\frac{2n+1}{2}\right) H^R(\omega)\right\} \times |H(\omega)|^2 \Phi_{QQ}(\omega)$$

$$= \{1 - 2N_m \omega_0^2 H^R(\omega)\} \ |H(\omega)|^2 \Phi_{QQ}(\omega)$$

which agrees with the corresponding result obtained by the perturbation method, provided that

$$\sigma_q^{2n-2} = \sigma_{q0}^{2n-2}.$$

Since

$$\sigma_q^2 = \int_0^\infty \Phi_{qq}(\omega) d\omega$$

a value of $\sigma_q^2$ may be obtained by solving the following algebraic equation:

$$\sigma_q^2 = \int_0^\infty \left\{1 - 2\omega_0^2 \sum_{n=2,3} d_{2n-1} \frac{2^n}{\sqrt{\pi}} \sigma_q^{2n-2} \Gamma\left(\frac{2n+1}{2}\right) H^R(\omega)\right\}$$

$$\times |H(\omega)|^2 \Phi_{QQ}(\omega) d\omega.$$

If the nonlinearities are neglected, the period of the response of the linear system is simply

$$T = 2\pi/\omega_0.$$

According to the theory for the equivalent linear equation, an approximation to the period of the nonlinear system is

$$T' = \frac{2\pi}{\sqrt{c_e}} = \frac{T}{\sqrt{(1+N_m)}}$$

which indicates that $T' < T$ for $N_m > 0$, and $T' > T$ for $N_m < 0$.

In the linear system, resonance occurs when $\omega = \omega_0$. It is now seen that this frequency is shifted to $\omega = \omega_0\sqrt{(1+N_m)}$ when the nonlinearity is admitted.

Kaplan[6] considered a roll damping moment of the form

$$r(\dot{q}, q) = \beta\dot{q}(t)|\dot{q}(t)|.$$

The appropriate linearised coefficient is

$$b_e = 2k\omega_0 + 2\sqrt{\frac{2}{\pi}}\,\beta\sigma_{\dot{q}}$$

where $\sigma_{\dot{q}}^2$ is the mean square value of the angular velocity $\dot{q}(t)$, which is assumed to be a random process with zero mean value. The relationship between the input and output spectral densities is

$$\Phi_{qq}(\omega) = |H_e(\omega)|^2 \Phi_{QQ}(\omega)$$

and the equivalent linearised factor

$$|H_e(\omega)|^2 = \{(\omega_0^2 - \omega^2)^2 + b_e^2\}^{-1}.$$

For small nonlinearities, this may be expressed as

$$|H_e(\omega)|^2 = |H(\omega)|^2 \{1 - N_d |H(\omega)|^2 + N_d^2 |H(\omega)|^4 \ldots\}$$

where

$$N_d = 4\beta\sqrt{\frac{2}{\pi}}\,\sigma_{\dot{q}}\left(2k\omega_0 + \beta\sqrt{\frac{2}{\pi}}\,\sigma_{\dot{q}}\right)^2$$

and the square of the modulus of the response amplitude operator of the linear system is

$$|H(\omega)|^2 = \{(\omega_0^2 - \omega^2)^2 + 4k^2\omega_0^2\}^{-1}.$$

Thus, if the mean square spectral density of the excitation is known, the spectral density of the response is found to be

$$\Phi_{qq}(\omega) = \{1 - N_d|H(\omega)|^2 + \ldots\}\,|H(\omega)|^2\Phi_{QQ}(\omega)$$

and the mean square $\sigma_{\dot{q}}^2$ is given by the algebraic equation

$$\sigma_{\dot{q}}^2 = \int_0^\infty \omega^2\Phi_{qq}(\omega)d\omega$$

$$= \int_0^\infty \omega^2\{1 - N_d|H(\omega)|^2 + \ldots\}|H(\omega)|^2\Phi_{QQ}(\omega)d\omega.$$

For $N_d \geqslant 0$ it is seen that the equivalent linearised factor $|H_e(\omega)|^2 \leqslant |H(\omega)|^2$ which indicates that the nonlinear contributions to the roll damping moment decrease the roll motion of the ship.

## References

[1] FROUDE, W. 1861, 'On the rolling of ships'. Trans INA, **11**, 180–229.

[2] CRANDALL, S. H. 1964, 'The spectrum of random vibration of a nonlinear oscillator'. MIT Report AFSOR 64–1057.

[3] YAMANOUCHI, Y. 1963, 'The change of damping and the nonlinearity of rolling by speed'. *Proc. 10th International Towing Tank Conference*, National Physical Laboratory, London.

[4] VOLTERRA, V. 1930, *Theory of Functional and Integral and Integro-Differential Equations*, Blackie, London.

[5] CAUGHEY, T. K. 1963, 'Equivalent linearisation techniques'. *J. Acoustic Soc. Am.*, **35**, 1706–1711.

[6] KAPLAN, P. 1966, 'Lecture notes on nonlinear theory of ship roll motion in a random seaway'. *Proc. 11th International Towing Tank Conference*, Tokyo, Japan.

# Index